Springer Texts in Business and Economics

More information about this series at http://www.springer.com/series/10099

Thorsten Hens • Marc Oliver Rieger

Financial Economics

A Concise Introduction to Classical and Behavioral Finance

Second Edition

Springer

Thorsten Hens
University of Zurich
Zurich, Switzerland

Marc Oliver Rieger
University of Trier
Trier, Germany

ISSN 2192-4333 ISSN 2192-4341 (electronic)
Springer Texts in Business and Economics
ISBN 978-3-662-49686-2 ISBN 978-3-662-49688-6 (eBook)
DOI 10.1007/978-3-662-49688-6

Library of Congress Control Number: 2016939949

Printed on acid-free paper

This Springer imprint is published by Springer Nature
The registered company is Springer-Verlag GmbH Berlin Heidelberg

Preface

In quiet times, most people do not pay too much attention to financial markets. Just a few years back, however, such quiet times came to an abrupt end with the onset of the financial crisis of the years 2007–2009. Before that crisis, everybody took it for granted that we can borrow money from a bank or get to save on interest payments on deposits, but all these fundamental beliefs were shaken in the wake of the financial crisis and – not much later – during the Euro crisis. Although times are now much quieter again, these crises left some lasting impact on how we trust the functioning of financial markets.

When the man on the street loses his faith in systems which he believed to function as steadily as the rotation of the earth, how much more have the beliefs of financial economists been shattered? But the good news is that, in recent years, the theory of financial economics has incorporated many aspects that now help to understand many of the bizarre market phenomena that we could observe during the financial crisis. In the early days of financial economics, the fundamental assumption was that markets are always efficient and market participants perfectly rational. These assumptions allowed to build an impressive theoretical model that was indeed useful to understand quite a few characteristics of financial markets. Nevertheless, a major financial crisis was not necessary to realize that the assumptions of perfectly efficient markets with perfectly rational investors did not hold – often not even "on average." The observation of systematic deviations gave birth to a new theory, or rather a set of new theories, *behavioral finance theories*.

While classical finance remains the cornerstone of financial theory – and be it only as a benchmark that helps us to judge how much real markets deviate from efficiency and rationality – behavioral finance enriches the view on the real market and helps to explain many of the more detailed phenomena that might be small on sunny days, but decisive in rough weather.

Often, behavioral finance is introduced as something independent of financial economics. It is assumed that behavioral finance is something students may learn after they have mastered and understood all of the classical financial economics.

In this book, we would like to follow a different approach. As market behavior can only be fully understood when behavioral effects are linked to classic models, this book integrates both views from the very beginning. There is no separate chapter on behavioral finance in this book. Instead, all classic topics (such as decisions on markets, the capital asset pricing model, market equilibria, etc.) are immediately connected with behavioral views. Thus, we will never stay in a purely theoretical world, but look at the "real" one. This is supported with many case studies on market phenomena, both during the financial crisis and before.

How this book works and how it can be used for teaching or self-study is explained in detail in the introduction (Chap. 1). Please notice that there is an accompanying book with exercises and solutions. With this, students can check their understanding of the material and prepare for exams.

For now, we would like to take the opportunity to thank all those people who helped us write this book. First of all, we would like to thank many of our colleagues for their valuable input, in particular Anke Gerber, Bjørn Sandvik, Mei Wang, and Peter Wöhrmann.

Parts of this book are based on scripts and other teaching materials that were initially composed by former and present students of ours, in particular by Berno Büchel, Nilüfer Caliskan, Christian Reichlin, Marc Sommer, and Andreas Tupak.

Many people contributed to the book by means of corrections or proofreading. We would like to thank especially Amelie Brune, Julia Buge, Marius Costeniuc, Michal Dzielinski, Mihnea Constantinescu, Mustafa Karama, R. Vijay Krishna, Urs Schweri, Vedran Stankovic, Christoph Steikert, Sven-Christian Steude, Laura Oehen, and the best secretary of the world, Martine Baumgartner.

That this book is not only an idea but a real printed book with hundreds of pages and thousands of formulas is entirely due to the fact that we had two tremendously efficient LaTeX professionals working for us. A big "thank you" goes therefore to Thomas Rast and Eveline Hardmeier. We are grateful to Simone Fuchs and Johannes Baltruschat for their great support in finalizing the second edition.

We also want to thank our publishers for their support and especially Martina Bihn for her patience in coping with the inevitable delays of finishing this book.

Finally, we thank our families for their even larger patience with their book-writing husbands and fathers.

We hope that you, dear reader, will have a good time with this book and that we can transmit some of our fascination for financial economics and its interplay with behavioral finance to you.

Have fun!

Zurich, Switzerland Thorsten Hens
Trier, Germany Marc Oliver Rieger

Contents

Part III Advanced Topics

Part I

Foundations

Introduction 1

"Advice is the only commodity on the market where the supply always exceeds the demand." ANONYMOUS

This first chapter provides an overview on financial economics and how to study it: you will learn how we have designed this textbook and how you can use it efficiently; we will give you an overview of the essence of financial economics and some of its central ideas; we will finally summarize how research in financial economics is done, what methods are used and how they interact with each other.

If you are new to the field of financial economics, we hope that at the end of this introduction your appetite to learn more about it has been sufficiently stimulated to enjoy reading the rest (or at least the main parts) of this book, and maybe even to immerse yourself deeper in this fascinating research area. If you are already working in this field, you can lean back and relax while reading the introduction and then pick the topics of this book that are interesting to you. Since financial economics is a very active area of research into which we have incorporated a number of very recent results, be assured that you will find something new as well.

1.1 An Introduction to This Book

This book integrates classical and behavioral approaches to financial economics and contains results that have been found only recently. It can serve several aims:

- as a textbook for a master or PhD course. Some parts can also be used on an advanced bachelor level,
- for self-study,

T. Hens, M.O. Rieger, *Financial Economics*, Springer Texts in Business and Economics, DOI 10.1007/978-3-662-49688-6_1

- as a reference to various topics and as an overview on current results in financial economics and behavioral finance.

In the following we want to give you some recommendations on how to use this book as a textbook and for self-studying.

The book has three parts: the foundations part consists of this introduction and a chapter on decision theory. The second part on financial markets builds a sophisticated model of financial markets step by step and is also the core of this book. Finally, the third part presents advanced topics that sketch some of the connections between financial economics and other fields in finance. In the first two parts, every chapter is accompanied by a number of tests (solutions can be found in the appendix). Tests are included in order to enable self-studying and as an assessment of the progress made in a chapter. Moreover, in the accompanying exercise book one can find many problem sets with their solutions.

The level of difficulty usually increases gradually within a chapter. Difficult parts not needed in the subsequent chapters are marked with an asterisk. The content of this book is enough for two semesters. For a one-semester class there are therefore various possible routes. A reasonable suggestion for a bachelor class could be to cover Chap. 1, excerpts of Chap. 2, Sects. 3.1 and 3.2. They may be spiced with some applications. A one-semester master course could be based on Chap. 1, main ideas of Chaps. 2, 3 and 4 and some parts of Chap. 5. A two-semester course could follow the whole book in order of presentation. For a one-semester PhD course for students who have already taken a class in financial economics, one could choose some of the advanced topics (especially Chaps. 5, 6, 7 and 8) and provide necessary material from previous chapters as needed (e.g., the behavioral decision theory from Sects. 2.4 and 2.5). The interdependence of the chapters in this book is illustrated in Fig. 1.1.

1.2 An Introduction to Financial Economics

Finance is composed of many different topics. These include public finance, international finance, corporate finance, derivatives, risk management, portfolio theory, asset pricing, and financial economics.

Financial economics is the interface that connects finance to economics. This means that different research questions, methods and languages meet, which can be very fruitful, but also sometimes confusing. To mitigate the confusion, we will present common topics from both points of view, the economics and the finance perspective. In doing so, we hope to reduce potential misunderstandings and help to explore the synergies of the subfields.

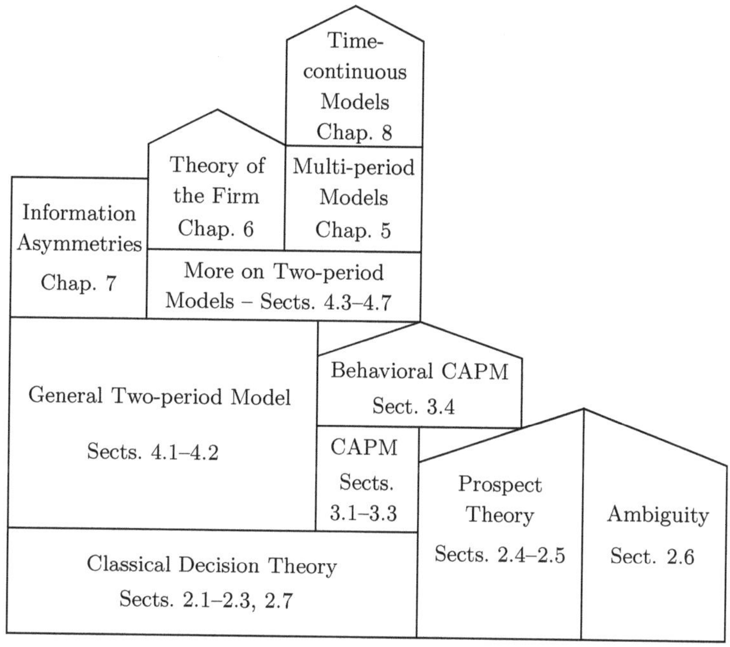

Fig. 1.1 An overview on the interdependence of the chapters in this book. If you want to build up your course on this book, be careful that the "bricks do not fall down"!

Most topics in finance are in some way or the other connected to financial economics. We will discuss several of these connections and the relation to neighboring disciplines in detail, see Fig. 1.2.

Having located financial economics on the scientific map, we are now ready to start our expedition by an overview of the key ideas and research methods. The central point is hereby the transfer of the concept of trade from economics (where tangible goods are traded) to the concept of valuation used in finance.

1.2.1 Trade and Valuation in Financial Markets

Financial economics is about trade among agents, trading in well functioning financial markets. At first sight, agents trade interest bearing or dividends paying assets (bonds or stocks) as well as derivatives thereof in financial markets. But from an economic perspective, on financial markets, agents trade time, risks and beliefs. Of course, agents are heterogeneous, i.e., they have different valuations of time, risks and beliefs. One of the main topics of financial economics is therefore the aggregation of those different valuations at a market equilibrium into market prices for time, risks and beliefs.

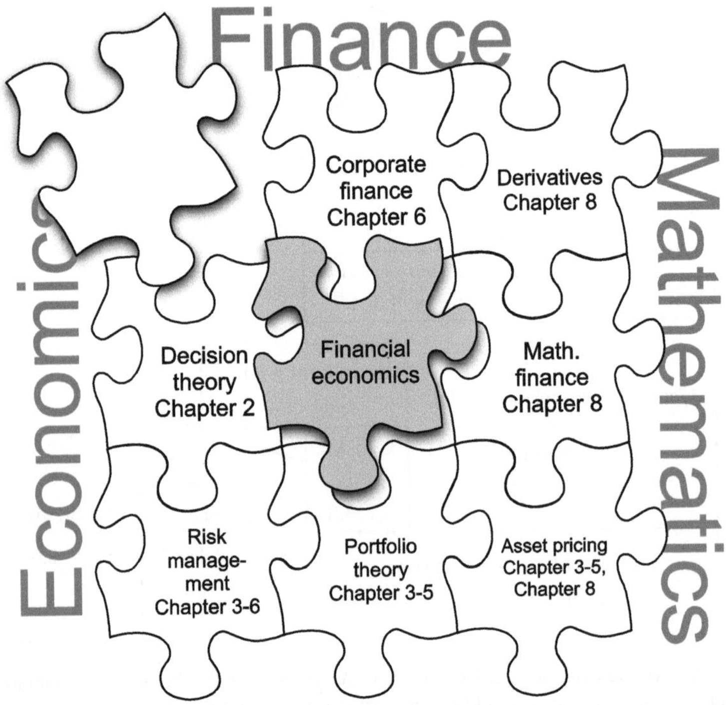

Fig. 1.2 Connections of financial economics with other subfields of finance and other disciplines

For a long time, researchers believed that the aggregation approach would be sufficient to describe financial markets. Recently, however, this classical view has been challenged by new theories (behavioral and evolutionary finance) as well as by the emergence of new trading strategies (as implemented, e.g., by hedge funds). One of the goals of this book is to describe to what degree these new views on financial markets can be integrated into the classical concepts and how they give rise to new insights into financial economics. In this way, we lay the foundations to understand practitioner's buzzwords like "Alpha", "Alternative Beta" and "Pure Alpha".

What do we mean by saying that agents trade risks, time and beliefs? Let us explain this idea with some examples. The trading of risks can be explained easily if we look at commodities. For example, a farmer is naturally exposed to the risk of falling prices, whereas a food company is exposed to the risk of increasing prices. Using forwards, both can agree in advance on a price for the commodity, and thus trade risk in a way that reduces both parties' risks.

There are other situations where one party might not reduce its risk, but is willing to buy the risk from another party for a certain price: hedge funds and insurance companies, although very different in their risk appetite, both work by this fundamental principle.

How to trade "time" on financial markets? Here the difference between investment horizons plays a role. If I want to buy a house, I prefer to do this rather earlier than later, since I get a benefit from owning the house. A bank will lend me money and wants to be paid for that with a certain interest. We can find the same mechanism on financial markets when companies and states issue bonds. Sometimes the loan issued by the bank is bundled and sold as some of these now infamous CDOs that were at the epicenter of the financial crisis.

We can also trade "beliefs" on financial markets. In fact, this is likely to be the most frequent reason to trade: two agents differ in their opinion about certain assets. If Investor A believes Asset 1 to be more promising and Investor B believes Asset 2 to be the better choice, then there is obviously some reason for both to trade. Is there really? Well, from their perspectives there is, but of course only one of them can be right, so contrary to the first two reasons for a trade (risk and time), where both parties will profit, here only one of them (the smarter or luckier) will profit. We will discuss the consequences of this observation in a simple model as "the hunt for Alpha" in Sect. 3.3.

But in all of these cases what does limit the amount of trading? If trading is good for both parties (or at least they believe so), why do they not trade infinite amounts? The reason is in all cases the decreasing marginal utility of the agents: eventually, the benefit from more trades will be outweighed by other factors. For instance, if agents trade because of different beliefs, they will still have the same differences in beliefs after their trade but they won't trade unlimited amounts due to their decreasing marginal utility in the states.

1.2.2 No Arbitrage and No Excess Returns

Financial markets are complex, and moreover practitioners and researchers tend to use the same word for different concepts, so sometimes these concepts get mixed-up. An example of this is the frequent confusion between no-arbitrage and no gains from trades. An efficient financial market is arbitrage-free. An arbitrage opportunity is a self-financing trading strategy that does not incur losses but gives positive returns. Many researchers and practitioners agree that arbitrage strategies are so rare that one can assume they do not exist.

This simple idea has far reaching conclusions for the valuation of derivatives. Derivatives are assets whose payoffs depend on the payoff of other assets, the underlying, the assets from which the derivative is derived. In the simple case where the payoff of the derivative can be duplicated by a portfolio of the underlying and e.g., a risk-free asset, the price of the derivative must be the same as the value of the duplicating portfolio. Why? Suppose the derivative's price is actually higher than the value of the duplicating portfolio. In that case, one can build an arbitrage strategy by shorting the asset and hedging the payoff by holding the duplicating portfolio. If the price of the derivative were less than that of the duplicating portfolio, one would trade the other way round. Hence the principle of no-arbitrage ties asset prices to each other. As we will see later, the absence of arbitrage also implies nice mathematical properties for asset prices which allow one to describe them by methods from stochastics, for example by martingales.

Often, however, the term "arbitrage" is used for a likely, but uncertain gain by an investment strategy. Now, forgetting about the motivations for trading like risk sharing and different time preferences, many people believe that the only reason to trade on financial markets would be to gain more than others, more precisely: to generate excess returns or "a positive Alpha".

Given that efficient markets are arbitrage-free, it is often argued that therefore such gains are not possible and hence trading on a financial market is useless: in any point of time the market has already incorporated all future opportunities. Thus, instead of cleverly weighing the pros and cons of various assets, one could also choose the assets at random, like in the famous monkey test, where a monkey throws darts on the Wall Street Journal to pick stocks and competes with investment professionals (see [Mal90]).

However, this point of view is wrong in two ways: first, it completely ignores the two other reasons for trading on financial markets, namely risk and time. Secondly, there is a distinction between an arbitrage-free market and one without any further opportunities for gains from trade returns. An efficient market, i.e. a market without any further gains from trade, must be arbitrage-free since arbitrage opportunities certainly give gains from trades. However, the converse is not true. Absence of arbitrage does not mean that you should not try to position yourself on the markets reflecting on your beliefs, time preferences and risk aversion.

Saying that investments could be chosen at random just because markets are *arbitrage-free* is like saying that when you go shopping in a shop without bargains, you can pick your goods at random. Just try to buy the ingredients for a tasty dinner in this way, and you will discover that this is not true.

There is another way of looking at this problem: If you consider the return distribution of your portfolio, forming asset allocations means to construct the return distribution that is most suitable for you. One motive for this may simply be controlling the risk of your initial portfolio, which could, e.g., be achieved by buying capital protection. Even though all possible portfolios would be arbitrage-free, the precise choice nevertheless matters to you.

Before we conclude this extremely important section we should mention how the notion of excess returns is related to the concepts of absence of arbitrage and no gains from trade. An excess return is a return higher than the risk-free rate. An excess return is usually no arbitrage opportunity since it carries some risks. Does it indicate gains from trade? In other words, should you buy assets that have excess returns? If you ought to buy or not depends on your risk preference relative to the risk the asset carries. For example, a positive alpha is an excess return that is attractive if your risk preference is to avoid variance and if your beliefs coincide with the average beliefs in the market. However, if one of these conditions is not met, an asset with positive alpha may not be a good choice, as we will see later.

1.2.3 Market Efficiency

The word "efficiency" has a double meaning in financial economics. One meaning – put forward by Fama – is that markets are efficient if prices incorporate all information. For example, paying analysts to research the opportunities and the risks of certain companies is worthless because the market has already priced the company reflecting all available information. To illustrate this view consider Fama and a pedestrian walking on the street. The pedestrian spots a 100 Dollar Bill and wants to pick it up. Fama, however, stops by saying if the 100 Dollar Bill were real, someone would have picked it up before.

The second meaning of efficiency is that efficient markets do not have any unexploited gains from trade. Thus the allocation obtained on efficient markets cannot be improved by raising the utility of one agent without lowering the utility of some other agent. This notion of efficiency is called Pareto-efficiency. Thus, whenever we refer to "efficiency" in our book, we will mean Pareto-efficiency.

1.2.4 Equilibrium

Economics is based on the idea of understanding markets from the interaction of optimizing agents. In a competitive equilibrium all agents trade in such a way as to achieve the most desirable consumption pattern, and market prices are such that all markets clear, i.e., in all markets demand is equal to supply.

Obviously, in a competitive equilibrium there cannot be arbitrage opportunities since otherwise no agent would find an optimal action. Exploiting the arbitrage more would drive the agent's utility to infinity and he would like to trade infinite amounts of the assets involved, which conflicts with market clearing. Note that the notion of equilibrium puts more restrictions on asset prices than mere no-arbitrage. Equilibrium prices reflect the relative scarcity of consumption in different states, the agents' beliefs of the occurrence of the states and their risk preferences. Moreover, in a complete market, at equilibrium there are no further gains from trade.

As a final remark on equilibrium one should note that for one given initial allocation there can be multiple equilibria. Which one is actually obtained may be a matter of exogenous factors like market sentiment or conventions. For example, stock returns could be high or low when the weather is extremely nice. Supposing that every trader believes in high stock returns when the weather is extremely nice, stock returns will turn out to be high because the agents' trades make this belief self-fulfilling. However it could also be the other way round, i.e., low returns when the weather is extremely nice.

In a financial market equilibrium the agents' beliefs determine the market reality and the market reality confirms agents' beliefs. In the words of George Soros [Sor98, page xxiii]:

> Financial markets attempt to predict a future that is contingent on the decisions people make in the present. Instead of just passively reflecting reality, financial markets are actively creating the reality that they, in turn, reflect. There is a two way connection between present decisions and the future events, which I call reflexivity.

1.2.5 Aggregation and Comparative Statics

Do we really need to know all agents' beliefs, risk attitudes and initial endowments in order to determine asset prices at equilibrium? The answer is "No", fortunately! If equilibrium prices are arbitrage free then they can be supported by a single decision problem in which one so-called "representative agent" optimizes his utility supposing he had access to all endowments. The equilibrium prices found in the competitive equilibrium can also be thought of as prices that induce a representative agent to demand total endowments.

For this trick to be useful one then needs to understand how the individual beliefs and risk attitudes aggregate into those of the representative agent. In the case of complete markets such aggregation rules can be found.

A final warning on the use of the representative agent methodology is in order. This method describes asset prices by some as-if decision problem. Hence it is constructed given the knowledge of the asset prices. It is not able to predict asset prices "out-of-sample", e.g., after some exogenous shock to the economy.

1.2.6 Time Scale of Investment Decisions

Investors differ in their time horizon, information processing and reaction time. Day traders for example make many investment decisions per day requiring fast information processing. Their reaction time is seconds long. Other investors have longer investment horizons (e.g., one or more years). Their investment decisions do not have to be made "just in time". A popular investment advice for investors with a longer investment horizon is: "Buy stocks and take a good long (20 years) sleep". Investors following this advice are expected to have a different perception of stocks as Benartzi and Thaler [BT95] make pretty clear with the following example:

> Compare two investors, Nick who calculates the gains and losses in his portfolio every day, and Dick who only looks at his portfolio once per decade. Since, on a daily basis, stocks go down in value almost as often as they go up, Nick's loss aversion will make stocks appear very unattractive to him. In contrast, loss aversion will not have much effect on Dick's perception of stocks since at ten year horizons stocks offer only a small risk of losing money.

Particularly important for an investment decision is the perception of the situation. In the words of a day trader, interviewed by the Wall Street Journal [Mos98], the situation is like this:

> Ninety percent of what we do is based on perception. It doesn't matter if that perception is right or wrong or real. It only matters that other people in the market believe it. I may know it's crazy, I may think it's wrong. But I lose my shirt by ignoring it. This business turns on decisions made in seconds. If you wait a minute to reflect on things, you're lost. I can't afford to be five steps ahead of everybody else in the market. That's suicide.

Thus, intraday price movements reflect how the average investor perceives incoming news. In the very long run price movements are determined by trends in fundamental data – like earnings, dividend growth and cash flows. A famous observation called excess volatility first made by Shiller [Shi81] is that stock prices fluctuate around the long term trend by more than economic fundamentals indicate. How the short run aspects get washed out in the long run, i.e., how aggregation of fluctuations over time can be modelled is rather unclear.

In this course we will consider three time scales: The short run (intraday market clearing of demand and supply orders), the medium run (monthly equalization of expectations) and the long run (yearly wealth dynamics).

1.2.7 Behavioral Finance

A rational investor should follow expected utility theory. However, it is often observed that agents do not behave according to this rational decision model. Since often important to understand actual investment behavior, the concepts of classical (rational) decision theory have often been replaced with a more descriptive approach that is labeled as "behavioral decision theory".

Its application to finance led to the emergence of "behavioral finance" as a subdiscipline. Richard Thaler once nicely defined what behavioral finance is all about [Tha93]:

> Behavioral finance is simply open-minded finance. [...] Sometimes in order to find a solution to an [financial] empirical puzzle it is necessary to entertain the possibility that some of the agents in the economy behave less than fully rational some of the time.

Whenever there is need to study deviations from perfectly rational behavior, we are already in the realm of behavioral finance. It is therefore quite obvious that a clear distinction of problems inside and outside behavioral finance is impossible: we will often be in situations where agents behave mostly rational, but not always, so that a simple model might be successful with only considering rational behavior, but behavioral "corrections" have to be made as soon as we take a closer look.

In this book we therefore aim to integrate behavioral views into classical theories to show how they can enhance our understanding of financial markets.

One particularly interesting behavioral model is Prospect Theory. It was developed by Daniel Kahneman and Amos Tversky [KT79] to describe decisions between risky alternatives. Prospect Theory departs from expected utility by showing the sensitivity of actual decisions to biases like framing, by using a valuation function that is defined on gains and losses instead of final wealth and by using non-linear probability when weighing the utility values obtained in various states. In particular Prospect Theory investors are loss averse, and they are risk averse when comparing two gains but risk seeking when comparing two losses. The question then is whether Prospect Theory is relevant for market prices. And indeed it is: many so-called asset pricing puzzles can be resolved with Prospect Theory. An example is the equity premium puzzle, i.e., the observation that stock returns are on average 6–7 % above the bond returns. This high excess return is hard to explain with plausible values for risk aversion, if one sticks to the expected utility paradigm. The idea of myopic loss aversion (Benartzi and Thaler [BT95]), the observation that investors have short horizons and are loss averse, can resolve the equity premium puzzle.

1.3 An Introduction to the Research Methods

We want to conclude this chapter by taking a look at the *research methods* that are used in financial economics. After all, we want to know where the results we are studying come from and how we can possibly add new results.

Albert Einstein is known to have said that "there is nothing more practical than a good theory." But what is a good theory? First of all, a good theory is based on observable assumptions. Moreover, a good theory should have testable implications – otherwise it is a religion which cannot be falsified. This falsification aspect cannot be stressed enough.[1] Finally, a good theory is a broad generalization of reality that captures its essential features. Note that a theory does not become better if it becomes more complicated.

But what are our observations and implications? There are essentially two ways to gather empirical evidence to support (or falsify) a theory on financial markets: one way is to study financial market data. Some of this data (e.g., stock prices) is readily available, some is difficult to obtain for reasons such as privacy issues or time constraints. The second way is to conduct surveys and laboratory experiments, i.e., to expose subjects to controlled conditions under which they have to perform financial decisions.

Both approaches have their advantages and limitations: market data is often noisy, depends on many uncontrollable factors and might not be available for a specific purpose, but by definition always comes from real life situations. Experimental

[1] Steve Ross, the founder of the econometric Arbitrage Pricing Theory (APT), for example, claims that "every financial disaster begins with a theory!" By saying this, he means that those who start trading based on a theory are less likely to react to disturbing facts because they are typically in love with their ideas. Falsification of their beloved theory is certainly not their goal!

The expected value is the average outcome of a lottery if played iteratively. It seems natural to use this value to decide when faced with a choice between two or more lotteries. In fact, this idea is so natural, that it was the only well-accepted theory for decisions under risk until the middle of the twentieth century. Even nowadays it is still the only one which is typically taught at high school, leaving many a student puzzled about the fact that "mathematics says that buying insurances would be irrational, although we all know it's a good thing". (In fact, a person who decides only based on the expected value would not buy an insurance, since insurances have negative expected values due to the simple fact that the insurance company has to cover its costs and usually wants to earn money and hence has to ask for a higher premium than the expected value of the insurance.)

But not only in high schools the idea of the expected value as the sole criterion for rational decision is still astonishingly widespread: when newspapers compare the performance of different pension funds, they usually only report the average return p.a. But what if you have enrolled into a pension fund with the highest average return over the past 100 years, but the average return over your working period was low? More general, what does the average return of the last year tell you about the average return in the next year?

The idea that rational decisions should only be made depending on the expected return was first criticized by Daniel Bernoulli in 1738 [Ber38]. He studied, following an idea of his cousin, Nicolas Bernoulli, a hypothetical lottery A set in a hypothetical casino in St. Petersburg which became therefore known as the "St. Petersburg Paradox". The lottery can be described as follows: After paying a fixed entrance fee, a fair coin is tossed repeatedly until a "tails" first appears. This ends the game. If the number of times the coin is tossed until this point is k, you win 2^{k-1} ducats (compare Fig. 2.2). The question is now: how much would you be willing to pay as an entrance fee to play this lottery?

If we follow the idea of using the expected value as criterion, we should be willing to pay an entrance fee up to this expected value. We compute the probability p_k that the coin will show "tail" after exactly k times:

$$\begin{aligned} p_k &= P(\text{"head" on 1st toss}) \cdot P(\text{"head" on 2nd toss}) \cdots \\ &\quad \cdots P(\text{"tail" on } k\text{-th toss}) \\ &= \left(\tfrac{1}{2}\right)^k. \end{aligned}$$

Now we can easily compute the expected return:

$$\mathbb{E}(A) = \sum_{k=1}^{\infty} x_k p_k = \sum_{k=1}^{\infty} 2^{k-1} \left(\frac{1}{2}\right)^k = \sum_{k=1}^{\infty} \frac{1}{2} = +\infty.$$

In other words, following the expected value criterion, you should be willing to pay an arbitrarily large amount of money to take part in the lottery. However, the probability that you win $1024 = 2^{10}$ ducats or more is less than one in a thousand

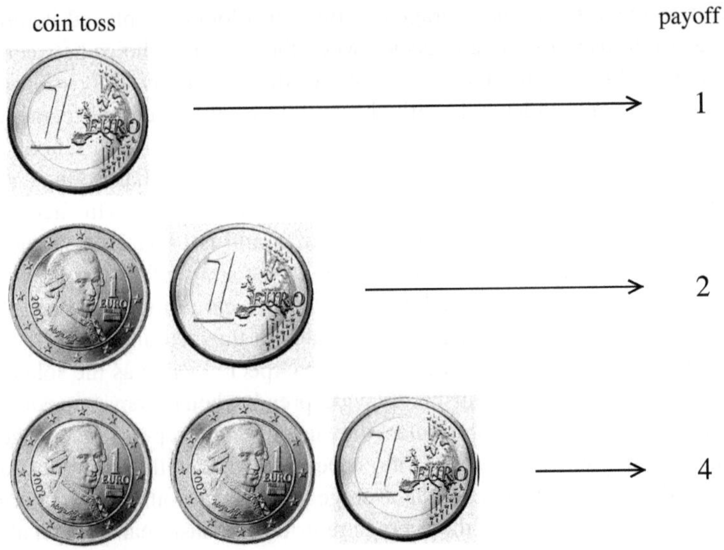

Fig. 2.2 The "St. Petersburg Lottery"

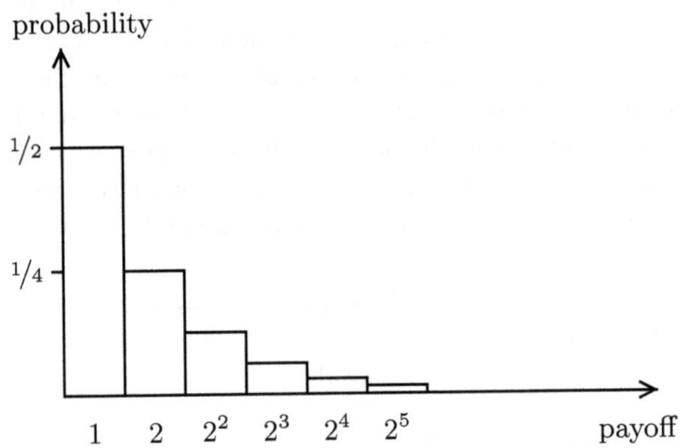

Fig. 2.3 The outcome distribution of the St. Petersburg Lottery

and the infinite expected value only results from the tiny possibility of extremely large outcomes. (See Fig. 2.3 for a sketch of the outcome distribution.) Therefore most people would be willing to pay not more than a couple of ducats to play the lottery. This seemingly paradoxical difference led to the name "St. Petersburg Paradox".

But is this really so paradoxical? If your car does not drive, this is not paradoxical (although cars are constructed in order to drive), but it needs to be checked, and probably repaired. If you use a model and encounter an application where it produces paradoxical or even plainly wrong results, then this model needs to be checked, and probably repaired. In the case of the St. Petersburg Paradox, the model was structured to decide according to the expected return. Now, Daniel Bernoulli noticed that this expected return might not be the right guideline for your choice, since it neglects that the same amount of money gained or lost might mean something very different to a person depending on his wealth (and other factors). To put it simple, it is not at all clear why twice the money should always be twice as good: imagine you win one billion dollars. I assume you would be happy. But would you be as happy about then winning another billion dollars? I do not think so. In Bernoulli's own words:

> There is no doubt that a gain of one thousand ducats is more significant to the pauper than to a rich man though both gain the same amount.

Therefore, it makes no sense to compute the expected value in terms of monetary units. Instead, we have to use units which reflect the usefulness of a given wealth. This concept leads to the *utility theory*, in the words of Bernoulli:

> The determination of the value of an item must not be based on the price, but rather on the utility [*"moral value"*] it yields.

In other words, every level of wealth corresponds to a certain numerical value for the person's utility. A utility function u assigns to every wealth level (in monetary units) the corresponding utility, see Fig. 2.4.[4] What we now want to maximize is the expected value of the utility, in other words, our utility functional becomes

$$U(p) = \mathbb{E}(u) = \sum_i u(x_i)p_i,$$

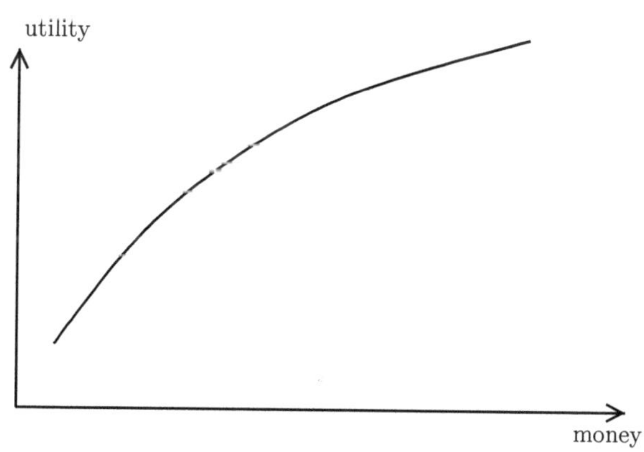

Fig. 2.4 A utility function

[4] We will see later, how to measure utility functions in laboratory experiments (Sect. 2.2.4), and how it is possible to deduce utility functions from financial market data (Sect. 4.6).

or in the continuum case

$$U(p) = \mathbb{E}(u) = \int_{-\infty}^{+\infty} u(x)\,\mathrm{d}p.$$

Since we will define other decision theories later on, we denote the Expected Utility Theory functional from now on by *EUT*.

Why does this resolve the St. Petersburg Paradox? Let us assume, as Bernoulli did, that the utility function is given by $u(x) := \ln(x)$, then the expected utility of the St. Petersburg lottery is

$$\begin{aligned} EUT(\text{Lottery}) = \sum_k u(x_k)p_k &= \sum_k \ln(2^{k-1})\left(\frac{1}{2}\right)^k \\ &= (\ln 2)\sum_k \frac{k-1}{2^k} < +\infty. \end{aligned}$$

This is caused by the "diminishing marginal utility of money", i.e., by the fact that $\ln(x)$ grows slower and slower for large x.

What other consequences do we get by changing from the classical decision theory (expected return) to the Expected Utility Theory (EUT)?[5]

Example 2.5 Let us consider a decision about buying a home insurance. There are basically two possible outcomes: either nothing bad happens to our house, in which case our wealth is diminished by the price of the insurance (if we decide to buy one), or disaster strikes, our house is destroyed (by fire, earthquake etc.) and our wealth gets diminished by the value of the house (if we do not buy an insurance) or only by the price of the insurance (if we buy one).

We can formulate this decision problem as a decision between the following two alternative lotteries A and B, where p is the probability that the house is destroyed, w is our initial wealth, v is the value of the house and r is the price of the insurance:

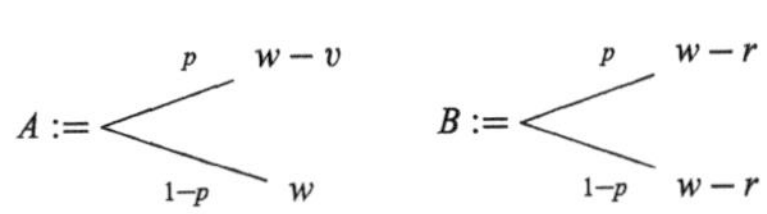

We can also display these lotteries as a table like this:

$A =$

Probability	$1-p$	p
Final wealth	w	$w-v$

$B =$

Probability	$1-p$	p
Final wealth	$w-r$	$w-r$

[5]EUT is sometimes called *Subjective Expected Utility Theory* to stress cases where the probabilities are subjective estimates rather than objective quantities. This is frequently abbreviated by SEU or SEUT.

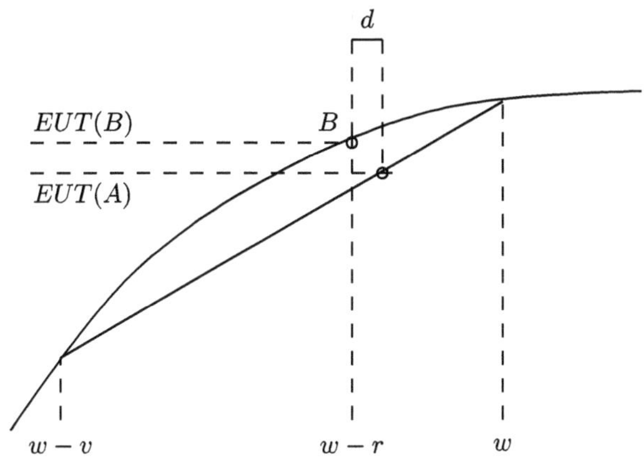

Fig. 2.5 The insurance problem

A is the case where we do not buy an insurance, in B if we buy one. Since the insurance wants to make money, we can be quite sure that $\mathbb{E}(A) > \mathbb{E}(B)$. The expected return as criterion would therefore suggest not to buy an insurance. Let us compute the expected utility for both lotteries:

$$EUT(A) = (1-p)u(w) + pu(w-v),$$
$$EUT(B) = (1-p)u(w-r) + pu(w-r) = u(w-r).$$

We can now illustrate the utilities of the two lotteries (compare Fig. 2.5) if we notice that $EUT(A)$ can be constructed as the value at $(1-p)v$ of the line connecting the points $(w-v, u(w-v))$ and $(w, u(w))$, since

$$EUT(A) = u(w-v) + (1-p)v\frac{u(w) - u(w-v)}{v}.$$

The expected profit of the insurance d is the difference of price and expected return, hence $d = r - pv$. We can graphically construct and compare the utilities for the two lotteries (see Fig. 2.5). We see in particular, that a strong enough concavity of u makes it advantageous to buy an insurance, but also other factors have an influence on the decision:

- If d is too large, the insurance becomes too expensive and is not bought.
- If w becomes large, the concavity of u decreases and therefore buying the insurance at some point becomes unattractive (assuming that v and d are still the same).
- If the value of the house v is large relative to the wealth, an insurance becomes more attractive.

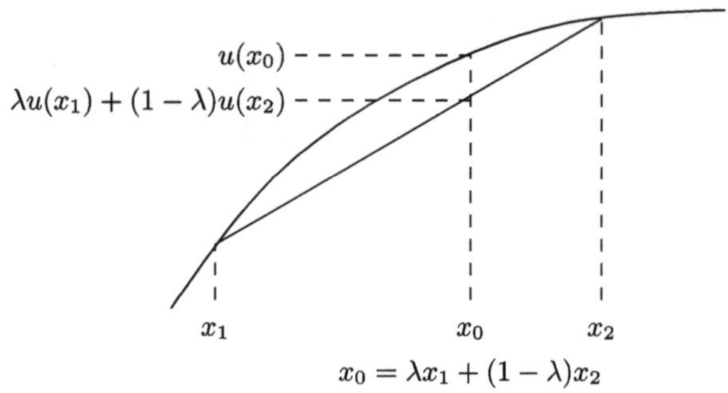

Fig. 2.6 A strictly concave function

We see that the application of Expected Utility Theory leads to quite realistic results. We also see that a crucial factor for the explanation of the attractiveness of insurances and the solution of the St. Petersburg Paradox is the concavity of the utility function. Roughly spoken, concavity corresponds to risk-averse behavior. We formalize this in the following way:

Definition 2.6 (Concavity) We call a function $u : \mathbb{R} \to \mathbb{R}$ *concave* on the interval (a, b) (which might be $\mathbb{R}$) if for all $x_1, x_2 \in (a, b)$ and $\lambda \in (0, 1)$ the following inequality holds:

$$\lambda u(x_1) + (1-\lambda)u(x_2) \le u\left(\lambda x_1 + (1-\lambda)x_2\right). \tag{2.1}$$

We call u *strictly concave* if the above inequality is always strict (for $x_1 \neq x_2$).

Definition 2.7 (Risk-averse behavior) We call a person *risk-averse* if he prefers the expected value of every lottery over the lottery itself.[6]

Formula (2.1) looks a little complicated, but follows with a small computation from Fig. 2.6. Analogously, we can define convexity and risk-seeking behavior:

Definition 2.8 (Convexity) We call a function $u : \mathbb{R} \to \mathbb{R}$ *convex* on the interval (a, b) if for all $x_1, x_2 \in (a, b)$ and $\lambda \in (0, 1)$ the following inequality holds:

$$\lambda u(x_1) + (1-\lambda)u(x_2) \ge u(\lambda x_1 + (1-\lambda)x_2). \tag{2.2}$$

We call u *strictly convex* if the above inequality is always strict (for $x_1 \neq x_2$).

[6]Sometimes this property is called "strictly risk-averse". "Risk-averse" then also allows for indifference between a lottery and its expected value. The same remark applies to risk-seeking behavior, compare Definition 2.9.

Definition 2.9 (Risk-seeking behavior) We call a person *risk-seeking* if he prefers every lottery over its expected value.

We have some simple statements on concavity and its connection to risk aversion.

Proposition 2.10 *The following statements hold:*

(i) If u is twice continuously differentiable, then u is strictly concave if and only if $u'' < 0$ and it is strictly convex if and only if $u'' > 0$. If u is (strictly) concave, then $-u$ is (strictly) convex.

(ii) If u is strictly concave, then a person described by the Expected Utility Theory with the utility function u is risk-averse. If u is strictly convex, then a person described by the Expected Utility Theory with the utility function u is risk-seeking.

To complete the terminology, we mention that a person which has an affine (and hence convex *and* concave) utility function is called *risk-neutral*, i.e., indifferent between lotteries and their expected return.

As we have already seen, risk aversion is the most common property, but one should not assume that it is necessarily satisfied throughout the range of possible outcomes. We will discuss these questions in more detail in Sect. 2.2.3.

An important property of utility functions is, that they can always be rescaled without changing the underlying preference relations. We recall that

$$U(x_1, \ldots, x_S) = \sum_{s=1}^{S} p_s u(x_s).$$

Then, U is fixed only up to monotone transformations and u only up to positive affine transformations:

Proposition 2.11 *Let $\lambda > 0$ and $c \in \mathbb{R}$. If u is a utility function that corresponds to the preference relation $\succ$, i.e., $A \succ B$ implies $U(A) \geq U(B)$, then $v(x) := \lambda u(x) + c$ is also a utility function corresponding to $\succeq$.*

For this reason it is possible to fix u at two points, e.g., $u(0) = 0$ and $u(1) = 1$, without changing the preferences. And for the same reason it is not meaningful to compare absolute values of utility functions across individuals, since only their preference relations can be observed, and they define the utility function only up to affine transformations. This is an important point that is worth having in mind when applying Expected Utility Theory to problems where several individuals are involved.

We have learned that Expected Utility Theory was already introduced by Bernoulli in the eighteenth century, but has only been accepted in the middle of the twentieth century. One might wonder, why this took so long, and why this mathematically simple method has not quickly found fruitful applications. We can only speculate what might have happened: mathematicians at that time felt a certain dismay to the muddy waters of applications: they did not like utility functions whose precise form could not be derived from theoretical considerations. Instead they believed in the unique validity of clear and tidy theories. And the mean value was such a theory.

Whatever the reason, even in 1950 the statistician Feller could still write in an influential textbook [Fel50] on Bernoulli's approach to the St. Petersburg Paradox that he "tried in vain to solve it by the concept of moral expectation." Instead Feller attempted a solution using only the mean value, but could ultimately only show that the *repeated* St. Petersburg Lottery is asymptotically fair (i.e., fair in the limit of infinite repetitions) if the entrance fee is $k \log k$ at the k-th repetition. This implies of course that the entrance fee (although finite) is unbounded and tends to infinity in the limit which seems not to be much less paradoxical than the St. Petersburg Paradox itself. Feller was not alone with his criticism: W. Hirsch writes about the St. Petersburg Paradox in a review on Feller's book:

> Various mystifying "explanations" of this paradox had been offered in the past, involving, for example, the concept of moral expectation... These explanations are hardly understandable to the modern student of probability.

The discussion in the 1960s even became at times a dispute with slight "patriotic" undertones; for an entertaining reading on this, we refer to [JB03, Chapter 13].

At that time, however, the ideas of von Neumann and Morgenstern (that originated in their book written in 1944 [vNM53]) finally gained popularity and the Expected Utility Theory became widely accepted.

The previous discussions seem to us nowadays more amusing than comprehensible. We will speculate later on some reasons why the time was ripe for the full development of the EUT at that time, but first we will present the key insights of von Neumann and Morgenstern, the axiomatic approach to EUT.

2.2.2 Axiomatic Definition

When we talk about "rational decisions under risk", we usually mean that a person decides according to Expected Utility Theory. Why is there such a strong link between rationality and EUT? However convincing the arguments of Bernoulli are, the main reason is a very different one: we can derive EUT from a set of much simpler assumptions on an individual's decisions. Let us start to compose such a list:

First, we assume that a person should always have *some* opinion when deciding between two alternatives. Whether the person prefers A over B or B over A or whether the person is indecisive, does not matter. But one of these should always be the case. Although this sounds trivial, it might well be that in some context this condition is violated, in particular when moral issues are involved. Generally, and

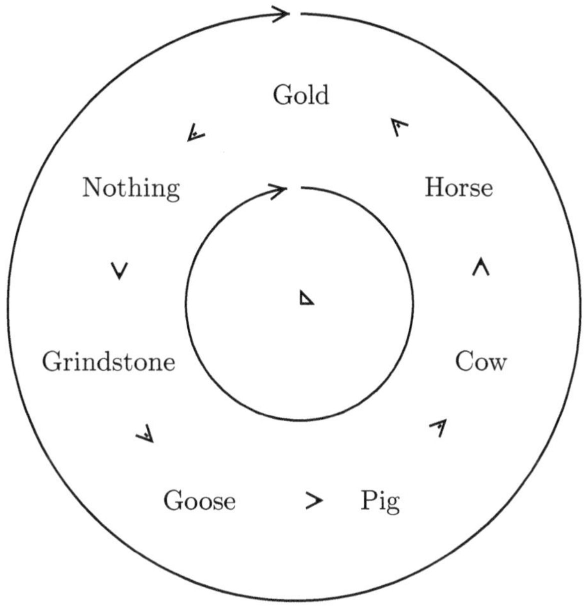

Fig. 2.7 The cycle of the "Lucky Hans", violating transitivity

in particular when only financial matters are involved, this condition is indeed very natural. We formulate it as our first *axiom*, i.e., a fundamental assumption on which our later analysis can be based:

Axiom 2.12 (Completeness) *For every pair of possible alternatives, A, B, either* $A \prec B$, $A \sim B$ *or* $A \succ B$ *holds.*

It is easy to see that EUT satisfies this axiom as long as the utility functional has a finite value.

The next idea is that we should have consistent decisions in the following sense: If we prefer B over A and C over B, then we should prefer C over A. This idea is called "transitivity". In the fairy tale "Lucky Hans" by the Brothers Grimm, this property is violated, as Lucky Hans happily exchanges a lump of solid gold, that he had earned for 7 years of hard work, for a horse, because the gold is so heavy to carry. Afterwards he exchanges the horse for a cow, the cow for a pig, the pig for a goose, and the goose finally for two knife grinder stones which he then accidentally throws into a well. But he is very happy about this accident, since the stones were so heavy to carry… At the end of the tale he has therefore the same that he had 7 years before – nothing. But nevertheless each exchange seemed to make him happy (Fig. 2.7).

In mathematical terms, "Lucky Hans" preferred B over A, C over B and A over C. Although we might not be blessed with such a cheerful nature, we have to accept that the behavior of some people can be very strange indeed and that the assumption of transitivity might be already too much to describe individuals. However, persons like "Lucky Hans" are probably quite an exception, and the fairy tale would not have its humorous effect if the audience considered such a transitivity-violating behavior normal. We can therefore feel quite safe by applying this principle, in particular in a prescriptive context.

Axiom 2.13 (Transitivity) *For every A,B,C with $A \preceq B$ and $B \preceq C$, we have $A \preceq C$.*

Transitivity is satisfied by EUT and by all other theories that are based on a utility functional, since for these decision theories, transitivity translates into transitivity of real numbers which is always satisfied.

The properties up to now could have been stated for preferences between apples and pears or for whatever one might wish to decide about. It was by no means necessary that the objects under considerations were lotteries. We will now focus on decision under risk, since the following axioms require more detailed properties of the items we wish to compare.

The next axiom is more controversial than the first two. We argue as follows: if we have to choose between two lotteries which are partially identical, then our decision should only depend on the difference between the two lotteries, not on the identical part. We illustrate this with an example:

Example 2.14 Let us assume that we decide about buying a home insurance.There are two insurances on the market that cost the same amount of money and pay out the same amount in case of a damage, but one of them excludes damages by floods and the other one excludes damages by storm. Moreover both insurances exclude damages induced by earthquakes.

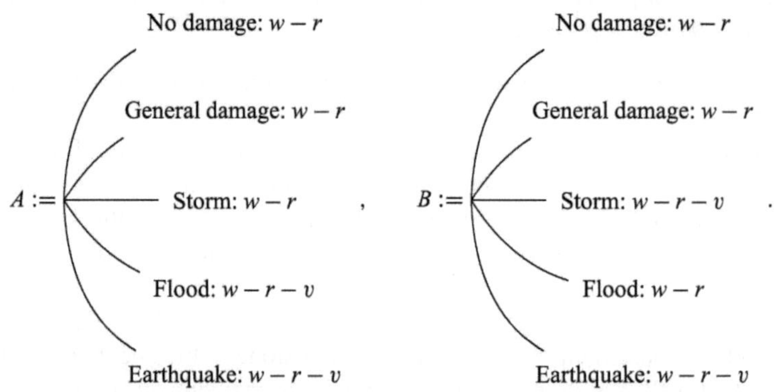

If we decide which insurance to buy, we should make our decision without considering the case of an earthquake, since this case (probability and costs) is identical for both alternatives and hence irrelevant for our decision.

Although the idea to ignore irrelevant alternatives sounds reasonable, it turns out not to be very consistent with experimental findings. We will discuss this when we study descriptive approaches like Prospect Theory in Sect. 2.4. For now, we can happily live with this assumption, since we are more interested in rational decisions, in other words we follow a prescriptive approach.

To formulate this axiom mathematically correctly, we need to understand what it means when we combine lotteries.

Definition 2.15 Let A and B be lotteries and $\lambda \in [0,1]$, then $\lambda A + (1-\lambda)B$ denotes a new combined lottery where with probability λ the lottery A is played, and with probability $1-\lambda$ the lottery B is played.[7]

Example 2.16 Let A and B be the following lotteries:

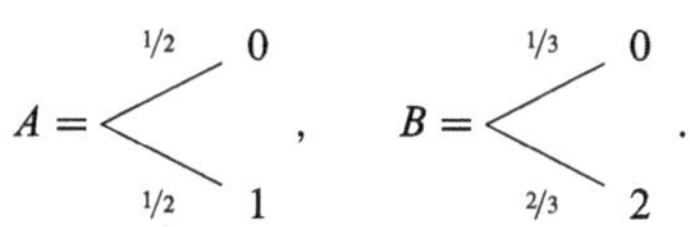

Then the lottery $C := \lambda A + (1-\lambda)B$ can be calculated as

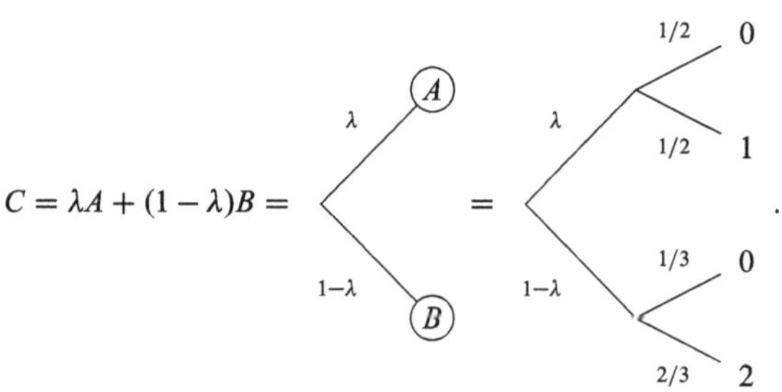

Alternatively, we can do the same calculation by representing the lottery in a table:

$$A = \begin{array}{lcc} \hline \text{Probability} & 1/2 & 1/2 \\ \text{Outcome} & 0 & 1 \\ \hline \end{array}, \quad B = \begin{array}{lcc} \hline \text{Probability} & 1/3 & 2/3 \\ \text{Outcome} & 0 & 2 \\ \hline \end{array}.$$

Then the lottery $C := \lambda A + (1-\lambda)B$ is

$$C = \lambda A + (1-\lambda)B \;=\; \begin{array}{cc} \hline \lambda & 1-\lambda \\ A & B \\ \hline \end{array} \;=\; \begin{array}{cccc} \hline \multicolumn{2}{c}{\lambda} & \multicolumn{2}{c}{1-\lambda} \\ \hline 1/2 & 1/2 & 1/3 & 2/3 \\ 0 & 1 & 0 & 2 \\ \hline \end{array}$$

Both formulations lead to the same result, it is basically a matter of taste whether we write lotteries as tree diagrams or tables. The Independence Axiom allows us now

[7]If the lotteries are given as probability measures, then the notation coincides with the usual algebraic manipulations of probability measures.

to collect compound lotteries into a single lottery, i.e.

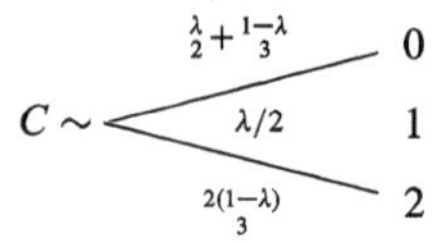

$$\text{or} \quad C \sim \begin{array}{ccc} \frac{\lambda}{2}+\frac{1-\lambda}{3} & \frac{\lambda}{2} & \frac{2(1-\lambda)}{3} \\ 0 & 1 & 2 \end{array}$$

A mathematically precise formulation of the Independence Axiom reads as follows:

Axiom 2.17 (Independence) *Let A and B be two lotteries with $A \succ B$, and let $\lambda \in (0, 1]$ then for any lottery C, it must hold*

$$\lambda A + (1 - \lambda)C \succ \lambda B + (1 - \lambda)C.$$

To see that EUT satisfies the Independence Axiom is not so obvious anymore, but the proof is not very difficult. To keep things simple, we assume that the lotteries A, B and C have only finitely many outcomes $x_1, \ldots, x_n$. (A general proof is given in Appendix A.6.) The probability to get the outcome x_i in lottery A is denoted by p_i^A. Analogously, we write p_i^B and p_i^C. We compute

$$\begin{aligned} U(\lambda A + (1-\lambda)C) &= \sum_{i=1}^n \left(\lambda p_i^A + (1-\lambda)p_i^C\right) u(x_i) \\ &= \lambda \sum_{i=1}^n p_i^A u(x_i) + (1-\lambda) \sum_{i=1}^n p_i^C u(x_i) \\ &= \lambda U(A) + (1-\lambda)U(C) \\ &> \lambda U(B) + (1-\lambda)U(C) \\ &= \lambda \sum_{i=1}^n p_i^B u(x_i) + (1-\lambda) \sum_{i=1}^n p_i^C u(x_i) \\ &= U(\lambda B + (1-\lambda)C). \end{aligned}$$

The last axiom we want to present is the so-called "Continuity Axiom"[8]: let us consider three lotteries A, B, C, where we prefer A over B and B over C. Then there should be a way to mix A and C such that we are indifferent between this mix and B. In a precise formulation, valid for finite lotteries[9]:

Axiom 2.18 (Continuity) *Let A, B, C be lotteries with $A \succeq B \succeq C$ then there exists a probability p such that $B \sim pA + (1 - p)C$.*

[8]Sometimes this is also called "Archimedian Axiom".

[9]In order to make this concept work for non-discrete lotteries, one needs to take a slightly more complicated approach. We give this general definition in Appendix A.6.

One might argue whether this axiom is natural or not, but at least for financial decisions this seems to be a very reasonable assumption. Again, it is not very difficult to see that EUT satisfies the Continuity Axiom. The proof for this is left as an exercise.

Why did we define all these axioms? We have seen that EUT satisfies them (sometimes under little additional conditions like continuity of u), but the reason why they are interesting is a different one: if we don't know anything about a system of preferences, besides that it satisfies these axioms, then they can be described by Expected Utility Theory! This is quite a surprise, since at first glance the definition of EUT as given by Bernoulli seemed to be a very special and concrete concept, but preference relations and the axioms we studied seem to be very general and abstract. Now, both approaches – the direct definition based on economic intuition and the careful, very general approach based only on a small list of natural axioms – lead exactly to the same concept. This was the key insight by Morgenstern and von Neumann [vNM53]. Therefore, utility functions in EUT are often called "von Neumann-Morgenstern utility functions".

We formulate this central result in the following Theorem that does not follow precisely the original formulation by von Neumann and Morgenstern, but is nowadays the most commonly used version of their result.

Theorem 2.19 (Expected Utility Theory) *A preference relation that satisfies the Completeness Axiom 2.12, the Transitivity Axiom 2.13, the Independence Axiom 2.17 and the Continuity Axiom 2.18, can be represented by an EUT functional. EUT always satisfies these axioms.*

Proof Since the result is so central, we give a sketch of its proof. However, the mathematically inclined reader might want to venture into the realms of Appendix A.6, where the complete proof together with some generalizations (in particular to lotteries with infinite outcomes) is presented.

First, we notice that the (simpler) half of the proof is already done: We have already checked that preference relations which are described by the Expected Utility Theory satisfy all of the listed axioms. What remains is to prove that if these axioms are satisfied, a von Neumann-Morgenstern utility function exists.

Let us consider lotteries with finitely many outcomes $x_1, \dots, x_n$ with $x_1 > x_2 > \dots > x_n$. A sure outcome of x_i can be replaced by a lottery having only the two outcomes x_1 and x_n with some probability q_i and $(1 - q_i)$, as we know from the Continuity Axiom. In other words:

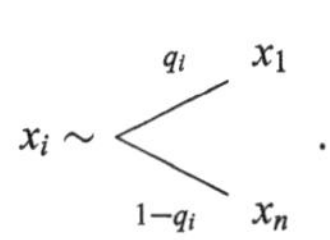

If we have an arbitrary lottery A with outcomes $x_1, \dots, x_n$, each of probability $p_1^A, \dots, p_n^A$, then we can use the Independence Axiom to substitute first the single

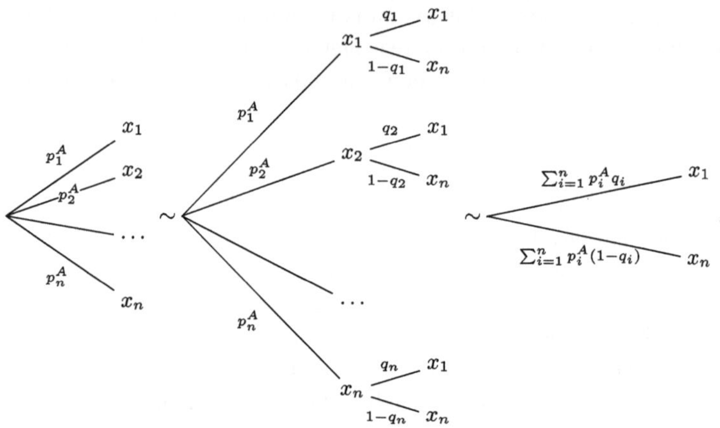

Fig. 2.8 Compound lottery

outcomes by lotteries in x_1 and x_n (using the above equivalence) and then collecting the new lottery into a compound lottery, shown in Fig. 2.8.

If we want to compare two lotteries A and B, we transform them both in this way to get equivalent lotteries A' and B'. Then it becomes very easy for us to decide which lottery is the best: we simply prefer A' over B' if the probability of A' having the better outcome (x_1 or x_n) is larger. To fix ideas, let us assume that x_1 is preferred over x_n, then we just need to compare $U(A) := \sum_{i=1}^n p_i^A q_i$ with $U(B) := \sum_{i=1}^n p_i^B q_i$: if $U(A) > U(B)$, than we prefer A over B; if $U(B) > U(A)$, then the other way around. Now we can define a utility function u in such a way, that its Expected Utility for any lottery A becomes $U(A)$: simply define $u(x_i) := q_i$, then

$$EUT(A) = \sum_{i=1}^n p_i^A u(x_i) = \sum_{i=1}^n p_i^A q_i. \qquad \square$$

Since we convinced ourselves that the listed axioms are all very reasonable, and we tend to say that a *rational* person should obey them, we can conclude that EUT is in fact a good *prescriptive* theory for decisions under risk. However, we have to assume that the utility function considers all relevant effects. Not in all situations the monetary amounts involved are the only relevant effect. Other effects could be based on moral standards, social acceptance etc. EUT as a prescriptive model will work the better the smaller the influence of such factors are that cannot readily be included into the definition of the utility function.

Whether it is also adequate to model behavior of people in real life is an entirely different question, and it will turn out that there are some discrepancies that lead to the development of new *descriptive* theories.

Coming back for a moment to the question, why it took more than 200 years for the development of Expected Utility Theory, a look at other sciences, and in particular mathematics can help us. In fact, the approach by von Neumann and Morgenstern follows a concept that had been used in mathematics intensely at the beginning of the twentieth century and can be summarized as the "axiomatic method": starting from some fundamental and simple axioms one tries to derive complex theories. Mathematicians stopped accepting objects like the real numbers and merely working with them, but instead developed methods to construct them from simple basic axioms: the natural numbers from some axioms on sets, the rational numbers as fractions of natural numbers, the real numbers as limits of rational numbers and so forth. This was the method that was waiting to be applied to the problems in decision theory under risk. There was also a strong input from psychology which understood at this time that the elementary object of decisions is the *preference* between objects. Von Neumann and Morgenstern (and together with them other scientists who, around the same time, derived similar models) took this as their starting point and used the axiomatic method from mathematics to derive a solid foundation for rational decisions under risk.

We can now even go a step further and say that the results of von Neumann and Morgenstern enable us to avoid any interpretation of the meaning of "utility". We may not have means to measure a person's utility, but we do not need to, since it just provides a useful mathematical concept of capturing the person's preference (which we can observe quite well). We don't even have to feel bad about using this mathematically convenient framework, since we have proved that it is not so much of an extra assumption, but a natural consequence of reasonable behavior.

To phrase this idea differently: we have at hands two complementary ways of understanding what the Expected Utility Theory is. Summarizing them will help to remember the core ideas of the theory much more than remembering the formula:

- First, we can use Bernoulli's idea of the utility function that assigns a "real" value to a given amount of money.[10] If we are faced with a decision under risk, we should use the expected value *of this utility* as a natural method to find the more advantageous alternative. This leads to the formula

$$EUT(A) = \mathbb{E}(u(A))$$

  for the expected utility of a lottery A.
- Second, we can neglect any potential deep meaning of the utility functions and consider them merely as a convenient and feasible (in realistic situations as defined by the axioms of this section) way of describing the preferences of a rational person.

[10]This approach has recently found a revival in the works of Kahneman and others, compare [KDS99].

The precise definition is made in a way that the utility of a lottery A can be computed as convex combination of the utilities of the various outcomes, weighted by their respective probabilities. If these outcomes are x_i and their probabilities are p_i, then this leads to the formula

$$EUT(A) = \sum_{i=1}^{n} u(x_i)p_i,$$

respectively the generalization to non-discrete probability measures

$$EUT(A) = \int u(x)\,\mathrm{d}p.$$

As we have seen, both approaches lead to the same result.

Looking back on the theory we have derived so far, we are now left with a different, very practical question: we know that we should use EUT with a monotone and continuous utility function u to model rational decisions under risk, but there are plenty of monotone and continuous functions – actually infinitely many. So, which one should we choose? Are there any further axioms that could guide us to select the right one?

2.2.3 Which Utility Functions Are "Suitable"?

We have seen that Expected Utility Theory describes a rational person's decisions under risk. However, we still have to choose the utility function u in an appropriate way. In this section we will discuss some typical forms of the utility function which have specific properties.

We have already seen that a reasonable utility function should be continuous and monotone increasing, in order to satisfy all axioms introduced in the last section. We have also already discussed that the concavity respectively convexity of the utility function corresponds to risk-averse respectively risk-seeking behavior. It would be nice if one could derive a quantitative measurement for the degree of risk aversion (or risk-seeking) of a person. Since convexity and concavity are characterized by the second derivative of a function (Proposition 2.10), a naive indicator would be this second derivative itself. However, we have seen that utility functions are only characterized up to an affine transformation (Proposition 2.11) which would change the value of u''. A way to avoid this problem is the standard risk aversion measure, $r(x)$, first introduced by J.W. Pratt [Pra64], which is defined as

$$r(x) := -\frac{u''(x)}{u'(x)}.$$

The larger r, the more a person is risk-averse. Assuming that u is monotone increasing, values of r smaller than zero correspond to risk-seeking behavior, values above zero correspond to risk-averse behavior.

What is the interpretation of r? The most useful property of r is that it measures how much a person would pay for an insurance against a fair bet. We formulate this as a proposition and give a proof for the mathematical inclined reader:

Proposition 2.20 *Let p be the outcome distribution of a lottery with $\mathbb{E}(p) = 0$, in other words, p is a fair bet. Let w be the wealth level of the person, then, neglecting higher order terms in $r(w)$ and p,*

$$EUT(w + p) = u\left(w - \frac{1}{2}\operatorname{var}(p)r(w)\right),$$

where $\operatorname{var}(p)$ denotes the variance of p. We could say that the "risk premium", i.e., the amount the person is willing to pay for an insurance against a fair bet, is proportional to $r(w)$.

Proof We denote the risk premium by a and get $EUT(w + p) = u(w - a)$. Using $EUT(w + p) = \mathbb{E}(u(w + p))$ and a Taylor expansion on both sides, we obtain

$$\mathbb{E}(u(w)) + \mathbb{E}(pu'(w)) + \mathbb{E}\left(\tfrac{1}{2}p^2u''(w)\right) + \mathbb{E}\left(O(p^3)\right)$$
$$= u(w) - au'(w) + O(a^2).$$

(Here O is the so-called *Landau symbol*, this means that $O(f(x))$ is a term which is asymptotically less or equal to $f(x)$.)

Using $\mathbb{E}(p) = 0$, we get

$$-\tfrac{1}{2}\operatorname{var}(p)u''(w) = au'(w) - O(\mathbb{E}(p^3)) - O(a^2)$$

and finally $\frac{1}{2}\operatorname{var}(p)r(w) = -a - O(\mathbb{E}(p^3)) + O(a^2)$. □

This result is particularly of interest, since it connects insurance premiums with a risk aversion measure, and the former can easily be measured from real life data.

What values can we expect for r? Looking at the problems we have studied so far – the St. Petersburg Paradox and insurances – it is natural to assume that risk aversion is the predominating property. However, there are situations in which people behave in a risk-seeking way:

Example 2.21 Lotteries are popular throughout the world. A typical example is the biggest German lottery, the "Lotto" with a turnover of about 25 Million Euro per draw. A lottery ticket of this lottery costs 0.75€ and the chances of winning a major prize (typically in the one million Euro range) are just 0.0000071 %. The chances of not getting any prize are 98.1 %. Only 50 % of the money spent by the participants is redistributed, the other half goes to the state and to welfare organizations.

Without knowing any more details, it is possible to deduce that a risk-averse or risk-neutral person should not participate in this lottery. Why? To prove our claim, we use the *Jensen inequality*:

Theorem 2.22 (Jensen inequality) *Let* $f: [a, b] \to \mathbb{R}$ *be a convex function, let* $x_1, \ldots, x_n \in [a, b]$ *and let* $a_1, \ldots, a_n \geq 0$ *with* $a_1 + \cdots + a_n = 1$. *Then*

$$f\left(\sum_{i=1}^{n} a_i x_i\right) \leq \sum_{i=1}^{n} a_i f(x_i).$$

If f *is instead concave, the inequality is flipped.*

We assume that you have encountered a proof of this inequality before, otherwise you may have a look into a calculus textbook. We refer the advanced reader to Appendix A.4 where we give a general form of Jensen's inequality that allows to generalize our results to non-discrete outcome distributions.

Let us now see, how this inequality can help us prove our statement on lotteries:

We choose as function f the utility function u of a person and assume that u is concave, corresponding to a risk-averse or at least risk-neutral behavior. We denote the lottery with L. The outcomes of L (prizes *plus* the initial wealth of the person *minus* the price of the lottery ticket) are denoted by x_i, their corresponding probabilities by a_i.

Jensen's inequality now tells us that

$$u(\mathbb{E}(L)) = u\left(\sum_{i=1}^{n} a_i x_i\right) \geq \sum_{i=1}^{n} a_i u(x_i) \quad = \quad EUT(L).$$

In other words: the utility of the expected return of the lottery is at least as good as the expected utility of the lottery. Now we know that only 50 % of the raised money are redistributed to the participants, in other words, to participate we have to pay twice the expected value of the lottery. Now since $u(2\mathbb{E}(L)) > u(\mathbb{E}(L))$, we conclude that a rational risk-averse or risk-neutral person should not participate in the lottery.

The fact that many people are nevertheless participating is a phenomenon that cannot be too easily explained, in particular since the same persons typically own insurances against various risks (which can only be explained by assuming risk-averse preferences).

A possible explanation is that their utility functions are concave for low values of money, but become convex for larger amounts. This could also explain why other games of chance, like roulette, that allow only for limited prizes, are by far less popular than big lotteries. One could argue that the marginal utility a person derives from a loss or gain of one Euro is not very high, but by increasing the wealth above a certain threshold, the marginal utility could grow. For instance, by winning one million Euro, a person could be free to stop working or move to a nice and otherwise never affordable house. Although we will see more convincing non-rational explanations of this kind of behavior later, we see that assuming that risk attitudes should follow a standard normalized pattern may not be a very convincing interpretation. We could also think of a more extreme example, taken from a movie:

Example 2.23 In the movie "Run Lola Run", Mannie, a wanna-be criminal, is supposed to deliver 100, 000 Deutsch Marks (50, 000€) to his new boss, but loses them on the way. Mannie and his girlfriend Lola have 20 min left to get the money somehow from somewhere, otherwise the boss is going to end Mannie's career, probably in a fatal way. Unfortunately, they are more or less broke.

The utility function for them will obviously be quite special: above a wealth level of 50, 000€ everything is fine (large utility), below that, everything is bad (low utility). It is therefore very likely that their utility function will not be concave. In the movie they are faced with the possibilities of robbing a grocery store, robbing a bank, or gambling in roulette in a casino to earn their money quickly. All three options are obviously very risky and reveal their highly risk-seeking preferences. However, advising them to put the little money they have on a bank account does not seem to be a very rational and helpful suggestion.

We conclude that there are no convincing arguments in favor of a specific risk attitude, other than that risk-averse behavior seems to be reasonable for very large amounts of money, as the St. Petersburg Paradox has taught us. Nevertheless, it is often convenient to do so, and one might argue that "on average" one or the other form could be a reasonable assumption.

One such standard assumption is that the risk aversion measure r is constant for all wealth levels. This is called *Constant Absolute Risk Aversion*, short: CARA. An example for such a CARA utility function is

$$u(x) :- -\mathrm{e}^{-Ax}.$$

We can verify this by computing $r(x)$ for this function:

$$r(x) = -\frac{u''(x)}{u'(x)} = \frac{A^2\mathrm{e}^{-Ax}}{A\mathrm{e}^{-Ax}} = A.$$

Realistic values of A would be in the magnitude of $A \approx 0.0001$.

Since it seems unlikely that risk attitudes are independent of a person's wealth, another standard approach suggests that $r(x)$ should be proportional to x. In other words, the *relative risk aversion*

$$rr(x) := xr(x) = -x\frac{u''(x)}{u'(x)}$$

is assumed to be constant for all x. We call such function *constant relative risk averse*, short: CRRA. Examples for such functions are

$$u(x) := \frac{x^R}{R}, \quad \text{where } R < 1,\ R \neq 0,$$

and

$$u(x) := \ln x.$$

Setting $R := 0$ for $\ln x$, we get $rr(x) = 1 - R$ for all of these functions. Typical values for R that have been measured are between -1 and -3, i.e., an appropriate utility function could be

$$u(x) := -\frac{1}{2}x^{-2}.$$

A subclass of these functions are probably the most widely used utility functions $u(x) := x^\alpha$ with $\alpha \in (0,1)$. These functions seem to be popular mostly for the sake of mathematical convenience: everybody knows their derivatives and how to integrate them. They are also strictly concave and correspond therefore to risk averse behavior which is often the only condition that one needs for a given application. – In other words, they are the perfect pragmatic solution to define a utility function. But please do not walk away with the idea that these functions are the *only natural* or the *only reasonable* or the *only rational* choice for a utility function! We have seen that things are not as easy and there is in fact no good reason other than convenience to recommend the utility function $u(x) = x^\alpha$.

A generalization of the classes of utility functions introduced so far are utility functions with *hyperbolic absolute risk aversion* (HARA). This class is defined as all functions where the *reciprocal of absolute risk aversion*, $T := 1/r(x)$, is an affine function of x. In other words: u is a HARA function if $T := -u'(x)/u''(x) = a + bx$ for some constants a, b. There is a classification of HARA functions by Merton [MS92]:

Proposition 2.24 *A function $u\colon \mathbb{R} \to \mathbb{R}$ is HARA if and only if it is an affine transformation of one of these functions:*

$$v_1(x) := \ln(x+a), \quad v_2(x) := -a\mathrm{e}^{-x/a}, \quad v_3(x) := \frac{(a+bx)^{(b-1)/b}}{b-1},$$

where a and b are arbitrary constants ($b \notin \{0,1\}$ for v_3). If we define $b := 1$ for v_1 and $b := 0$ for v_2, we have in all three cases $T = a + bx$.

It is now easy to see that HARA utilities include logarithmic, exponential and power utility functions. (we give an overview in Table 2.1.) Of course, by definition,

Table 2.1 Important classes of utility functions and some of their properties. All belong to the class of HARA functions

Class of utilities	Definition	ARA $r(x)$	RRA $rr(x)$	Special properties
Logarithmic	$\ln(x)$	decr.	const.	"Bernoulli utility"
Power	$\frac{1}{\alpha}x^{\alpha}, \alpha \neq 0$	decr.	const.	Risk-averse if $\alpha < 1$, bounded if $\alpha < 0$
Quadratic	$x - \alpha x^2, \alpha > 0$	incr.	incr.	Bounded, monotone only up to $x = \frac{1}{2\alpha}$
Exponential	$-e^{-\alpha x}, \alpha > 0$	const.	incr.	Bounded

they contain all CARA and CRRA functions. (v_2 is CARA and v_1 and v_3 for $a = 0$ give all CRRA functions.) To assume that a utility function has to belong to the HARA class is therefore certainly an improvement over more specific ad hoc assumptions, like risk-neutrality. It is, however, only a mathematically convenient simplification. We should not forget this fact, when we use EUT.

Unfortunately, it is not uncommon to read of one or the other class of utility functions as being the only reasonable class. Be careful when encountering such statements! Big minds have erred in such questions: take Bernoulli as an example, who suggested a particular CRRA function (the logarithm) as utility function. He argued that it would be reasonable to assume that the marginal utility of a person is inversely proportional to his wealth level. In modern mathematical terminology $u'(x) \sim 1/x$. Integrating this differential equation, we arrive at the logarithmic function that Bernoulli used to explain the St. Petersburg Paradox. However, is this utility function really so reasonable?

Let us go back to the St. Petersburg Paradox and see whether the solution Bernoulli suggested is really sufficient. Can we make the paradox reappear if we change the lottery? Yes, we can: we just need to change the payoffs to the (even larger) value of e^{2^k}. Then with $u(x) := \ln(x)$ (Bernoulli's suggestion), we get $u(x_k) = \ln(e^{2^{k-1}}) = 2^{k-1}$ and the same computation as in the case of the original paradox now proves that the expected utility of the new lottery is infinite:

$$EUT = \sum_k u(x_k)p_k = \sum_k 2^{k-1}\left(\frac{1}{2}\right)^k = \sum_k \frac{1}{2} = +\infty.$$

More generally, one can find a lottery that allows for a variant of the St. Petersburg paradox for *every unbounded utility function*, as was first pointed out by Menger [Men34].

There are basically two ways of solving this new paradox, which is sometimes called the "Super St. Petersburg Paradox". We can understand them, like in the case of the original St. Petersburg Paradox, by comparing the decision theory with a car. If your car does not drive, this might basically be due to two factors: either something is wrong with the car (e.g., no fuel, engine broken…) or something is wrong with the place where you try to drive it (e.g., you are stuck on an icy road).

In the case of a model that could mean that there is either something wrong with the model that needs to be fixed or that you try to apply it at a wrong place, in other words you encountered a restriction to its applicability. In the case of the "Super St. Petersburg Paradox" that leaves us with two ways out:

- We can assume an upper bound on the utility function, take for example $u(x) = 1 - \mathrm{e}^{-x}$ which is bounded by 1. In this case, every lottery has an expected utility of less than 1, and therefore there is a finite amount of money that corresponds to this utility value.
- We can try to be a little bit more realistic in the setting of our original paradox, and take into account that a casino would only offer lotteries with a finite expected value, in order to be able to earn money by asking for an entrance fee above this value. Under this restriction, one can prove that the St. Petersburg paradox disappears as long as the utility function is asymptotically concave (i.e., concave above a certain value) [Arr74].

In the second case, we restricted the range of applicable situations ("a car does not drive well on icy roads, so avoid them"). In the first case, we fixed our model to cover even these extreme situations ("always have snow chains with you").

We formulate this as a theorem:

Theorem 2.25 (St. Petersburg Lottery) *Let p be the outcome distribution of a lottery. Let $u\colon \mathbb{R} \to \mathbb{R}$ be a utility function.*

(i) If u is bounded, then $EUT(p) := \int u(x)\,\mathrm{d}p < \infty$.
(ii) Assume that $\mathbb{E}(p) < \infty$. If u is asymptotically concave, i.e., there is a $C > 0$ such that u is concave on the interval $[C, +\infty)$, then $EUT(p) < \infty$.

It is difficult to decide which of the two solutions is more appropriate, an interesting discussion on this can be found in [Aum77]. Considerations in the context of Cumulative Prospect Theory seem to favor a bounded utility function, compare Sect. 2.4.4.

There is another interesting idea that tries to select a certain shape of utility function via an evolutionary approach by Blume and Easley [BE92], see also [Sin02]. There are many experiments for decisions under risk on animals which show that phenomena like risk aversion are much older than humankind. Therefore it makes sense to study their evolutionary development. If the number of offspring of an animal is linearly correlated to the resources it obtained, and if the animal is faced with decisions under risk on these resources, then it can be shown that the only evolutionary stable strategy is to decide by EUT with a *logarithmic utility function*. This is a quite surprising and strong result. In particular, all other possible decision criteria will eventually become marginalized. In this sense EUT with logarithmic utility function would be the only decision model we would expect to observe.

One could also try to apply this idea to financial markets and argue that in the long run all investment strategies that do not follow the EUT maximization with logarithmic utility function will be marginalized and their market share will be negligible. Hence to model a financial market, we only need to consider EUT maximizer with a logarithmic utility function. – This would certainly be a very interesting insight!

However, there are a couple of problems with this line of argument. First, in the original evolutionary setting, the assumption that the number of offspring is proportional to the resources is a light oversimplification. There is, for instance, certainly a lower bound on the resources below which the animal will simply die and the average number of offspring will therefore be zero, on the other hand, there is some upper bound for the number of offspring. Second, the application to financial markets (as suggested, e.g., in [Len04]) is questionable: under-performance on the stock market does not have to lead to marginalization, since it may be counteracted by adding external resources and the investment time might just not be sufficiently long. New investors will moreover not necessarily implement the same strategies as their predecessors which prevents the market from converging to the theoretically evolutionary stable solution. The idea of using evolutionary concepts in the description of financial markets per se is very interesting, and we will come back to this starting in Sect. 5.7.1, but this concept does not seem to have strong implications for the shape of utility functions.

We have seen that there are plenty of ideas how to choose "suitable" utility functions. We have also found a list of properties (continuous, monotone increasing, either bounded or at least asymptotically concave) that rational utility functions should satisfy. Moreover, we have seen various suggestions for suitable utility functions that are frequently used. However, it is important to understand that there is no single class of functions that can claim to be the "right one". Therefore the choice of a functional form follows to some extent rather convenience than necessity.

2.2.4 Measuring the Utility Function

When we want to elicit a person's utility function, we have several possible methods to do so. First, we can rely on real-life data, e.g., from investment or insurance decisions. Second, we can perform laboratory experiments with test subjects. In the latter case, there are various possible procedures, which measure points of the utility function. Using these points, a fit of a function can be made, where usually a specific functional form (for instance x^α) is assumed.

We present here just one of the many methods, the *midpoint certainty equivalent method*. In this method, a subject is asked to state a monetary equivalent to a lottery with two outcomes that each occur with probability $1/2$, compare Fig. 2.9. Such a monetary equivalent ("the price of a lottery") is called a *Certainty Equivalent (CE)*.

If we set $u(x_0) := 0$ and $u(x_1) := 1$ (which we can do, since u is only determined up to affine transformations), then $u(CE) = 0.5$. We set $x_{0.5} := CE$ and iterate this method by comparing a lottery with the outcomes x_0 and $x_{0.5}$ and probabilities $1/2$ each etc.

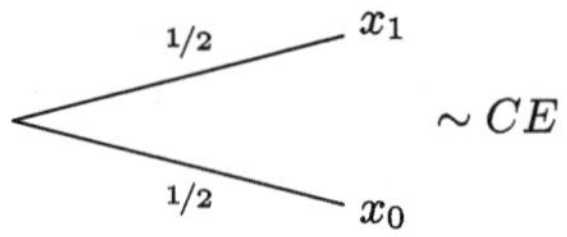

Fig. 2.9 A typical question when measuring a utility function is to ask for a certainty equivalent (CE) for a simple lottery

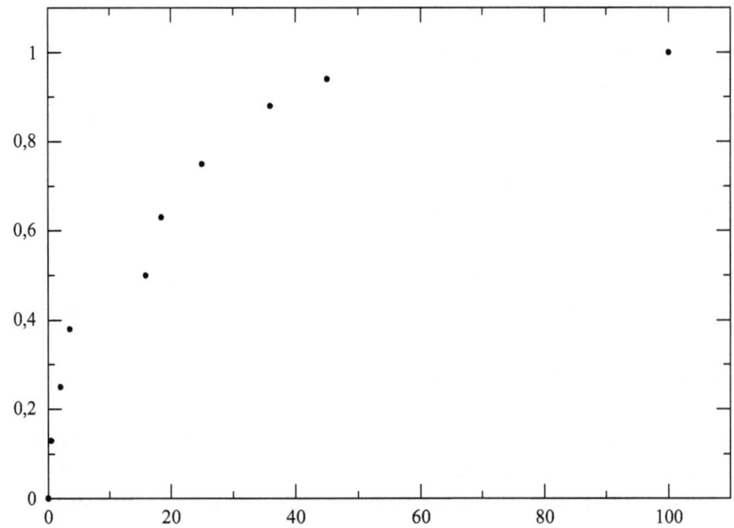

Fig. 2.10 Measured utility function of a test person. x-axis: return of a lottery, y-axis: utility

Let us try this in an example with wealth level w: we set $x_0 := w + 0€$ and $x_1 := w + 100€$. The certainty equivalent of a lottery with these outcomes is measured as, say, $w + 15€$. Thus $x_{0.5} = w + 15$. In the next step we determine the CE of a lottery with outcomes x_0 and $x_{0.5}$. The answer of our test person is 2€. We then ask for the CE of a lottery with outcomes $x_{0.5}$ and x_1 and get the answer 25€. Going on with this iteration, we can obtain more data points which ultimately leads to a sketch of the utility function, see Fig. 2.10.

This method has a couple of obvious advantages: it uses simple, transparent lotteries that do not involve complicated, unintuitive probabilities. Moreover, it only needs relatively few questions to elicit a utility function. However, it also has two drawbacks:

- It is not very easy to decide about the certainty equivalent. Pairwise preference decisions are much simpler to do. However, pairwise decisions reveal less information (only yes or no, rather than a numerical value), hence more questions have to be asked in order to get similar results.
- If the test person makes an error, it propagates through the whole experiment, and it is difficult to correct it later on.

There are other methods that avoid these problems, but typically have their own disadvantages. We do not want to discuss them here, but we hope that the example we have given is sufficient to give some ideas on how one can obtain information on this at first glance unascertainable object and what kind of problems this poses.

Assume now that we have measured in an experiment a utility function of a person. The next question we have to ask is, whether EUT is in fact a suitable theory to *describe* these experimental results, since only under this condition our measurements can be used to derive statements about real life situations, e.g., to give advice regarding investment decisions or to model financial markets.

In fact, this question is much more difficult than one might expect. One of the fundamental contributions to this problem has been made by M. Rabin [Rab00] who studied the following question: is it possible to explain the risk aversion that one measures in small stake experiments by means of the concavity of the utility function?

If we have a look on Fig. 2.10, we tend to answer the question affirmatively. The data resembles a function like x^α. However, the x-axis is not the final wealth of the person, but it is just the return of the lotteries, in other words we have to add the wealth w. (In the above example, the person's wealth was roughly $50,000$€). Rabin was analyzing such examples a little closer: If we assume a given risk-averse behavior (like rejecting a 50–50 gamble of gaining 105€ or losing 100€) below a certain, not too low wealth level, then it is possible to deduce that very advantageous lotteries would be rejected – regardless of the precise form of the utility function! One can prove, e.g., that if a 50–50 gamble of gaining 105€ or losing 100€ is rejected up to a wealth level of $300,000$€, then, at a wealth level of $290,000$€, a 50–50 gamble of losing 6000€ and gaining 1.5 Million Euro would still be rejected. This behavior seems to be quite unlikely and not very rational, hence we can conclude that a rational person would not reject the initial offer (lose 100€, gain 105€) up to such a large wealth level.

How does Rabin prove his strong, and somehow surprising result? Without going into the details, we can get an intuition of the result by considering a Taylor expansion of a utility function u at the wealth level w and compute the expected utility of a 50–50 gamble with loss l or gain g:

$$\frac{1}{2}u(w-l)+\frac{1}{2}u(w+g)$$
$$= u(w)+\frac{1}{2}\left(u'(w)(g-l)+\frac{1}{2}u''(w)(g^2-l^2)+O(l^3)-O(g^3)\right).$$

Here O is the Landau symbol (see Appendix). Comparing this with $u(w)$, the initial wealth utility, one sees that in order to reject the gamble for all wealth levels or at least for up to a substantial wealth level, $-u''(w)$ has to be sufficiently large. On the other hand $u'(w) > 0$ for all w. This leads to a quickly flattening utility function and to the paradoxical situations observed by Rabin.

The result indicates that EUT might not work well in explaining small stake experiments as illustrated in Fig. 2.10, since it has difficulties in explaining the strong risk aversion that individuals still show – even at relatively large wealth levels. The simplest way to explain this discrepancy is to use a different "frame", i.e., to compute the utility function in terms of the potential gains and losses in a given situation, instead of the final wealth. We will see later how this "framing effect" influences decisions and that it is an essential ingredient in modern descriptive theories, in particular in Prospect Theory. It is interesting to observe that this "change of frame" is often intuitively and unintentionally done in textbooks on expected utility theory, a brief search will surely provide the reader with some examples.

Although the paper by Rabin is suggesting to use an alternative approach to describe results of small and medium stake experiments, it has often been misunderstood, in particular in experimental economics, where it is frequently cited as a justification to assume risk-neutrality in experiments. Rabin himself, together with Richard Thaler, admits in a comment [RT02] that

> we can see … how our choice of emphasis could have made our point less clear to some readers

and goes on to remind that risk aversion has been observed in nearly all experiments:

> We refer the reader who believes in risk-neutrality to pick up virtually any experimental test of risk attitudes. Dozens of laboratory experiments show that people are averse to far more favorable bets for smaller stakes. The idea that people are not risk neutral in playing for modest stakes is uncontroversial.

He underlines the fact that

> because most people are *not* risk neutral over modest stakes, expected utility should be rejected by economists as a *descriptive* theory of decision-making.

Alas, it seems that these clarifications were not heard by everybody.

We will see in Sect. 2.4 what kind of theories are superior as a descriptive model for decisions under risk. Nevertheless it is important to keep in mind that Expected Utility Theory as a *prescriptive* model for *rational* decisions under risk is still largely undisputed. In the next section we will turn our attention to the widely used Mean-Variance Theory which is popular for its "ease of use" that allows fruitful applications where the more complicated EUT is too difficult to apply.

2.3 Mean-Variance Theory

2.3.1 Definition and Fundamental Properties

Mean-Variance Theory was introduced in 1952 by Markowitz [Mar52, Mar91] as a decision criterion for portfolio selection. His key idea was to measure the risk of an asset by only one parameter, the variance σ. Together with the mean μ, these are the only two parameters that are used in this decision model. Harry Markowitz was awarded the Nobel Prize in 1990 for his pioneering work in financial economics.

In order to make precise what we mean with the "mean-variance approach", we start with a formal definition:

Definition 2.26 (Mean-Variance approach) A *mean-variance utility function u* is a utility function $u: \mathbb{R} \times \mathbb{R}_+ \to \mathbb{R}$ which corresponds to a utility functional $U: \mathcal{P} \to \mathbb{R}$ that only depends on the mean and the variance of a probability measure p, i.e., $U(p) = u(\mathbb{E}(p), \mathrm{var}(p))$.

This definition means that for two lotteries A, B described by the probability measures p_A and p_B, the lottery A is preferred over B if and only if $u(\mathbb{E}(p_A), \mathrm{var}(p_A)) > u(\mathbb{E}(p_B), \mathrm{var}(p_B))$.

The mean is usually denoted by μ, the variance by σ^2. We can hence express a mean-variance utility functional by writing down the function $u(\mu, \sigma)$.

Of course not every mean-variance utility function is reasonable. – We have already seen in the case of EUT utility functions that for theoretical and practical reasons some properties should be assumed. Most commonly one expects the utility function to be strictly increasing in μ, which corresponds to the "more money is better" maxim. Since σ reflects the risk of a lottery, one usually also assumes that the utility decreases when σ increases. Let us define this precisely:

Definition 2.27 A mean-variance utility function $u: \mathbb{R} \times \mathbb{R}_+ \to \mathbb{R}$ is called *monotone* if $u(\mu, \sigma) \geq u(\nu, \sigma)$ for all μ, ν, σ with $\mu > \nu$. It is called *strictly monotone* if even $u(\mu, \sigma) > u(\nu, \sigma)$.

We will always assume that u is strictly monotone.

Definition 2.28 A mean-variance utility function $u: \mathbb{R} \times \mathbb{R}_+ \to \mathbb{R}$ is called *variance-averse* if $u(\mu, \sigma) \geq u(\mu, \tau)$ for all μ, τ, σ with $\tau > \sigma$. It is called *strictly variance-averse*[11] if $u(\mu, \sigma) > u(\mu, \tau)$ for all μ, τ, σ with $\tau > \sigma$.

Often this is assumed as well, but we will not turn this into a general condition. Instead we expect the preference to be risk-averse, i.e., that the expected value of a lottery is always preferred over the lottery itself, compare Definition 2.7. This leads to the following trivial observation:

Remark 2.29 Let u be a mean-variance function. Then the preference induced by u is risk-averse if and only if $u(\mu, \sigma) < u(\mu, 0)$ for all μ, σ. The preference is risk-seeking if and only if $u(\mu, \sigma) > u(\mu, 0)$.

[11]We use these standard names, although they are not coherent with the use of the similar term "risk-averse" where a *strict* inequality occurs.

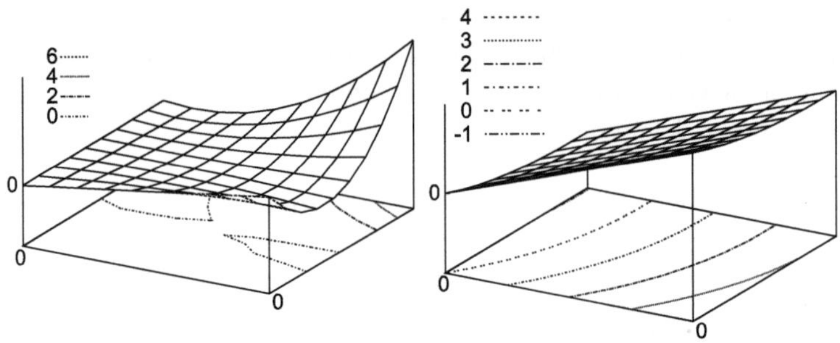

Fig. 2.11 Two mean-variance diagrams

We have found ample evidence for risk-averse behavior in the last section, therefore we consider only mean-variance functions which describe risk-averse behavior.

There is a very convenient way to deduce information on a given mean-variance utility function and the preferences induced by it: the mean-variance diagram, also known as (μ, σ)-diagram. It corresponds to the indifference curves of the utility function on the set of all μ and σ. As an example take the two utility functions[12]

$$u_1(\mu, \sigma) := \mu - \sigma^2, \qquad u_2(\mu, \sigma) := 2\mu - 1.3\sigma + 0.5\sigma^2 - 0.054\sigma^3.$$

Their corresponding (μ, σ)-diagrams can be found in Fig. 2.11.

2.3.2 Success and Limitation

The main advantage of the mean-variance approach is its simplicity that reduces the complexity of decisions under risk to only two parameters, the mean μ and the variance σ. This allows us to use (μ, σ)-diagrams in order to characterize the key properties (average return and risk) of an asset. It also allows us to handle complicated models much more easily than with the clumsy EUT. It is therefore no surprise that the Mean-Variance Theory is the most frequently used decision theory by theorists, but also by practitioners in finance, both as a descriptive and prescriptive tool.

On the theoretical side, we will see in the next chapter how this approach can be used to derive the famous Capital Asset Pricing Model which provides easy formulas for price of an asset. We will also see how (μ, σ)-diagrams can be used to derive the Two-Fund-Separation Theorem which states that if everybody is a mean-variance investor and the market is complete and efficient, it is best to hold

[12] The function u_2 is, by the way, an unorthodox suggestion to resolve Allais' Paradox which we will meet in the next section.

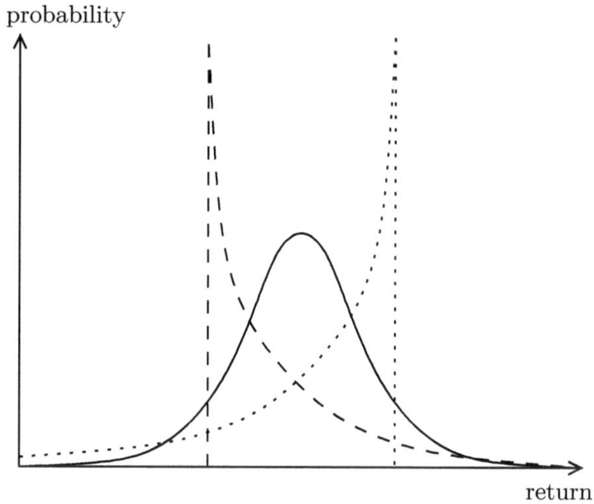

Fig. 2.12 Three outcome distributions that are indistinguishable for the Mean-Variance Theory, since their means and variances each agree

a portfolio composed out of a risk-free asset and a representative market portfolio. This outlook highlights the efficiency of the mean-variance approach as a tool in financial economics. However, it also shows its limitation, since practitioners are obviously not following this result, and we may assume that they have reasons.

On the practical side, we mention that banks are usually providing clients with two main informations on assets: the average return and the risk, the latter usually measured as variance.

Although the practical use of an easy method to solve complex problems is surely valuable, there are nevertheless certain problems and limitations of the Mean-Variance Theory. Practitioners sometimes raise the question whether the variance is really an appropriate tool to measure risk. As a simple – albeit more academic – example take, e.g., the following two assets which have identical mean and variance and are hence considered to be equal by the mean-variance criterion:

$A :=$

payoff	0€	1010€
probability	99.505 %	0.495 %

$B :=$

payoff	−1000€	10€
probability	0.495 %	99.505 %

There are obviously important reasons why one would like to prefer either A or B, but it seems worthwhile to distinguish both assets! This also holds true for more realistic distributions, compare Fig. 2.12.

A practical application is that banks who are only reducing their risk as measured by the variance might still accept a *very small* probability of an *enormous* loss that might even lead to a bankruptcy. A modern risk management, however, would rather accept a *substantial* risk of a *small* loss which can still be mitigated, even if the associated variance is larger. Another typical practical problem occurs with options which tend to have very skewed payoff distributions. Here the variance as risk measure does not distinguish the upside risk of making a large profit with low probability from the downside risk of losing a lot with low probability.

There are several other methods to measure risk, like *value at risk* which measures the value of which the payoff falls short in only $n\,\%$ of the cases or the *expected tail loss* which measures the expected loss that occurs in the worst $n\,\%$ of the cases. All these practical modifications have in common that they aim to measure the risk with a single quantity, but that they replace the variance by a more sophisticated measure.

Besides these practical problems there are also strong theoretical limitations of the mean-variance approach. The strongest is the so-called "Mean-Variance Paradox". We formulate it as a theorem:

Theorem 2.30 (Mean-Variance Paradox) *For every continuous mean-variance utility function $u(\mu, \sigma)$ which corresponds to a risk-averse preference, there exist two assets A and B where A state dominates B, but B is preferred over A.*

Proof Let us construct an explicit example, where for simplicity we assume that u is strictly monotone. Consider for $N \geq 1$ the following lottery

$$A_N := \begin{array}{lcc} \hline \text{payoff in €} & 0 & \text{N} \\ \text{probability} & 1-\frac{1}{N^2} & \frac{1}{N^2} \\ \hline \end{array}.$$

The expected value of A_N is

$$\mathbb{E}(A_N) = \left(1 - \frac{1}{N^2}\right) \cdot 0 + \frac{1}{N^2} N = \frac{1}{N}.$$

The variance can now be easily computed as

$$\begin{aligned} \text{var}(A_N) &= \left(1 - \frac{1}{N^2}\right) \frac{1}{N^2} + \frac{1}{N^2} \left(N - \frac{1}{N}\right)^2 \\ &= \frac{1}{N^2} - \frac{1}{N^4} + 1 - \frac{2}{N^2} + \frac{1}{N^4} = 1 - \frac{1}{N^2}. \end{aligned}$$

Now we compare this with the mean and variance of the lottery A_0 that always gives a payoff of zero:

$$\mathbb{E}(A_0) = 0, \quad \text{var}(A_0) = 0.$$

If N becomes large, its mean value converges to zero, whereas its variance converges to 1. Since u is continuous and risk-averse, this implies that

$$U(A_N) = u\left(\frac{1}{N}, 1 - \frac{1}{N^2}\right) \to u(0, 1) < u(0, 0) = u(A_0).$$

Therefore, we can choose N large enough, such that the inequality

$$u\left(\frac{1}{N}, 1 - \frac{1}{N^2}\right) < u(0,0)$$

holds. Since A_N gives a payoff of zero in states with the total probability $1 - \frac{1}{N^2}$, but a positive payoff N in states with a total probability $\frac{1}{N^2}$ which is strictly larger than zero, but A_0 gives in both cases a zero payoff, A_N state dominates A_0. However, we have just proved that *any* mean-variance utility (which satisfies the initial assumptions) would prefer A_0 over A_N. Setting $A := A_N$ and $B := A_0$ we have proved the theorem. □

Let us stop here for a moment and think about what we have proved right now: there are two assets, one which never gives any profit, and the other one which does, although with a small probability, and besides that never loses you any money. Sure, you would prefer the latter one! After all, it poses no risk for you. But this is wrong, if you define "risk" as variance. We learn from this that the variance is *really* not a particularly good measure for risk.

Another interesting fact about Mean-Variance Theory which follows directly from the Mean-Variance Paradox is that it does not satisfy the Independence Axiom (compare Definition 2.17):

Corollary 2.31 *Every strictly monotone and risk-averse Mean-Variance Utility violates the Independence Axiom.*

Proof Take the lottery A_N as constructed in the last proof, such that A_0 is preferred over A_N. Both lotteries have a common part: with a probability of $1-1/N^2$ they both yield an outcome of zero. Only in the remaining cases (with probability $1/N^2$) they differ: whereas A_N gives an outcome of N, A_0 still gives only 0. If the Independence Axiom were satisfied, we could neglect the common part, and the preference relation would carry over to the remaining cases. However, these lotteries correspond to a sure gain of N or a sure gain of zero, and according to strict monotonicity the gain of N would be preferred. □

We remark that both assumptions (strict monotonicity and risk averseness) are indeed necessary requirements for the corollary, since they exclude the special cases of risk-neutral EUT and indifference to mean.

It is also possible to illustrate the violation of the Independence Axiom on a simple example, sometimes referred to as "common ratio effect":

Example 2.32 Consider four investment alternatives A, B, C and D that yield returns of 2 %, 4 % or 6 % with the probabilities given in Table 2.2. We can list mean and variance of the four investments and then compute their Mean-Variance utility which we choose for simplicity as $U(\mu, \sigma) := \mu - \sigma^2$. In this way we obtain Table 2.2.

Table 2.2 Payoff probabilities for the four hypothetic investments A, B, C and D

Investment	2 %	4 %	6 %
A	0.2	0	0.8
B	0	1	0
C	0.8	0	0.2
D	0.75	0.25	0

Table 2.3 Mean μ, variance σ^2 and $U(\mu, \sigma) = \mu - \sigma^2$ for the four investments from Table 2.2

Investment	Mean μ	Variance σ^2	$\mu - \sigma^2$
A	5.2	2.56	2.64
B	4	0	4
C	2.8	2.56	0.24
D	2.5	2.75	0

This implies that Mean-Variance Theory with the utility $U(\mu, \sigma^2) = \mu - \sigma^2$ predicts the preference pattern B $\succ$ A $\succ$ C $\succ$ D (Table 2.3).

Let us take a closer look at the investments. Then we see that the lottery C is equivalent to playing lottery A with a probability of $1/4$ and getting 2 % with a probability of $3/4$. Similarly, lottery D is equivalent to playing lottery B with a probability of $1/4$ and getting 2 % with a probability of $3/4$. Thus, the Independence Axiom would imply that if C is preferred over A, then B had to be preferred over C. The Mean-Variance utility from above, however, shows a different pattern of preferences, thus the Independence Axiom is violated.

That the pattern of preferences predicted by Mean-Variance Theory contradicts Expected Utility Theory, can also be seen directly by a short computation: denote $x := u(2\,\%)$, $y := u(4\,\%)$, $z := u(6\,\%)$. Then B $\succ A$ implies $0.2x + 0.8z < y$ and C $\succ D$ implies $0.8x + 0.2z > 0.75x + 0.25y$ or $0.05x + 0.2z > 0.25y$. Multiplying the last inequality by four gives a contradiction, thus the preference pattern cannot be explained by Expected Utility Theory.

In Sect. 2.5 we compare EUT and Mean-Variance Theory and we will see that there are in fact certain cases, where the problems we have encountered cannot occur and Mean-Variance Theory even becomes a special instance of EUT. In general, however, we need to keep in mind that there is always a risk to apply the mean-variance approach to general situations: beware of being too credulous when applying Mean-Variance Theory!

2.4 Prospect Theory

So, how do people *really* decide? As if they were maximizing their expected utility? Or as if they were following the mean-variance approach? Or do they deviate from both models and decide in a random manner that makes it completely impossible to predict their decisions beforehand? – It turns out that none of these is the case. In this section we will present models that describe actual decisions quite well.

2.4.1 Origins of Behavioral Decision Theory

Although the axioms of Expected Utility Theory were so convincing that we refer to a behavior described by this model as "rational", it is nevertheless possible to observe people deviating systematically from this rational behavior. One of the most striking examples is the following (often called "Asian disease"):

Example 2.33 Imagine that your country is preparing for the outbreak of an unusual disease, which is expected to kill 600 people. Two alternative programs to combat the disease have been proposed. Assume that the exact scientific estimates of the consequences of the programs are as follows: If program A is adopted, 200 people will be saved. If program B is adopted, there is a one-third probability that 600 people will be saved and a two-thirds probability that no people will be saved. Which of the two programs would you choose?

The majority (72 %) of a representative sample of physicians preferred program A, the "safe" strategy. Now, consider the following, slightly different problem:

Example 2.34 In the same situation as in Example 2.33 there are now instead of A and B two different programs C and D: If program C is adopted, 400 people will die. If program D is adopted, there is a one-third probability that nobody will die and a two-thirds probability that 600 people will die. Which of the two programs would you favor?

In this case, the large majority (78 %) of an equivalent sample preferred the program D. – Obviously, it would be cruel to abandon the lives of 400 people by choosing program C!

You might have noticed already that both decision problems are exactly identical in contents. The only difference between them is how they are formulated, or more precisely how they are *framed*. Applying EUT cannot explain this observation, neither can Mean-Variance Theory. Moreover, it would not help to modify our notion of a rational decider to capture this "framing effect", since any rational person should definitely *not* make a difference between the two identical situations. Let us have a look on another classical example of a deviation from rational behavior.[13]

Example 2.35 In the so-called "Allais paradox" we consider four lotteries (A, B, C and D). In each lottery a random number is drawn from the set $\{1, 2, \ldots, 100\}$ where each number occurs with the same probability of 1 %. The lotteries assign outcomes to every of these 100 possible numbers (states), according to Table 2.4.

[13]This example might remind the reader of Example 2.32 that demonstrated how Mean-Variance Theory can lead to violations of the Independence Axiom.

Table 2.4 The four lotteries of Allais' Paradox

Lottery A	State	1–33	34–99	100
	Outcome	2500	2400	0
Lottery B	State	1–100		
	Outcome	2400		
Lottery C	State	1–33	34–100	
	Outcome	2500	0	
Lottery D	State	1–33	34–99	100
	Outcome	2400	0	2400

The test persons are asked to decide between the two lotteries A and B and then between C and D. Most people prefer B over A and C over D.

This behavior is not rational, although this time it might be less obvious. The axiom that most people violate in this case is the Independence Axiom. We can see this by neglecting in both decisions the states 34–99, since they give the same result each. What is left (the states 1–33 and the state 100) are the same for both decision problems. In other words, the part of our decisions which is independent of irrelevant alternatives, is the same when deciding between A and B and when deciding between C and D. Hence, if we prefer B over A we should also prefer D over C, and if we prefer C over D, we should also prefer A over B.

We have already encountered other observed facts that can be explained with EUT only under quite delicate and even painstaking assumptions on the utility function:

- People tend to buy insurances (risk-averse behavior) and take part in lotteries (risk-seeking behavior) at the same time.
- People are usually risk-averse even for small-stake gambles and large initial wealth. This would predict a degree of risk aversion for high-stake gambles that is far away from standard behavior.

Other experimental evidence for systematic deviation from rational behavior has been accumulated over the last decades. One could joke that there is quite an industry for producing more and more such examples.

Does this mean, as is often heard, that the "homo economicus" is dead and that all models of humans as rational decision makers are obsolete? And does this mean that the excoriating judgment that we quoted at the beginning of this chapter holds in a certain way and that "science is at a loss" when it comes to people's decisions?

Probably none of these fears is appropriate: the "homo economicus" as a rationally behaving subject is still a central concept, and on the other hand there are modifications of the rational theories that describe the irrational deviations from the rational norm in a systematic way which leads to surprisingly good descriptions of human decisions. In the following we will introduce some of the most important concepts that such behavioral decision theories try to encompass.

The first example has already shown us one very important effect, the "framing effect". People decide by comparing the alternatives to a certain "frame", a point of reference. The choice of the frame can be influenced by phrasing a problem in a certain way. In Example 2.33 the problem was phrased in a way that made people frame it as a decision between *saving* 200 people for sure or saving 600 people with a probability of 1/3. In other words, the decision was framed in *positive* terms, in *gains*. It turns out that people behave risk-averse in such situations. This does not come as a surprise, since we have encountered this effect already several times, e.g., when we measured the utility function of a test person (see Sect. 2.2.4). In Example 2.34 the frame is inverted: now it is a decision about *letting people die*, in other words it is a decision about *losses*. Here, people tend to behave risk-seeking. They would rather take a 1/3 chance of letting all 600 persons die than choosing to let 200 people die.

But let us think about this for a moment. Doesn't this contradict the observation that people buy insurances and that people buy lottery tickets? An insurance is surely about losses (and their prevention), whereas a lottery is definitely about gains, but still people behave risk-averse when it comes to insurances and risk-seeking when it comes to lotteries.

The puzzle can be solved by looking on the probabilities involved in these situations: In the two initial examples the probabilities were in the mid-range (1/3 and 2/3), whereas in the cases of insurances and lotteries the probabilities involved can be very small. In fact, we have already observed that lotteries which attract the largest number of participants typically have the smallest probabilities to win a prize, compare Example 2.21. If we assume that people tend to systematically *overweight* these small probabilities, then we can explain why they buy insurances against small probability risks and at the same time lottery tickets (with a small probability to win). Summarizing this idea we get a four-fold pattern of risk-attitudes[14]:

Can we explain Allais' Paradox with this idea? Indeed, we can: When choosing between the lotteries A and B the small probability not to win anything when choosing A is perceived much larger than the difference in the probabilities of not winning anything when deciding between the lotteries C and D. This predicts the observed decision pattern (Table 2.5).

The fact that people overweight small probabilities should be distinguished from the fact that they often *overestimate* small probabilities: if you ask a layman for the

[14]It is historically interesting to notice, that a certain variant of the key ideas of Kahneman and Tversky have already been found 250 years earlier in the discussion on the St. Petersburg paradox: Nicolas Bernoulli had the idea to resolve the paradox by assuming that people *underweight* very small probabilities, whereas Gabriel Cramer, yet another Swiss mathematician, tried to resolve the paradox with an idea that resembles the value function of Prospect Theory.

Table 2.5 Risk attitudes depending on probability and frame

	Losses	Gains
Medium probabilities	Risk-seeking	Risk-averse
Low probabilities	Risk-averse	Risk-seeking

probability to die in an airplane accident or to get shot in the streets of New York, he will probably overestimate the probability, however, the effect we are interested in is a different one, namely that people even when they *know* the precise probability of an event still behave *as if* this probability were higher. This effect seems in fact to be quite universal, whereas the overestimating of small probabilities is not as universal as one might think. Indeed, small probabilities can also be underestimated. This is typically the case when a person neither experienced nor heard that a certain small probability event happened before. If you, for instance, let a person sample a lottery with an outcome with unknown, but low probability, then the person will likely not experience any such outcome and hence *underestimate* the low probability. Such a sampling will nowadays (in times of excessive media coverage) not be our only possibility to estimate the probabilities of events that we haven't experienced by ourselves before. But what about events that are too unimportant to be reported? Such events might nevertheless surprise us, since in these situations we have to rely on our own experience and tend to underestimate the probability of such events before we experience them. – Surely everybody can remember an "extremely unlikely" coincidence that happened to him, but it couldn't have been *that* unlikely if everybody experiences such "unlikely" coincidences, could it?

In the next section we formalize the ideas of framing and probability weighting and study the so-called "Prospect Theory" introduced by Kahneman and Tversky [KT79].

2.4.2 Original Prospect Theory

Framing effect and probability overweighting, these are the two central properties we want to include into a behavioral decision theory. We follow here the ideas of Kahneman and Tversky and use as starting point for this theory the Expected Utility Theory. Instead of the final wealth we consider the gain and loss induced by a given outcome (framing effect) and instead of the real probabilities we consider weighted probabilities that take into account the overweighting of small probabilities. This *Prospect Theory* (PT) leads us to the following definition of a "subjective utility" of a lottery A with n outcomes $x_1, \ldots, x_n$ (relative to a frame) and probabilities $p_1, \ldots, p_n$:

$$PT(A) := \sum_{i=1}^{n} v(x_i)w(p_i), \tag{2.3}$$

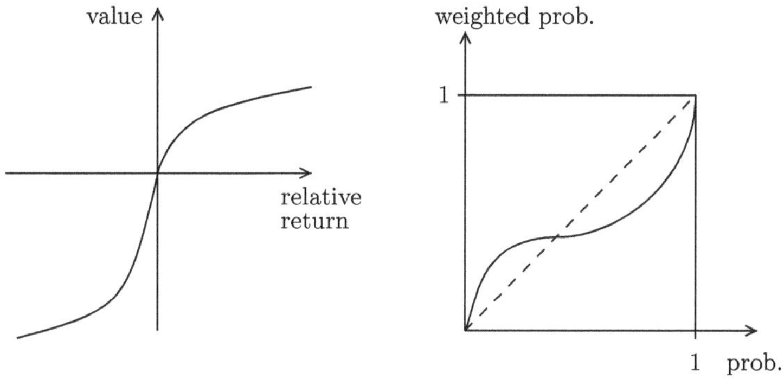

Fig. 2.13 A rough sketch of the typical features of value function (*left*) and weighting function (*right*) in Prospect Theory

where $v: \mathbb{R} \to \mathbb{R}$ is the *value function*, a certain kind of utility function, but defined on losses and gains rather than on final wealth, and $w: [0, 1] \to [0, 1]$ is the *probability weighting function* which transforms real probabilities into subjective probabilities. The key features of the value function are the following:

- v is continuous and monotone increasing.
- The function v is strictly concave for values larger than zero, i.e., in gains, but strictly convex for values less than zero, i.e., in losses.
- At zero, the function v is "steeper" in losses than in gains, i.e., its slope at $-x$ is bigger than its slope at x.

The weighting function satisfies the following properties:

- The function w is continuous and monotone increasing.
- $w(p) > p$ for small values of $p > 0$ (probability overweighting) and $w(p) < p$ for large values of $p < 1$ (probability underweighting), $w(0) = 0$, $w(1) = 1$ (no weighting for sure outcomes).

Typical shapes for v and w are sketched in Fig. 2.13.

If we have many events, all of them will probably be overweighted and the sum of the weighted probabilities will be large. There is an alternative formulation of Prospect Theory in [Kar78] that fixes the problem by a simple normalization:

$$PT(A) = \frac{\sum_{i=1}^{n} v(x_i)w(p_i)}{\sum_{i=1}^{n} w(p_i)}. \tag{2.4}$$

For didactical reasons it is easier to consider (2.3), hence we will mainly concentrate on this formulation. Equation (2.4) shares many of the common features with (2.3) and has some technical advantages that we will discuss later.

Can this new theory predict the four-fold pattern of risk-attitudes observed in the examples of the Sect. 2.4.1? Yes, it can. If we have two outcomes of similar probability, their weighted probability is approximately identical to their real probability, hence the concavity of the value function in gains, leads to risk-averse behavior, and the convexity of the value function in losses, leads to risk-seeking behavior. We know this already from EUT and do not need to compute anything new. (This explains the results of Examples 2.33 and 2.34.) Now if one of the probabilities is very small, then it is strongly overweighted ($w(p) > p$). In the case of losses this means that the overall utility is reduced even more. This effect can cancel the convexity of the value function and lead to a risk-averse behavior. On the other hand an overweighting of a gain might increase the value of the utility so much that it outperforms a sure option, even though the concavity of the value function would predict a risk-averse behavior.

Prospect Theory in this general form can only give a rough explanation of the experimental evidence, but is not useful for computations. To make precise predictions and to classify people's attitude towards risk, we need to make the functional forms of v and w precise. Nowadays the most commonly used functional forms are the ones introduced for *Cumulative Prospect Theory*(CPT), and we will discuss them in the next section. For the moment, we just note that Prospect Theory seems to be a good candidate for a descriptive model of decisions under risk. However, there are a couple of limitations to this theory that led to further developments.

We know that PT does not satisfy the Independence Axiom. This is a feature, not a bug, since otherwise we could not explain Allais' Paradox. There are some other axioms we are not so eager to give up in a descriptive theory. One of them is stochastic dominance: we have already briefly mentioned this concept which is essentially a "state-independent version" of state dominance:

Definition 2.36 (Stochastic dominance) A lottery A is *stochastically dominant* over a lottery B if, for every payoff x, the probability to obtain more than x is larger or equal for A than for B and there is at least some payoff x such that this probability is strictly larger.

This notion is quite natural: if we set our goal to get at least x€ as payoff, we will choose A, since the probability to reach our goal with A is at least as large and sometimes strictly larger than with B. If this holds for all x, then A is in this sense "better" than B. If a decision criterion always prefers A over B when A is stochastic dominant over B, we say it *satisfies* or *respects stochastic dominance.*

Let us have a look at the following example: if we compare the two lotteries

$$A := \begin{array}{lcccccc} \hline \text{Outcome} & 0.95 & 0.96 & 0.97 & 0.98 & 0.99 & 1 \\ \text{Probability} & 1/6 & 1/6 & 1/6 & 1/6 & 1/6 & 1/6 \\ \hline \end{array}, \qquad B := \begin{array}{lc} \hline \text{Outcome} & 1 \\ \text{Probability} & 1 \\ \hline \end{array},$$

then it is obvious that B is stochastic dominant over A (e.g., the probability to gain at least 0.97 is $1/2$ for A, but 1 for B) and should hence be preferred by a reasonable decision criterion. (Or would *you* prefer lottery A?) In fact, it is easy to prove that EUT always satisfies stochastic dominance as long as the utility function is strictly increasing. Nevertheless, this does not need to be the case in Prospect Theory: the probability of $1/6$ is quite small, thus we expect $w(1/6) > 1/6$. On the other hand, the outcomes $0.95, \ldots, 1$ are all quite close to 1, therefore

$$PT(A) = \sum_{i=1}^{6} w(1/6)v(x_i) \approx \sum_{i=1}^{6} w(1/6)v(1) > v(1). \tag{2.5}$$

This argument can easily be made rigorous to show that for every weighting function w that overweights at least *some* small probabilities, two lotteries can be constructed that show that PT violates stochastic dominance. In other words, if we want to have small probabilities being overweighted, there is no way we can at the same time rescue stochastic dominance. The alternative formulation (2.4) somehow reduces this problem such that stochastic dominance is not violated for lotteries with at most two outcomes, for lotteries with more outcomes, however, the problem persists. – This seems like bad news for the theory.

There is another problem involved in this example, namely a lack of continuity in this model. Roughly spoken, two very similar lotteries can have very different subjective utilities. We will discuss this problem in Sect. 2.4.5 more in detail.

Already Kahneman and Tversky knew about these problems and that their theory violates stochastic dominance and continuity. They suggested as "fix" a so-called "editing phase": before a person evaluates the PT-functional (or rather behaves *as if* he evaluates this functional, since of course nobody assumes that people actually do these computations when deciding), this person would check a couple of things on the lotteries under consideration. In particular, the frame would be chosen, very similar outcomes would be collected to one, and stochastically dominating lotteries would automatically be preferred, regardless of any subjective utility.

The procedure is unfortunately not very well defined and leaves a lot of space for interpretations. (When are outcomes "close"? How does a person set the frame?) This causes problems when applying the theory and limits its usability.

Another limitation is that PT can only be applied for finitely many outcomes. In particular in finance, however, we are interested in situations with infinitely many outcomes. (An asset yields typically a return which can potentially be any amount, not just one out of a small list.)

We will discuss later why it is so difficult to extend PT to infinitely many outcomes and how one can improve PT regarding stochastic dominance and continuity (compare Sect. 2.4.6). Historically, however, these problems led first to the nowadays most important theory of behavioral decisions, the *Cumulative Prospect Theory*.

2.4.3 Cumulative Prospect Theory

We have seen that many problems in Prospect Theory were caused by the overweighting of small probabilities. In a certain sense, our example for violation of stochastic dominance was based on the fact that a large number of small probability events added up to a "subjective" probability larger than one. The key idea of [TK92] was to replace the probabilities by *differences of cumulative probabilities*. In other words, we replace in the definition of Prospect Theory the probabilities p_i with the expression $F_i - F_{i-1}$, where $F_i := \sum_{j=1}^{i} p_j$ are the cumulative probabilities. (We set $F_0 := 0$.) Of course, the order of the events is now important, and we order them in the natural way, i.e., by the amount of their outcomes.

We write down the formula of Cumulative Prospect Theory precisely:

Definition 2.37 (Cumulative Prospect Theory[15]) For a lottery A with n outcomes $x_1, \dots, x_n$ and the probabilities $p_1, \dots, p_n$ where $x_1 < x_2 < \dots < x_n$ and $\sum_{i=1}^{n} p_i = 1$ we define

$$CPT(A) := \sum_{i=1}^{n} \left(w(F_i) - w(F_{i-1})\right) v(x_i), \tag{2.6}$$

where $F_0 := 0$ and $F_i := \sum_{j=1}^{i} p_j$ for $i = 1, \dots, n$.

There exist slightly different definitions of the CPT functional. In particular the original formulation in [TK92] differed in that it used the above formula only for losses, but a *de*-cumulative probability (i.e., $F_i := \sum_{j=i+1}^{n} p_j$) for gains. In finance, however, the above formula is more frequently used, since it is structurally simpler and essentially equivalent with the original formulation if one allows for changes in the weighting function.

How is this formula connected to Prospect Theory? Let us have a look on the case of a three-outcome lottery A (with outcomes x_1, x_2, x_3 with respective probabilities p_1, p_2, p_3) to see a little clearer, here the formulae reduce to

$$\begin{aligned} CPT(A) &= w(p_1)v(x_1) + (w(p_1 + p_2) - w(p_1))v(x_2) \\ &\quad + (1 - w(p_1 + p_2))v(x_3), \\ PT(A) &= w(p_1)v(x_1) + w(p_2)v(x_2) + w(p_3)v(x_3). \end{aligned}$$

We see that both formulae slightly differ, but not much. The difference between both models is essentially that in PT every probability is, regardless of their outcome, over- or underweighted, whereas in CPT, usually only probabilities that reflect

[15]The definition of *CPT* can be generalized if we use different weighting function w_- and w_+ for negative resp. positive outcomes. To keep things simple, we assume that $w_- = w_+ = w$.

extreme outcomes tend to be overweighted and probabilities that reflect outcomes in the middle are in general underweighted: if we compare the three terms in the formula for CPT, we see that the middle term indeed is likely to be the smallest, since the slope of w is typically small in the mid-range (compare Fig. 2.13). On *average*, events are neither over- nor underweighted in CPT[16]:

$$\sum_{i=1}^{n} (F_i - F_{i-1}) = F_n - F_0 = 1.$$

In many applied problems, probability distributions look similar to a normal distribution: extremely low and extremely high outcomes are rare, mid-range outcomes are frequent. This explains why often the difference between PT and CPT is small. Whereas PT overweights small probabilities which are associated with extreme outcomes, CPT overweights extreme outcomes which have small probabilities. Nevertheless, there can be situations where both theories deviate substantially, namely whenever small probability events in the mid-range of outcomes play a significant role.

There is another related theory, *Rank Dependent Utility* (RDU), which predates CPT and shares the cumulative probability weighting with CPT. However, it does not use the framing of PT and CPT, but uses a standard utility function in units of finite wealth, compare [Qui82].

In order to use CPT for applications, in particular in financial economics, we need to choose specific forms for v and w.

The prototypical example of a value function v has been given in [TK92] for $\alpha, \beta \in (0, 1)$ and $\lambda > 1$:

$$v(x) := \begin{cases} x^\alpha & , x \geq 0, \\ -\lambda(-x)^\beta & , x < 0, \end{cases} \tag{2.7}$$

compare Fig. 2.14. The parameter λ reflects the experimentally observed fact that people react to losses stronger than to gains: the resulting function v has a "kink" at zero, a marginal loss is considered a lot more important than a marginal gain. λ is usually assumed to be somewhere between 2 and 2.5.[17]

The probability weighting function w has been given by

$$w(p) := \frac{p^\gamma}{(p^\gamma + (1-p)^\gamma)^{1/\gamma}}, \tag{2.8}$$

[16]This is not the case in the original formulation of CPT when applying the weighting function on cumulative probabilities in losses and de-cumulative probabilities in gains.

[17]If $\alpha < \beta$, however, even a value of $\lambda < 1$ can lead to loss aversion. – In fact, when measuring λ on experimental data one often gets values substantially smaller than 2.

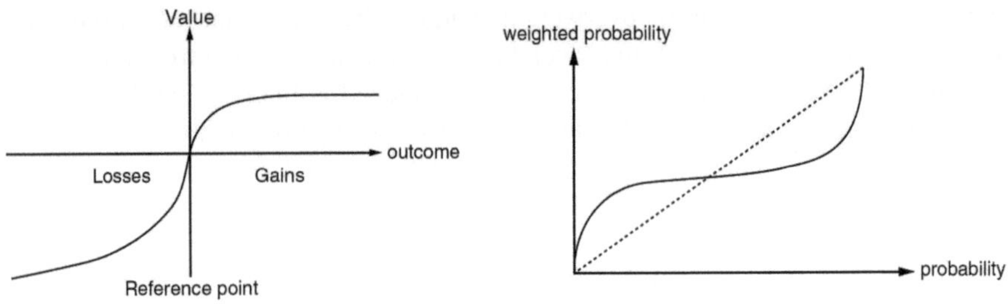

Fig. 2.14 Value and weighting function suggested by [TK92]

Table 2.6 Experimental values of α, β and γ, δ from various studies, compare (2.7) and (2.8) for the definition of $\alpha, \beta, \gamma, \delta$

Study	Estimate for α,β	Estimate for γ, δ
Tversky and Kahneman [TK92]		
Gains:	0.88	0.61
Losses:	0.88	0.69
Camerer and Ho [CH94]	0.37	0.56
Tversky and Fox [TF95]	0.88	0.69
Wu and Gonzalez [WG96]		
Gains:	0.52	0.71
Abdellaoui [Abd00]		
Gains:	0.89	0.60
Losses:	0.92	0.70
Bleichrodt and Pinto [BP00]	0.77	0.67/0.55
Kilka and Weber [KW01]	0.76–1.00	0.30–0.51

compare Fig. 2.14. It is possible to assign different weighting functions for gains and losses (denoted by w_+ and w_-, where in the loss region the constant γ is replaced by δ). There are also different suggestions how to choose v and w. We discuss in Sect. 2.4.4 which types of value and weighting functions are advantageous and which restrictions we have to take into account. For the moment we work for simplicity with the original suggestions by Tversky and Kahneman [TK92], although they are not the best choice. (For instance, w is not monotone increasing for $\gamma \leq 0.279$ [CH94, RW06, Ing07].) The parameters of their model have been measured experimentally in several studies, compare Table 2.6.

We see from this table that the results sometimes differ which might depend on the selection of the test sample or on the choice of the kind of experiment done to elicit these numbers. The overall impression, however, is that the values are typically in the range of $\alpha \approx \beta \approx 0.75 \pm 0.1$, $\gamma \approx \delta \approx 0.65 \pm 0.1$. Risk preferences also depend on economical and cultural factors, see [HW07a] for parameter estimates for some countries.

To fix ideas, we will in the following choose $\lambda := 2.25$, $\alpha := \beta := 0.8$ and $\gamma := \delta := 0.7$.

By the way, Prospect Theory (and also CPT) coincides with risk-neutral EUT when $\alpha = \beta = \gamma = \delta = \lambda = 1$. As can be seen from the experimental numbers, there is a strong deviation from this.

There are now two questions, we need to answer. Does the modified theory solve the problems that PT had (only finitely many outcomes, violation of stochastic dominance, lack of continuity) and does it still provide a good descriptive model of behavior under risk?

Let us first extend CPT to arbitrary lotteries. Since we all the time assume state-independent preferences, we can describe lotteries by probability measures, see Appendix A.4 for details.

Definition 2.38 Let p be an arbitrary probability measure, then the generalized form of CPT[18] reads as

$$CPT(p) := \int_{-\infty}^{+\infty} v(x) \left(\frac{\mathrm{d}}{\mathrm{d}t} w(F(t))|_{t=x} \right) \mathrm{d}x, \tag{2.9}$$

where

$$F(t) := \int_{-\infty}^{t} \mathrm{d}p.$$

For the cognoscenti we remark that the formula (2.6) for lotteries with finitely many outcomes is just a special case of (2.9) when choosing p as a finite sum of Diracs.

Definition 2.38 paths the way to applications of CPT in financial economics and other areas where models require more than just a couple of potential outcomes. Although it looks at first glance much more involved than its finite counterpart (compare Definition 2.37), a closer look reveals the similarity: the sum in the definition of the cumulative probability is simply replaced by an integral, and the difference of weighted cumulative probabilities is replaced by a differential. Nothing special about this, it is just the usual process when proceeding from discrete to continuous situations.

We turn our attention now to stochastic dominance. Does CPT violate stochastic dominance? The answer is given by the following proposition:

Proposition 2.39 *CPT does not violate stochastic dominance, i.e., if A is stochastic dominant over B then $CPT(A) > CPT(B)$.*

[18]Here we consider again only the form defined in this book. In the original formulation we would need to write down two integrals for negative and positive outcomes and invert the direction of integration on the latter one. Compare the remark after Definition 2.37.

Proof We prove the case of finite outcomes. The general case is slightly tricky, in particular in the original formulation of CPT by Tversky and Kahneman, see, e.g., [Lév05].

Let x_i denote the potential outcomes of A and B. Let F_i denote the cumulative probabilities of A. Let G_i denote the cumulative probabilities of B. Then

$$\begin{aligned} CPT(A) &= \sum_{i=1}^{n} v(x_i)(w(F_i) - w(F_{i-1})) \\ &= \sum_{i=1}^{n} v(x_i)w(F_i) - \sum_{i=0}^{n-1} v(x_{i+1})w(F_i) \\ &= \sum_{i=1}^{n-1}(v(x_i) - v(x_{i+1}))w(F_i) + w(F_n)v(x_n). \end{aligned}$$

By Definition 2.36, we know that the probability to get a payoff of at most x_i with lottery A should be less or equal to the corresponding probability for lottery B. These probabilities are nothing else than F_i and G_i, and therefore we get $F_i \leq G_i$ for all $i = 1, \ldots, n$ and that there is at least one i such that $F_i < G_i$. Moreover, using the monotonicity of v, $v(x_i) - v(x_{i+1}) < 0$. Finally $F_n = 1 = G_n$, so we get

$$\begin{aligned} CPT(A) &= \sum_{i=1}^{n-1}(v(x_i) - v(x_{i+1}))w(F_i) + w(F_n)v(x_n) \\ &> \sum_{i=1}^{n-1}(v(x_i) - v(x_{i+1}))w(G_i) + w(G_n)v(x_n) = CPT(B). \end{aligned}$$

This concludes the proof. □

The final theoretical property that we hoped CPT to satisfy, since PT did not, is continuity. We expect that “similar” lotteries should have “similar” CPT values. The precise meaning of this will be explained in Sect. 2.4.5, for the moment we just convey that CPT is in fact continuous, compare Theorem 2.44. This excludes in particular any “event splitting effects”, in other words a lottery does not become more attractive if we partition an outcome into several very similar outcomes.

There is another attractive feature of CPT: it can be axiomatized. In other words, we can mimic the approach that von Neumann and Morgenstern used for Expected Utility Theory and define a set of axioms on preferences that describe Cumulative Prospect Theory. This has been observed first by Wakker and Tversky [Wak93]. Unfortunately, the axioms used are more complicated than in the case of Expected Utility Theory. The rough idea is first to replace the Independence Axiom with an equivalent set of (albeit less intuitive) axioms. This gives an alternative characterization of EUT. Then one weakens this assumption by restricting the validity of these axioms only to a certain subclass of prospects. This characterizes CPT.[19] By restricting the axioms to larger subclasses one also obtains two other decision models (Cumulative Utility and Sign-Dependent Expected Utility).

We have learned now that CPT is a conceptual adequate theory: it satisfies properties that we expect to hold for a behavioral theory for decisions under risk. Let us now take a look on the descriptive qualities of CPT. How well does CPT explain actual choices? Does it explain the phenomena we have encountered before as well as PT?

Let us first consider the Allais Paradox. If we choose v and w as the functions defined by Kahneman and Tversky (compare (2.7) and (2.8) for a definition) with the parameters $\lambda := 2.25$, $\alpha := \beta := 0.8$ and $\gamma := \delta := 0.7$, we can indeed explain the paradox by simply computing the CPT values of the four lotteries A, B, C and D. You may verify this as an exercise.

In general, we will also recover the four-fold pattern of risk-attitudes, but we have to change its definition slightly. Since we are not over– and underweighting solely depending on the size of the probabilities involved, things become a little bit more complicated. These complications, however, disappear as soon as we study the simple case of a lottery A with only two outcomes. The CPT functional in this case simply becomes

$$CPT(A) = w(p)v(x_1) + (1 - w(p))v(x_2).$$

Although this is not precisely the same formula as in PT,[20] it shares the same properties with it: small probabilities (either for x_1 or for x_2) are overweighted, large probabilities are underweighted. Since the value function has the same convex-concave shape in CPT as in PT, the four-fold pattern of risk-attitudes can be explained in exactly the same manner. – As long as we consider only two-outcome lotteries. This means in particular that we can explain the behavioral quirks that we encountered before: the life-death problem (Examples 2.33 and 2.34) and the fact that people both play lotteries and buy insurances.

We will see in the following chapters that CPT can also be used to explain several striking observations in finance, for instance the asset allocation puzzle. CPT has also been confirmed as a reasonable description of choices under risk by numerous quantitative studies.

After so much praise for this theory (which was a key reason for Daniel Kahneman to win the Nobel Prize in 2002 [TK92]), we also like to mention two limitations. To do so, we have to overcome a certain bias with which we have happily lived so far, namely that people are, if not fully, so at least partial, rational. We have tacitly assumed that people act according to the simple motto "more money is

[19]This characterization is mathematically quite involved. Brave readers might want to look into the original paper [Wak93].

[20]Unless the weighting function w satisfies the symmetry property $w(1-p) = 1 - w(p)$.

better" and apply the principle of stochastic dominance. Of course, one could always phrase a problem in a way that convinces people to make a wrong decision. (Some professions live from that.) But even if we provide clear, non-misleading conditions, this assumption, as natural as it seems, has been questioned severely in experiments. Let us have a look on the following example:

Example 2.40 There are two lotteries. In each case there are 100 marbles in total, one of which is drawn by chance. Every marble corresponds to a prize. The two lotteries have the following frequencies of marbles:

$A :=$

Number of marbles	Prize in €
90	96
5	14
5	12

,

$B :=$

Number of marbles	Prize in €
85	96
5	90
10	12

.

Which lottery do you prefer?

This example is taken from [BC97]. Which lottery did you choose? In several studies, a significant majority of persons preferred B over A. The percentage differed somehow with the educational background. (PhD students favored B only in around 50 % of the cases, whereas undergraduate students preferred in around 70 % of the cases.) What is wrong about this? You might have noticed that lottery A is stochastic dominant over B in the sense of Definition 2.36: the probability to win at least 96€ is larger for A, the probability to win at least 90€ is the same for both, the probability to win 14€ is again larger for A and in both cases you have the same probability (100 %) to win at least 12€, so in this sense, A really *is* better.

That A is stochastic dominant over B means in particular that not only EUT, but also CPT would predict a preference for A, since they both respect stochastic dominance. PT, however, can violate stochastic dominance, and in this particular case it can predict correctly that B is preferred over A. The reason for this difference is that PT overweights the intermediate outcome that occurs with only 5 % probability, but CPT does not. (Remember that CPT usually overweights only extreme events, not low probabilities in the mid-range.)

There are several other models for decision under risk that can predict such a behavior as well (e.g., RAM or TAX models, see [BN98, Bir05]), but since they are not used much in finance we refrain from describing them. Instead, we will give some information on a different way of extending PT that can describe this violation of stochastic dominance, but also allows for applications in finance (compare Sect. 2.4.6).

Important for us is to remember that we cannot expect people to follow *always* the stochastic dominance principle. Their decisions might deviate from this. This is not necessarily bad news, since deviations from rational behavior are, for instance, the key ingredients of active investment strategies! In *most* cases, however, assuming that preferences are compatible with stochastic dominance is a safe thing to do, and it is enough to consider the irrational behavioral patterns like overweighting of small probabilities and framing effect that can be described well with PT or CPT.

2.4.4 Choice of Value and Weighting Function

When we use CPT (or PT) to model decisions under risk, we need to decide what value and weighting functions to choose. There are, in principle, two methods to obtain information on their shape: one is to measure them directly in experiments, the other one is to derive them from principal considerations. The former is the way that Tversky and Kahneman originally went, the latter one mimics the ideas that Bernoulli went with the St. Petersburg Paradox in the case of EUT.

Measuring value functions in experiments follow the same ideas outlined in Sect. 2.2.4. The measurement of the weighting function is more difficult. Some information on this can be found in [TK92] or [WG96]. The original choice of Kahneman and Tversky seems reasonable in both cases, although different forms for the weighting function have been suggested, the most popular being

$$w(F) := \exp(-(-\ln(F))^{\gamma})$$

for $\gamma \in (0, 1)$, see [Pre98].

The measurement of these functions is of course limited to lotteries with relatively small outcomes. (Otherwise, laboratory experiments become too expensive.) This makes it also difficult to measure very small probabilities, since for small-stake lotteries, events with very small probability do not influence the decision much.

These are important restrictions if we want to apply behavioral decision theory to finance, since we will frequently deal with situations where large amounts of money are involved and where investment strategies may pose a risk connected to a very large loss occurring with a very small probability. We therefore are interested in finding at least some qualitative guidelines about the global behavior of value and weighting function based on theoretical considerations.

At this point it is helpful to go back to the St. Petersburg Paradox. We remember that the St. Petersburg Paradox in EUT was solved completely if we restricted ourselves to lotteries with *finite* expected value. Then the only structural assumption that we had to pose on the utility function was concavity above a fixed value.[21] Does this result also hold for CPT? A closer look at this reveals some subtle difficulty: the far-out events of the St. Petersburg Lottery are overweighted by CPT which leads to a more risk-seeking behavior. (Remember the four-fold pattern of risk-attitudes!) Therefore one might wonder whether it is not possible to construct lotteries that have a finite expected return, but nevertheless an infinite CPT value.

[21] Compare Theorem 2.25 (ii).

This observation has been done in [Bla05] and [RW06]. The following result gives a precise characterization of the cases where this happens. We formulate it for general probability measures, but its main conclusions holds of course also for discrete lotteries with infinitely many outcomes.

Theorem 2.41 (St. Petersburg Paradox in CPT [RW06]) *Let CPT be a CPT subjective utility given by*

$$CPT(p) := \int_{-\infty}^{+\infty} v(x) \frac{\mathrm{d}}{\mathrm{d}x}(w(F(x)))\,\mathrm{d}x,$$

where the value function v is continuous, monotone, convex for $x < 0$ and concave for $x > 0$. Assume that there exist constants $\alpha, \beta \geq 0$ such that

$$\lim_{x\to+\infty} \frac{u(x)}{x^\alpha} = v_1 \in (0, +\infty), \qquad \lim_{x\to-\infty} \frac{|u(x)|}{|x|^\beta} = v_2 \in (0, +\infty), \tag{2.10}$$

and that the weighting function w is a continuous, strictly increasing function from $[0, 1]$ to $[0, 1]$ such that $w(0) = 0$ and $w(1) = 1$. Moreover assume that w is continuously differentiable on $(0, 1)$ and that there is a constant γ such that

$$\lim_{y\to 0} \frac{w'(y)}{y^{\gamma-1}} = w_0 \in (0, +\infty). \tag{2.11}$$

Let p be a probability distribution with $\mathbb{E}(p) < \infty$ and $\mathrm{var}(p) < \infty$. Then $CPT(p)$ is finite if $\alpha < \gamma$ and $\beta < \gamma$. This condition is sharp.

In particular, the CPT value may be infinite for distributions with finite EV in the usual parameter range where $\alpha > \gamma$.

What does this tell us about CPT as a behavioral model? Did it fail, because it cannot describe this variant of the St. Petersburg Paradox? Fortunately, this is not the case: we can restrict the theory to a subclass of lotteries or we can change the shape of the value and/or weighting function. Roughly spoken, one can show that there are three ways to fix the problem [RW06]:

1. If we allow only for probability distributions with exponential decay at infinity (or even with bounded support), the problem does not occur. In many applications, this is the case, for instance if we study normal distributions or finite lotteries. However, in problems where we are interested in finding the optimal probability distribution (subject to some constraints), it might well happen that we obtain a "solution" with infinite subjective utility. This renders CPT useless for applications like portfolio optimization.
2. We could modify the weighting function w such that $w'(0)$ and $w'(1)$ are finite. This guarantees a finite subjective utility, independently of the choice of the value function (as long as it has a convex–concave structure).

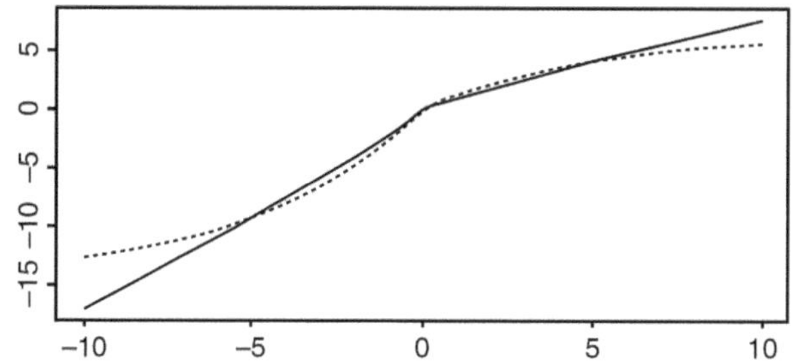

Fig. 2.15 Comparing classical (*solid line*) and exponential (*dashed line*) value function: they agree for small, but disagree for large outcomes

3. The value function can be modified for large gains and losses such that it is bounded. This again ensures a finite subjective utility. This is probably the best fix, since there are other theoretical reasons in favor of a bounded value function, compare Sect. 3.4.

There is of course a very strong reason in favor of keeping weighting and value function unchanged, namely that it has been introduced in a groundbreaking article and has subsequently used by many other people. Although this argument sounds strange at first, and arguments like this are often not fostering the scientific progress, there is in this case some grain of truth in it, namely that there is already a large amount of data on measuring CPT parameters, all based on the standard functional forms of value and weighting function. Changing the model means reanalyzing the data, estimating new parameters and generally making different studies less compatible.

How can we avoid such problems and still use functional forms that satisfy reasonable theoretical assumptions?

Fortunately, there are simple bounded value functions that are very close to the x^α-function used by Tversky and Kahneman, e.g. the exponential functions

$$v(x) := \begin{cases} \lambda^- e^{-\alpha x} - \lambda^- & , \text{ for } x < 0, \\ -\lambda^+ e^{-\alpha x} + \lambda^+ & , \text{ for } x \geq 0, \end{cases} \tag{2.12}$$

where the ratio λ^-/λ^+ corresponds to the loss aversion λ in PT and CPT, and α reflects the risk aversion (similar to PT and CPT). This function has been suggested in [DGHP05]. In Fig. 2.15 we compare the classical value function with the bounded variant. We see that the agreement for small values of x is very good. Since experiments are typically performed in this range, the descriptive behavior of both value functions should be very similar. For large values there is a strong disagreement which resolves the St. Petersburg Paradox and helps us applying CPT to problems in finance where we need a reasonable behavior of the CPT functional for lotteries involving the possibility of large gains and losses.

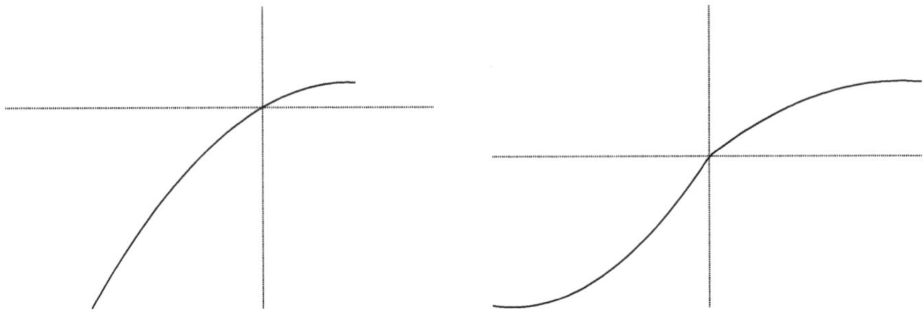

Fig. 2.16 A piecewise quadratic value function can describe MV- or PT-like preferences, depending on the parameters chosen

Another interesting example of an alternative value function has been introduced in [ZK09]: it makes an interesting connection between MV and PT by providing a common framework for both. Let us define the value function as

$$v(x) := \begin{cases} x - \alpha x^2 & , \text{for } x < 0, \\ \lambda(x + \beta x^2) & , \text{for } x \geq 0, \end{cases} \tag{2.13}$$

then for the case $\alpha = -\beta$ and $\lambda = 1$ we obtain a quadratic value function which implies that the corresponding decision model is the MV model – at least up to possible probability weighting and framing. By adjusting the parameters α, β and λ we can therefore generalize MV into the framework of PT which turns out particularly useful for applications in finance, compare Fig. 2.16. We will therefore use this functional form occasionally in later chapters.

There is, of course, the usual drawback in this specification that we inherit from PT and which is related to the mean-variance puzzle: the value function becomes decreasing for large values, thus we have to make sure that our outcomes do not become too large.

We have now developed the necessary tools to deal with decision problems in finance, from a rational and from a behavioral point of view. In the following section (which is intended for the advanced reader and is therefore marked with a *) we will discuss an interesting, but mathematically complicated concept in detail, namely continuity of decision theories. Afterwards, still marked with a * as warning to the nonspecialist, we introduce a different extension of PT that keeps more of the initial ideas of PT than CPT, but can nevertheless be extended to arbitrary lotteries. It might therefore be of some use for applications in finance, in particular in situations where the computation of CPT is computationally too difficult.

The nonspecialist might now turn his attention to Sect. 2.5, where we draw some connections between EUT, MV, PT and CPT. In particular, we will try to understand in which cases these theories agree, where they disagree, and in what situations we should apply them.

2.4.5 Continuity in Decision Theories*

We have already several times encountered the fundamental notion of *continuity*. This is a central property not only of decision theories, but virtually of all mathematical models, be it in economics, natural sciences or engineering. Its main insight is, that a model is only valuable if it allows for predictions that can be checked experimentally. In other words, for some given data, the model computes values for quantities that can be measured. Since the given data can in practical applications be never given with infinite precision, and it is also generally impossible to do computations with infinite accuracy, a fundamental property, which a reasonable model should satisfy, is that a slight change in the data only leads to a slight change in the predicted quantities. We call such behavior "continuous".

Of course many systems are not continuous under every circumstances: think about the movement of a pendulum (a mass, attached with a bar to a fixed point) which can be predicted by the laws of gravity with high accuracy, unless we put the mass directly above the fixed point, from which a movement to either sides is equally likely and determined by indiscernibly small changes in the initial position (i.e., the data). However, in reasonable models such non-continuous situations should be a rare exception.

At this point, we need to add a word of warning: unfortunately, the word "continuous" has two quite different meanings in the English language: first, *continuous* means *non-discrete*. We have already used this notion when talking about measures (or lotteries), and we have seen how to extend the notion of EUT and CPT to such continuous distributions. Second, *continuous* means *not discontinuous*. *This* is what we mean when we speak about continuity in this section. Historically, both ideas are related, but nowadays they are distinct properties that should not be mixed up.

If we want to know whether a decision theory is continuous, we need to find a mathematically precise definition of continuity. In order to define continuity, we need to define what it means if $U(A_n) \to U(A)$, i.e., when the sequence of lotteries A_n *converges* to the lottery A. We know from calculus what it means if a sequence of numbers converges to another number, but what does it mean when lotteries converge? Intuitively, we would say that for $n \to \infty$, the following sequence of lotteries A_n converges to the lottery A with the certain outcome of 1:

$$A_n := \begin{array}{lcc} \hline \text{Outcome} & 1-\frac{1}{n} & 1+\frac{1}{n} \\ \text{Probability} & 1/3 & 2/3 \\ \hline \end{array}.$$

But how can we formulate this in mathematical terms? Fortunately, we can describe lotteries (in the state-independent setting which we consider here) by probability measures. There is a well-developed mathematical concept for the convergence of probability measures, but before giving the mathematical definition, we want to motivate it a little: we could say that a sequence of probability measures p_n converges to a probability measure p if every expected utility of p_n converges to the expected utility of p. This would imply that no rational investor would see a difference between p and the limit of p_n. This idea leads to the mathematical concept of weak-$\star$-convergence:

Definition 2.42 (Weak-$\star$-convergence of probability measures) We say that a sequence $\{p_n\}$ of probability measures on $\mathbb{R}^N$ converges weakly-$\star$ to a probability measure p if for all bounded continuous functions f

$$\int_{\mathbb{R}^N} f(x)\, \mathrm{d}p_n(x) \to \int_{\mathbb{R}^N} f(x)\, \mathrm{d}p(x)$$

holds. We write this as $p_n \overset{\star}{\rightharpoonup} p$. The function f is sometimes called a *test function*.

To see the correspondence to the intuitive approach sketched above, we can consider $f(x)$ as a utility function.

In the above example, we can easily check that this definition is satisfied: first consider as f the indicator function[22] on some interval $[x_1, x_2]$: in fact, if $x_2 < 1$, then the integral of A_n becomes zero when n is large enough, and the integral of A over this interval is also zero. The same holds if $x_1 > 1$. If $x_1 \leq 1 \leq x_2$ then the integral of A_n becomes eventually 1 and the integral of A is 1 as well. We then can approximate an arbitrary continuous function f by sums of indicator functions.

We can now formulate the definition of continuity[23]:

Definition 2.43 (Continuity of a utility functional) We say that a utility functional U is *continuous*, if for all sequences of lotteries A_n with $A_n \overset{\star}{\rightharpoonup} A$ we have $U(A_n) \to U(A)$.

The concept of continuity, so natural it is in other situations, seems at first glance quite involved in the case of decision theory. However, having in mind that the mathematical formalism is just a way to clarify a quite intuitive concept (namely that "similar" lotteries should be evaluated in a "similar" way), is the main message we want you to remember.

Regarding the decision models we have encountered so far, we state that PT is discontinuous, whereas EUT, Mean-Variance Theory and CPT are continuous. We sketch a proof for the most complicated case, CPT. The other cases are left as an exercise for the mathematically inclined reader.

[22]The indicator function is of course not continuous, but one can work around this problem by approximating the indicator function by continuous functions – a quite useful little trick that works here.

[23]This definition is not related to the "Continuity Axiom" of von Neumann and Morgenstern (Axiom 2.18), even though the (unfortunate) name of the axiom suggests this.

Theorem 2.44 *If the weighting function w is continuously differentiable on* $(0, 1)$ *and the value function v is continuous, then CPT is weak-$\star$ continuous.*

Proof We assume for simplicity that p is absolutely continuous. If $p_n \stackrel{\star}{\rightharpoonup} p$, then, by definition, $\int f \, \mathrm{d}p_n \to \int f \, \mathrm{d}p$ for all bounded continuous functions f. Using that p_n and p are probability measures and that p is absolutely continuous, one can prove that $F_n(x) = \int_{-\infty}^{x} \mathrm{d}p_n \to \int_{-\infty}^{x} \mathrm{d}p = F(x)$ for all $x \in \mathbb{R}$. Since w' is continuous, also $w'(F_n) \to w'(F)$. We compute

$$\int_{-\infty}^{+\infty} v(x) \frac{\mathrm{d}}{\mathrm{d}y} \left(w(F_n(y)) \right) |_{y=x} \, \mathrm{d}x = \int_{-\infty}^{+\infty} v(x) w'(F_n(x)) \, \mathrm{d}p_n(x).$$

This is a product of a weak-$\star$ converging term and a pointwise converging term. Using a standard result from functional analysis, this converges to the desired expression. □

2.4.6 Other Extensions of Prospect Theory*

Since we have seen that not all properties of CPT correspond well with experimental data (in particular its lack of violations of stochastic dominance), there are some descriptive reasons favoring PT. There is another, practical argument in favor of PT: computations in finance often involve large data sets and involved optimizations. In this case, PT is the computationally simpler model, since it does not need outcomes to be sorted by their amounts. For these reasons it is useful to look for an extension of PT to arbitrary (not necessarily discrete) lotteries. This is in fact possible if we use the variant of PT introduced by [Kar78], i.e.

$$PT(p) := \frac{\sum_{i=1}^{n} v(x_i) w(p_i)}{\sum_{i=1}^{n} w(p_i)}.$$

We assume as before that the weighting function w behaves for p close to zero like p^{γ} (with some $\gamma > 0$), compare (2.11).

The result of [RW08] is now summarized in the following theorem:

Theorem 2.45 *Let p be a probability distribution on $\mathbb{R}$ with exponential decay at infinity and let p_n be a sequence of discrete probability measures with outcomes $x_{n,z}$ in equal distances of $1/n$ (each with probability $p_{n,z}$), i.e., $x_{n,z+1} = x_{n,z} + \frac{1}{n}$. Let $p_n \stackrel{\star}{\rightharpoonup} p$. Assume that the value function $v \in C^1(\mathbb{R})$ has at most polynomial growth and that the weighting function w: $[0, 1] \to [0, 1]$ satisfies the above condition. Then the normalized PT utility*

$$PT(p_n) := \frac{\sum_z w(p_{n,z}) v(x_{n,z})}{\sum_z w(p_{n,z})}$$

converges to

$$\lim_{n\to\infty} PT(p_n) = \frac{\int v(x)p(x)^\gamma \, dx}{\int p(x)^\gamma \, dx}.$$

This limit functional can therefore be considered as a version of PT for continuous distributions. A small problem is that we need to choose particular approximating sequences for p. Remark 2.50 shows how this can be fixed.

Theorem 2.45 can be generalized to lotteries that also contain singular parts. We summarize this in the following definition:

Definition 2.46 If p is a probability measure that can be written as a sum of finitely many weighted Dirac masses[24] $\pi_i\delta_{x_i}$ and an absolutely continuous measure p_a, i.e., $p = p_a + \sum_{i=1}^n \pi_i\delta_{x_i}$, then we can define

$$PT(p) := \frac{\sum_{i=1}^n v(x_i)\pi_i^\alpha + \int v(x)p_a(x)^\alpha \, dx}{\sum_{i=1}^n \pi_i^\alpha + \int p_a(x)^\alpha \, dx}.$$

Remark 2.47 The normalization is necessary, since otherwise the limit functional is either infinite (if $\gamma < 1$) or equivalent to a version of EUT (if $\gamma = 1$). Thus there would be no probability weighting in the limit.

Let us finally have a look at a related extension of PT [RW08]. Smooth Prospect Theory (SPT) encompasses parts of the editing phase of PT into the functional form, in that it collects "nearby" outcomes to one. This leads to a functional which is, unlike PT, continuous in the sense of the last section. We give here only its definition and some remarks on its properties:

Definition 2.48 Let p be a discrete outcome distribution. Then we define

$$SPT_\varepsilon(p) := \frac{\int w\left(\int_{x-\varepsilon}^{x+\varepsilon} dp\right) v(x) \, dx}{\int w\left(\int_{x-\varepsilon}^{x+\varepsilon} dp\right) dx}. \tag{2.14}$$

Remark 2.49 The parameter $\varepsilon > 0$ marks how small the distance between two outcomes can be until they are collected to one outcome. As long as $\varepsilon > 0$, SPT is continuous. It converges to PT when $\varepsilon \to 0$.

The definition of SPT allows us to generalize the convergence result of Theorem 2.45 to arbitrary approximating sequences:

[24] For a definition of Dirac masses, see Appendix A.4.

Remark 2.50 If $p^k \overset{\star}{\rightharpoonup} p$, then, for all sequences $k(\varepsilon) \to \infty$ that converge sufficiently slowly as $\varepsilon \to 0$, the SPT utility of p^k converges to $PT(p)$, i.e.:

$$\lim_{\varepsilon \to 0} SPT_\varepsilon(p^{k(\varepsilon)}) = PT(p) = \frac{\int v(x)p(x)^\alpha \, dx}{\int p(x)^\alpha \, dx}.$$

Proofs and further details on these results can be found in [RW08].

2.5 Connecting EUT, Mean-Variance Theory and PT

The main message of the last sections is that there are several different models for decisions under risk, the most important being EUT, Mean-Variance Theory and PT/CPT. The question we need to ask is: how important are the differences between these models? Maybe in "natural" cases all (or some) of these theories agree? In this section, we will check this idea. Moreover we will characterize the different approaches and their fields of applications. You should then be able to judge in a given situation which model is best to be applied.

First, we compare EUT and Mean-Variance Theory. Are they in general the same? Obviously not, since we have demonstrated in Theorem 2.30 that Mean-Variance Theory can violate state dominance, but we have seen in Sect. 2.2 that EUT does not, hence both theories cannot coincide. This shows that it is usually not possible to describe a *rational* person by Mean-Variance Theory.

This is certainly bad news if you still believed that Mean-Variance Theory is *the* way of modeling decisions under risk, but maybe we can rescue the theory by restricting the cases under consideration? This is in fact possible, and there are several important cases where Mean-Variance Theory can be interpreted as a special variant of EUT:

- If the von Neumann-Morgenstern utility function is quadratic.
- If the returns are all normally distributed.
- If the returns all follow some other special patterns, e.g., they are all lotteries with two outcomes of probability $1/2$ each.
- In certain time-continuous trading models.

We will state in the following a couple of theorems that make these cases precise and show how they lead to an equivalence between both theories. First we define:

Definition 2.51 Let $\succeq$ be an expected utility preference relation. We call EUT and Mean-Variance *compatible* if there exists a von Neumann-Morgenstern utility function $u(x)$ *and* a mean-variance utility function $v(\mu, \sigma)$ which both describe $\succeq$.

We have the following result:

Theorem 2.52 *Let $\succeq$ be a preference relation on probability measures.*

(i) If u is a quadratic von Neumann-Morgenstern utility function describing $\succeq$, then there exists a mean-variance utility function $v\,(\mu, \sigma)$ which also describes $\succeq$.
(ii) If $v\,(\mu, \sigma)$ describes $\succeq$ and there is a von Neumann-Morgenstern utility function u describing $\succeq$, then u must be quadratic.

Proof We prove (i): Let us write u as $u(x) = x - bx^2$. (We can always achieve this by an affine transformation.) The utility of a probability measure p is then

$$\begin{aligned} EUT(u) &= \mathbb{E}_p(u(x)) = \mathbb{E}_p(x - bx^2) = \mathbb{E}_p(x) - b\mathbb{E}_p(x^2) \\ &= \mathbb{E}(p) - b\mathbb{E}(p)^2 - b\,\mathrm{var}(p) = \mu - b\mu^2 - b\sigma^2 =: v(\mu, \sigma). \end{aligned}$$

The proof of (ii) is more difficult, see [Fel69] for details and further references. □

There is of course a problem with this result: a quadratic function is either affine (which would mean risk-neutrality and is not what we want) or its derivative is changing sign somewhere (which means that the marginal utility would be negative somewhere, violating the "more money is better" maxim) or that the function is strictly convex (but that would mean risk-seeking behavior for all wealth levels). None of these alternatives looks very appealing. The only case where this theorem can be usefully applied is when the returns are bounded. Then we do not have to care about a negative marginal utility above this level, since such returns just do not happen. The utility function looks then like $u(x) = x - bx^2$, $b > 0$, where $u'(x) > 0$ as long as we are below the bound. The minus sign ensures that $u'' < 0$, i.e., u is strictly concave. The drawback of this shape is that on the one hand it does not correspond well to experimental data and on the other hand there is no reason why this particular shape of a utility function should be considered as the only rational choice.

More important are cases where the compatibility is restricted to a certain subset of probability measures, e.g., when we consider only normal distributions:

Theorem 2.53 *Let $\succeq$ be an expected utility preference relation on all normal distributions. Then there exists a mean-variance utility function $v(\mu, \sigma)$ which describes $\succeq$ for all normal distributions.*

This means that, if we restrict ourselves to normal distributions, we can always represent an EUT preference by a mean-variance utility function.

Proof Let $N_{\mu,\sigma}$ be a normal distribution. Then using some straightforward computation and the substitution $z := (x - \mu)/\sigma$, we can define v:

$$EUT(u) = \mathbb{E}_p(u(x)) = \int_{-\infty}^{\infty} u(x)N_{\mu,\sigma}(x)\,\mathrm{d}x = \int_{-\infty}^{\infty} u(\mu + \sigma z)\frac{1}{\sqrt{2\pi}}\mathrm{e}^{-\frac{z^2}{2}}\,\mathrm{d}z$$
$$= \int_{-\infty}^{\infty} u(\mu + \sigma z)N_{0,1}(z)\,\mathrm{d}z =: v(\mu, \sigma).$$

□

This idea can be generalized: the crucial property of normal distributions is only that all normal distributions can be described as functions of their mean and their variance. There are many classes of probability measures, where we can do the same. In this way, we can modify the above result to such "two-parameter families" of probability measures, e.g., to the class of log-normal distributions or to lotteries with two outcomes of probability $1/2$ each.

After discussing the cases where Mean-Variance Theory and EUT are compatible, it is important to remind ourselves that these cases do *not* cover a lot of important applications. In particular, we want to apply our decision models to investment decisions. If we construct a portfolio based on a given set of available assets, the returns of the assets are usually assumed to follow a normal distribution. This allows for the application of Mean-Variance Theory as we have seen in Theorem 2.53. The assumption, however, is not necessarily true as we can invest into options and their returns are often not at all normally distributed. Given the manifold variants of options, it seems also quite hopeless to find a different two-parameter family to describe their return distributions.

We could also argue that the returns are bounded. Even if it is difficult to give a definite bound for the returns of an asset, we might still agree that there exists at least *some* bound. We could then apply Theorem 2.52, but this would mean that the utility function in the EUT model must be quadratic. Although theoretically acceptable, this seems not to fit well with experimental measurements of the utility function.

Finally, time-continuous trading is not the right framework in which to cast typical financial decisions of usual investors.

Therefore we see that there are many practical situations where Mean-Variance Theory does not work as a model for rational decisions. On the other hand, there are many situations where it is at least not too far from EUT (e.g., if the assets are not too far from being normally distributed etc.) and since Mean-Variance Theory is mathematically by far simpler than EUT, it is often for pragmatic reasons a good decision to use Mean-Variance Theory. However, results obtained in this way should always be watched with a critical eye, in particular if they seem to contradict our expectations.

How is it now with CPT (as prototypical representative of the PT family)? When does it reduce to a special case of EUT? How is its relation to Mean-Variance Theory?

Again, we see immediately, that CPT in general neither agrees with EUT nor with Mean-Variance Theory: it satisfies stochastic dominance, hence it cannot agree with Mean-Variance Theory, and it does not satisfy the Independence Axiom, thus it cannot agree with EUT.

How is it in the special case of normal distributions? In this case, the probability weighting does in fact not make a qualitative difference between CPT and Mean-Variance Theory, but the convex-concave structure of the value function can lead to risk-seeking behavior in losses, as we have seen. This implies that a person prefers a larger variance over a smaller variance, when the mean is fixed and contradicts classical Mean-Variance Theory.

We could also wonder how CPT relates to EUT if the probability weighting parameter becomes one, i.e., there is no over– and underweighting. In this case we arrive at some kind of EUT, but only with respect to a frame of gains and losses and not to final wealth. A person following this model, which is nothing else than the Rank-Dependent Utility (RDU) model, is therefore still not acting rationally in the sense of von Neumann and Morgenstern. We cannot see this from a single decision, but we can see this when we compare decisions of the same person for different wealth levels. There is only one case where CPT really coincides with a special case of EUT, namely when not only the weighting function parameter, but also the value function parameter and the loss aversion are one. In this case CPT coincides with a risk-neutral EUT maximizer, in other words a maximizer of the expected value.

On the other hand, we should not forget that CPT is only a modification of EUT. Therefore its predictions are often quite close to EUT. We might easily forget about this, since we have concentrated on the cases (like Allais' paradox) where both theories disagree. Nevertheless for many decisions under risk, neither framing effect nor probability weighting play a decisive role and therefore both models are in good agreement. We can illustrate this in a simple example:

Example 2.54 Consider lotteries with two outcomes. Let the low outcome be zero and the high outcome x million €. Denote the probability for the low outcome by p. Then we can compute the certainty equivalent (CE) for all lotteries with $x \geq 0$ and $p \in (0, 1)$ using EUT, Mean-Variance Theory, CPT. To fix ideas, we use for EUT the utility function $u(x) := x^{0.7}$ and an initial wealth level of 5 million €. For Mean-Variance Theory we fix the functional form $\mu - \sigma^2$ and for CPT we choose the usual function and parameters as in [TK92]. How do the predictions of the theories for the CE agree or disagree?

The result of this example is plotted in Fig. 2.17.

Summarizing we see that EUT and Mean-Variance Theory coincide in certain special situations; CPT usually disagrees with both models, but does often not deviate too much from EUT. We summarize the similarities and differences of EUT, Mean-Variance Theory and CPT in a diagram, see Fig. 2.18

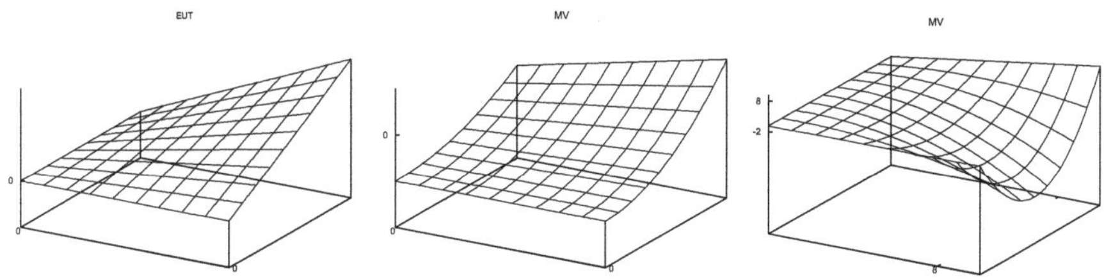

Fig. 2.17 Certainty equivalents for a set of two outcome lotteries for different decision models: EUT (*left*), CPT (*center*), Mean-Variance Theory (*right*). Small values for the high outcome x of the lottery are *left*, large values *right*. A small probability p to get the low outcome (zero) is on the back, a large probability on the front. The height of the function corresponds to its Certainty Equivalent

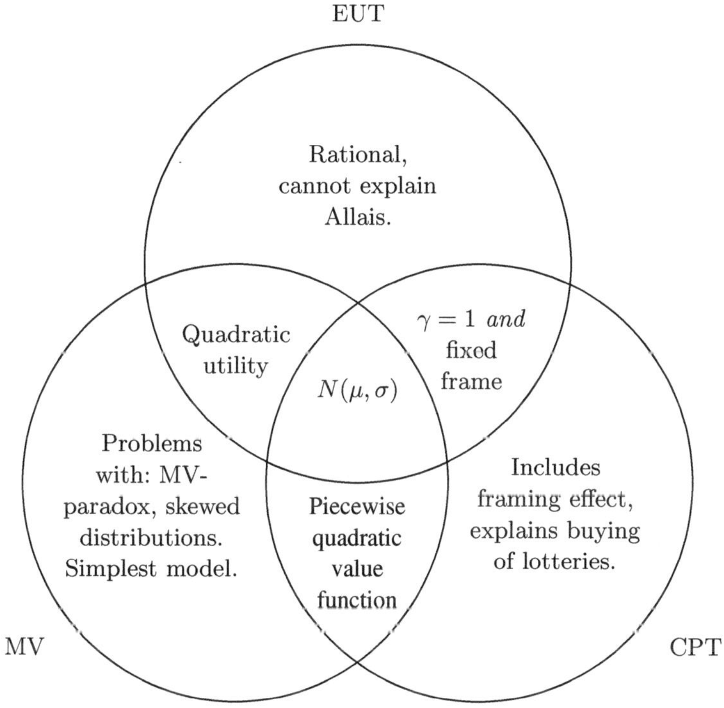

Fig. 2.18 Differences and agreements of EUT, PT and Mean-Variance

What does this tell us for practical applications? Let us sketch the main areas of problems where the three models excel:

- EUT is the "rational benchmark". We will use it as a reference of rational behavior and as a prescriptive theory when we want to find an objectively optimal decision.

- Mean-Variance Theory is the "pragmatic solution". We will use it whenever the other models are too complicated to be applied. Since the theory is widely used in finance, it can also serve as a benchmark and point of reference for more sophisticated approaches.
- CPT (and the whole PT family) model "real life behavior". We will use it to describe behavior patterns of investors. This can explain known market anomalies and can help us to find new ones. Ultimately this helps, e.g., to develop new financial products.

We will observe that often more than one theory needs to be applied in one problem. For instance, if we want to exploit market biases, we need to model the market with a behavioral (non-rational) model like CPT and then to construct a financial product based on the rational EUT. Or we might consider the market as dominated by Mean-Variance investors and model it accordingly, and then construct a financial product along some ideas from CPT that is taylor-made to the subjective (and not necessarily rational) preferences of our clients.

In the next chapters we will develop the foundations of financial markets and will use all of the three decision models to describe their various aspects.

2.6 Ambiguity and Uncertainty*

We have defined at the beginning of this chapter that *risk* corresponds to a multitude of possible outcomes whose probabilities are known. Often we deal with situations where the probabilities are not known, sometimes they cannot even be estimated in a reasonable way. (What is the probability that a surprising new scientific invention will render the product of a company we have invested in useless?) In other occasions, there are ways to quantify the probabilities, but a person might not be aware of these probabilities. (Somebody who has no idea of the stock market will have no idea how (un)likely it is to lose half of his wealth when investing into a market portfolio, although a professional investor will be able to quantify this probability.) We call this *ambiguity* or *uncertainty*.[25]

The difference between risk and uncertainty has first been pointed out by F. Knight in 1921, see [Kni21]. For the actual behavior of people, this difference is very important, as the famous Ellsberg Paradox [Ell61] shows:

Example 2.55 There is an urn with 300 balls. 100 of them are red, 200 are blue or green. You can pick red or blue and then take one ball (blindly, of course). If it is of the color you picked, you win 100€, else you don't win anything. Which color do you choose?

[25]Sometimes there are attempts in the literature to use both words for slightly different concepts, but so far there seems to be no commonly accepted definition, hence we take them as synonyms and will usually use the word "uncertainty".

Which color did you choose? Most people choose red. Let us go to the second experiment:

Example 2.56 Same situation, you pick again a color (either red or blue) and then take a ball. This time, if the ball is *not* of the color you picked, you win 100€, else you don't win anything. Which color do you choose?

Here the situation is different: if you pick red, you win if either blue or green is chosen, and although you do not know the number of the green or the number of the blue balls, you know that there are in total 200. Most people indeed pick red.

However, this seems a little strange: let us say, in the first experiment you must have estimated that there are fewer blue balls than red balls, and hence picked red. Then in the second experiment you should have chosen blue, since the estimated combined number of red and green balls would be larger than the combined number of blue and green balls.

What happens in this experiment is that people go both times for the "sure" option, the option where they know their probabilities to win. In a certain way, this is nothing else than risk-aversity, but of a "second order", since the "prizes" are now probabilities! One possible explanation of this experiment is therefore that people tend to apply their way of dealing with risky options, which works (more or less) well for decisions on lotteries,[26] also to situations where they have to decide between different probabilities. This is very natural, since these winning-probabilities can be seen as "prizes", and it is natural to apply the usual decision methods that one uses for other "prizes" (being it money, honor, love or chocolate). Unfortunately, probabilities are different, and so we run into the trap of the Ellsberg Paradox.

It is interesting to notice that the "uncertainty-aversity" that we observed in the Ellsberg Paradox occasionally reverts to an uncertainty-seeking behavior, in the same way, the four-fold pattern of risk-attitudes can lead to risk-averse behavior in some instances and to risk-seeking behavior in others.

This is, however, only one possible explanation, and the Ellsberg Paradox and its variants are still an active research area, which means that there are many open questions and not many definite answers yet.

The Ellsberg Paradox has of course interesting implications to financial economics. It yields, for instance, immediately a possible answer to the question why so many people are reluctant to invest into stocks or even bonds, but leave their money on a bank account: besides the problem of procrastination ("I will invest my money tomorrow, but today I am too busy.") which we will discuss in the next section, these people are often not very knowledgeable about the chances and risks of financial investments. It is therefore natural that when choosing between a known and an unknown risk, i.e., between a risk and an uncertain situation, they choose the safe option. This also explains why many people invest into very few stocks (that

[26]We have seen that CPT models such decisions quite well, and that the rational decisions modeled by EUT are not too far away from CPT.

they are familiar with) or even only into the stock of their own company (even if their company is not performing well).

2.7 Time Discounting

Often, financial decisions are also decisions about time. Up to now we have not considered effects on decisions induced by time. In this little section we will introduce the most important notion regarding time dependent decisions, the idea of *discounting*.

A classical example for financial decisions strongly involving the time component is retirement, where the consumption is reduced *today* in order to save for *later*.

If you are faced with a decision to either obtain 100€ now or 100€ in 1 year, you will surely choose the first alternative. Why this? According to the classical EUT both should be the same, at least at first glance. On a second look, one notices that investing the 100€ that you get today will yield an interest, thus providing you with more than 100€ after 1 year. There are other very rational reasons not to wait, e.g., you may simply die in the meanwhile not being able to enjoy the money after 1 year. In real life, you might also not be sure whether the offer will really still hold in 1 year, so you might prefer the "sure thing".

In all these cases, the second alternative is reduced in its value. In the simplest case, this reduction is "exponential" in nature, i.e., the reduction is proportional to the remaining utility at every time: if we assume that the proportion by which the utility u decreases is constant in time, we obtain the differential equation $u'(t) = -\delta u(t)$, where $\delta > 0$ is called *discounting factor*. This reduces the original utility $u(0)$ after a time $t > 0$ to

$$u(t) = u(0)\mathrm{e}^{-\delta t}, \tag{2.15}$$

as we can see by solving the differential equation. If we consider only discrete time steps $i = 1, 2, \ldots$, we can write the utility as $u(0)\delta^i$ (where the δ does not necessarily have the same value as before). To see this, set $t = 1, 2, \ldots$ in (2.15).

Classical time discounting is perfectly rational and leads to a time-consistent preference: if a person prefers A now over B after a time t, this person will also prefer A after a time s over B after a time $s + t$ and vice versa:

$$\begin{aligned} u_B(s+t) - u_A(s) &= u_B(0)\mathrm{e}^{-\delta(s+t)} - u_A(0)e^{-\delta(s)} \\ &= \mathrm{e}^{-\delta(s)}\left(u_B(0)\mathrm{e}^{-\delta(t)} - u_A(0)\right) \\ &= \mathrm{e}^{-\delta(s)}\left(u_B(t) - u_A(0)\right), \end{aligned}$$

where we use that $\mathrm{e}^{-\delta t}$ is a positive constant that does not influence the sign of the last expression.

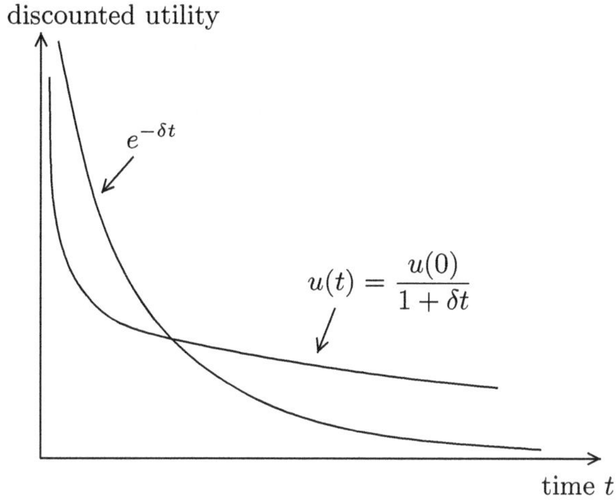

Fig. 2.19 Rational versus hyperbolic time discounting

Experience, however, shows that people do not behave according to the classical discounting theory: in a study test persons were asked to decide between 100hfl (former Dutch currency) now and 110hfl in 4 weeks [KR95]. Eighty-two percent decided that they preferred the money now. Another group, however, preferred 110hfl in 30 weeks over 100hfl in 26 weeks with a majority of 63 %. This is obviously not time-consistent and hence cannot be explained by the classical discounting theory. This phenomenon has been frequently confirmed in experiments. The extent of the effect varies with level of education, but also depends on the economic situation and cultural factors. For a large international survey on this topic see [WRH16].

The standard concept in economics and particularly in finance to model this behavior is the so-called "hyperbolic discounting". The utility at a time t is thereby modeled by a hyperbola, rather than an exponential function, following the equation

$$u(t) = \frac{u(0)}{1+\delta t}$$

where δ is the *hyperbolic discounting factor*, compare Fig. 2.19.

A similar definition is also often called hyperbolic discounting (or more accurately "quasi-hyperbolic" discounting), namely

$$u(t) = \begin{cases} u(0) & \text{, for } t = 0, \\ \frac{1}{1+\beta} u(0) e^{-\delta t} & \text{, for } t > 0, \end{cases} \qquad \text{where } \beta > 0.$$

Hyperbolic discounting explains the behavioral pattern observed in the experiment by Roelofsma and Keren [KR95] and similar ones. Nevertheless, there is also some serious criticism against this concept, notably by Rubinstein [Rub03] who points out that there are other inconsistencies in time-dependent decisions that cannot be explained by hyperbolic discounting, and that therefore the case for this model is not very strong. There is also recent work by Gerber [GR07] that demonstrates how uncertainties in the future development of a person's wealth can lead to effects that look like time-inconsistencies, but actually are not: in the classical experiment by Roelofsma and Keren, the results could e.g., be explained by classical time-discounting if people are nearly as unsure about their wealth level in the next week as in 30 weeks: the uncertainty of the wealth level reduces the expected utility of a risk-averse person at a given time. Although hyperbolic discounting is therefore not completely accepted, it is nevertheless a useful descriptive model for studying time-discounting.

A popular application of hyperbolic time-discounting is the explanation of undersaving for retirement. Here we give an example where hyperbolic discounting is combined with the framing effect:

Example 2.57 (Retirement) Assume a person has at time $t = 0$ a certain amount of money $w := 1$ which he could save for his retirement at time $t = 10$ yielding a fixed interest rate of $r := 0.05$. Alternatively, he can consume the interest rate of this amount immediately. The extra utility gained by consuming the interest rate wr is assumed to be wr and the utility gained by a total saving of x at the retirement age is $2x$, the factor 2 taking care of the presumably larger marginal utility at the retirement age, where the income, and hence the wealth level, shrinks. The hyperbolic discounting constant is $\delta = 0.25$. Does the person save or not?

We assume for simplicity that the person would either *always* or *never* save. A first approach would compare the discounted utility of the alternative "never saving" with the alternative "always saving". A short computation gives

$$u(\text{always saving}) = \frac{u(w(1+r)^t)}{1+\delta t} = \frac{2 \times 1.05^{10}}{3.5} \approx 0.9308,$$

$$u(\text{never saving}) = \frac{u(w)}{1+\delta t} + \sum_{s=0}^{t} \frac{u(rw)}{1+\delta s} \approx 0.8550.$$

This would imply that the person is indeed saving for his retirement. However, the decision whether or not to save might be framed differently: the person might decide on whether to start saving *now* or *tomorrow*. If he applies this frame[27] then his

[27]This framing seems at least to be used frequently enough to produce proverbs like "A stitch in time saves nine" and "Never put off till tomorrow what you can do today".

computation looks like this:

$$u(\text{start saving today}) = u(\text{always saving}) \approx 0.9308,$$

$$u(\text{start saving next year}) = \frac{u(w(1+r)^{t-1})}{1+\delta t} + u(wr) \approx 0.9365.$$

"Starting to save next year" is therefore the preferred choice – until next year, where the new alternative "starting to save yet another year later" suddenly becomes very appealing.

This theoretical explanation can also be verified empirically, e.g. by comparing data on time discounting from various countries with household saving rates [WRH16]: households in countries where people show stronger time discounting tend to save less.

The typical interaction of framing effect and hyperbolic discounting that we observe in retirement saving decisions can also be observed in other situations. Many students who start preparing for an examination in the last minute will know this all too well: one more day of procrastination seems much more preferable than the benefit from a day of hard work for the examination results, but of course everybody would still agree that it is preferable to start the preparation tomorrow (or at least *some* day) rather than to fail the exam...

2.8 Summary

Decisions under risk are decision between alternatives with certain outcomes which occur with given probabilities.

We have seen three models of decisions under risk: Expected Utility Theory (EUT) follows directly from the "rational" assumptions of completeness, transitivity (no "Lucky Hans"), continuity and independence of irrelevant alternatives (for a decision between A and B, only the differences between A and B matter). It is therefore the "rational benchmark" for decisions. The choice of the utility function allows to model risk-averse as well as risk-seeking behavior and can be used to explain rational financial decisions, e.g., on insurances or investments. The main purpose of EUT, however, is a *prescriptive* one: EUT helps to find the optimal choice from a rational point of view.

Sometimes EUT is too difficult to use. In particular when considering financial markets, it is often much easier to consider only two parameters: the expected return of an asset and its variance. This leads to the Mean-Variance Theory. We have seen that this theory has certain drawbacks, in particular it can violate state dominance. (This is called the "Mean-Variance paradox".) In certain cases, in particular when the returns are normally distributed, Mean-Variance Theory turns out to be a special case of EUT, and hence we can more confidently use it.

EUT is about how people *should* decide. But how *do* people decide? The pessimistic statement of Chomsky on the unpredictable nature of human decisions, which we had put at the beginning of this chapter, has been disproved to some extend in recent years: in particular Prospect Theory (PT) and Cumulative Prospect Theory (CPT) *describe* choices under risk quite well. Certain irrational effects like the violation of the "independence of irrelevant alternatives" make such approaches necessary to model actual behavior. Key features are the overweighting of small probabilities (respectively extreme events) and decision-making with respect to a reference point ("framing"). It is possible to explain the "four-fold pattern of risk-attitudes" and famous examples like Allais' Paradox with these models.

Finally, we had a look on the time-dimensions of decisions. Whereas a discounting of the utility of future events can be explained with rational reasons, the specific kind of time-discounting that is observed is clearly irrational, since it is not time-consistent. Such time-inconsistent behavior can be used to explain, e.g., undersaving for retirement.

After finishing this chapter, we have now a very solid foundation on which we can build our financial market theories in the next chapters.

Tests

The following tests should enable the reader to check whether he understood the key ideas of decision theory. Some of the multiple choice questions are tricky, but most should be answered correctly.

Tests

1. How do you define that a lottery A with finitely many outcomes *state dominates* a lottery B with finitely many outcomes?

   - ☐ If A gives a higher outcome than B in every state.
   - ☐ If A gives a higher or equal outcome than B in every state, and there is at least one outcome where A gives a higher outcome than B.
   - ☐ If the expected return of A is larger than the expected return of B.
   - ☐ If, for every x, the probability to get a return of more than x is larger for A than for B.

2. What is the expected utility (EUT) of a lottery A with outcomes x_1 and x_2 and probabilities p_1 and p_2?

   - ☐ $EUT(A) = x_1p_1 + x_2p_2$.
   - ☐ $EUT(A) = u(x_1p_1 + x_2p_2)$.
   - ☐ $EUT(A) = u(x_1)p_1 + u(x_2)p_2$.
   - ☐ $EUT(A) = u(p_1)x_1 + u(p_2)x_2$.

3. Let us assume that u is an EUT utility function describing a person's preference relation $\prec$, then:

   - ☐ $A \prec B$ if and only if $\mathbb{E}(u(A)) < \mathbb{E}(u(B))$.
   - ☐ $v(x) := u(2x + 42)$ is a utility function that describes the preference relation $\prec$.

- ☐ $v(x) := (u(x))^3$ is a utility function that describes $\prec$.
- ☐ If u is concave, then the person should not take part in any lottery that costs more than its expected value.
- ☐ If u is convex, then the person should take part in any lottery.
- ☐ If u is strictly convex on some interval then $\prec$ cannot be rational.

4. In which cases is a function $u\colon [a,b] \subset \mathbb{R} \to \mathbb{R}$ concave?

- ☐ If $\lambda u(x_1) + (1-\lambda)u(x_2) \leq u(\lambda x_1 + (1-\lambda)x_2)$ for every $x_1, x_2 \in [a,b]$ and $\lambda \in [0,1]$.
- ☐ If $\lambda u(x_1) + (1-\lambda)u(x_2) \geq u(\lambda x_1 + (1-\lambda)x_2)$ for every $x_1, x_2 \in [a,b]$ and $\lambda \in [0,1]$.
- ☐ If $\lambda u(x_1) + (1-\lambda)u(x_2) = u(\lambda x_1 + (1-\lambda)x_2)$ for every $x_1, x_2 \in [a,b]$ and $\lambda \in [0,1]$.
- ☐ If $u'' \leq 0$.
- ☐ If $u'' \geq 0$.

5. The absolute risk aversion is defined by

- ☐ $r(x) := -u''(x)$.
- ☐ $r(x) := -u''(x)/u'(x)$.
- ☐ $r(x) := -x\frac{u''(x)}{u'(x)}$.

6. Which of the following utility functions is the most rational choice:

- ☐ $u(x) := x^\alpha$. where $\alpha \in (0,1)$.
- ☐ $u(x) := x$.
- ☐ $u(x) := \ln x$.
- ☐ They are all equally rational.

7. What does Allais' Paradox tells us?

- ☐ It is irrational to follow Expected Utility Theory.
- ☐ Expected Utility Theory does not explain actual behavior of persons sufficiently well.
- ☐ People tend to violate the Independence Axiom.

8. Which are the key ideas of Prospect Theory (PT)?

- ☐ People frame their decisions in gains and losses rather than considering their potential final wealth.
- ☐ People tend to overweight small probabilities and underweight large probabilities. This can be modeled by a probability weighting function.
- ☐ People do not know probabilities exactly and hence overestimate small probabilities. This can be modeled by a probability weighting function.
- ☐ People compute the PT or CPT functional in order to make decisions.

9. How does PT explain why people gamble and buy insurances?

- ☐ People have a value function which is concave in gains (gamble) and convex in losses (insurance).
- ☐ People overweight small probabilities, like winning in a lottery or losing their home in a fire.

10. Why does PT violate stochastic dominance?

- ☐ Extreme events are overweighted, hence a small chance to lose a larger amount makes a lottery overly unattractive. This leads to a violation of stochastic dominance.
- ☐ Several small-probability events with similar outcome are overweighted relative to a single outcome with a slightly larger payoff, thus PT prefers the former to the latter, violating stochastic dominance.
- ☐ The convex shape of the value function in losses leads to risk-seeking behavior that makes people prefer risky lotteries over safe outcomes, violating stochastic dominance.

11. Which properties does Cumulative Prospect Theory (CPT) satisfy?

- ☐ Events with extremely low or high outcomes are overweighted.
- ☐ All small-probability events are overweighted.
- ☐ CPT does not violate stochastic dominance.
- ☐ CPT agrees with PT for lotteries with finitely many outcomes.
- ☐ CPT can be formulated for lotteries with finitely many outcomes as well as for arbitrary lotteries.

12. In which cases do Mean-Variance Theory and EUT coincide?

- ☐ When we consider only normal distributions of outcomes.
- ☐ When the utility function is concave.
- ☐ When the utility function is quadratic.
- ☐ When the utility function is linear.
- ☐ In lotteries with at most two outcomes.

13. Which axioms are satisfied by mean-variance theory?

- ☐ Completeness.
- ☐ Transitivity.
- ☐ Continuity.
- ☐ Independence.

14. *In-betweenness* says that the certainty equivalent of a lottery must be between its smallest and largest values.
    Do the following four theories satisfy in-betweenness?

- ☐ Expected utility theory, i.e. $U = \sum p_i u(x_i)$,
- ☐ Classical prospect theory, i.e. $U = \sum w(p_i)v(x_i)$,
- ☐ Cumulative prospect theory, i.e. $U = \sum (w(F_i) - w(F_{i-1}))v(x_i)$,
- ☐ Normalized prospect theory by Karmarkar, i.e. $U = (\sum w(p_i)v(x_i)) / \sum w(p_i)$.

15. Which of the following statements on decision models are correct?

- ☐ From the von Neumann-Morgenstern axioms can we derive the existence of a utility function.
- ☐ A concave von Neumann-Morgenstern utility function corresponds to risk averse behavior.
- ☐ From the independence axiom we can derive that the utility function must be concave.
- ☐ Mean-Variance Theory describes rational decisions.
- ☐ EUT describes rational decisions.

- ☐ A typical utility function with constant relative risk aversion is $u(x) = x^\alpha/\alpha$.
- ☐ A typical utility function with constant relative risk aversion is $u(x) = -e^{-\alpha x}$.
- ☐ CPT is the most widely used descriptive model for decision behavior.
- ☐ Mean-Variance Theory can violate stochastic dominance.
- ☐ CPT can violate stochastic dominance.

16. Which of the following statements on time discounting are correct?

- ☐ In the classical model, the discounted utility at time $t > 0$ is given by $u(t) := \frac{u(0)}{1+\delta t}$ for some $\delta > 0$.
- ☐ In the classical model, the discounted utility at time $t > 0$ is given by $u(t) := u(0)\mathrm{e}^{-\delta t}$ for some $\delta > 0$.
- ☐ Classical discounting is time-consistent, hyperbolic discounting is not.
- ☐ If somebody prefers 100€ now over 110€ tomorrow, this cannot be explained by classical discounting, but by hyperbolic discounting.
- ☐ If somebody prefers 100€ now over 110€ tomorrow, but 110€ in 101 days over 100€ in 100 days, then this cannot be explained by classical discounting, but by hyperbolic discounting.

Part II

Financial Markets

Two-Period Model: Mean-Variance Approach 3

千里之行始于足下

"A journey of a thousand miles starts with the first step."
CHINESE PROVERB

Indeed we will start our journey to financial markets with only one step: the step from one time period (in which we invest into assets) to another time period (in which the assets pay off). To make this two-period model even simpler, we assume in this chapter mean-variance preferences. We will see later that this model is a special case of two-period models with more general preferences (Chap. 4) and that we can extend the model to arbitrarily many time-periods (Chap. 5). Finally we generalize to continuous models, where the time does not any longer consists of discrete steps (Chap. 8). For now, the assumptions of two periods and mean-variance preferences allow us to get some intuition on financial markets without being overwhelmed by an overdose of mathematical formalism. Nevertheless, we want to point out that this simplicity comes at a price: we need to impose strong and not very natural assumptions. In Sect. 2.3, we have seen some of the potential problems of the mean-variance approach. In practical applications, however, this approach is still standard. We will use it to develop a first model of asset pricing, the so called "Capital Asset Pricing Model" (CAPM). This model has been praised by many researchers in finance, and in 1990 Markowitz and Sharpe were awarded the Nobel Prize in economics for its development.

As we have already mentioned in the last chapter, mean-variance analysis goes back to H. Markowitz (1952). In his work "Portfolio Theory Selection" [Mar52] he recommends the use of an expected return-variance of return rule,

T. Hens, M.O. Rieger, *Financial Economics*, Springer Texts in Business and Economics, DOI 10.1007/978-3-662-49688-6_3

> ... both as a hypothesis to explain well-established investment behavior and as a maxim to guide one's own action.

We have seen in Sect. 2.3 that both uses, the descriptive and the normative, have their limitations, nevertheless the mean-variance analysis and the Capital Asset Pricing Model have been recognized as "one of the major contributions of academic research in the post-war era" [JW96]. Campbell and Viceira [CV02] write:

> Most MBA courses, for example, still teach mean-variance analysis as if it were a universally accepted framework for portfolio choice.

And even top researchers in mathematical finance who have no difficulty to handle more complex models, like Duffie [Duf88] write on the CAPM:

> The CAPM is a rich source of intuition and also the basis for many practical decisions.

In short: finance without the CAPM is like Hamlet without the Prince.

3.1 Geometric Intuition for the CAPM

One nice feature about the CAPM is that it can be used to obtain some intuition for some of the more sophisticated models that we will encounter in the following chapters. Hence we start with an intuitive approach to its derivation, before we discuss more formal derivations that can be generalized in the sequel.

Let us describe the model in terms of returns. There are $k = 1, 2, \ldots, K$ assets. The gross return of asset k is denoted by $R_k := A_k / q_k$, where q_k is its first period market price and A_k its second period payoff. We write $\mu_k := \mu(R_k)$ for the expected return[1] and $\sigma_k^2 := \text{var}\,(R_k)$ for the variance of the gross returns. All assets can be represented in a two-dimensional diagram with expected return μ as a reward measure and standard deviation σ as a risk measure on the axes (Fig. 3.1).

The attractiveness of a single asset k can be characterized by the mean and standard deviation of its returns. The risk-free-asset for example has an expected return of R_f with a zero standard deviation. An investor who puts all of his money into one risky asset expects to achieve a return of μ_k with a standard deviation σ_k.

3.1.1 Diversification

It is nowadays difficult to imagine that there was a time when diversification as a means of reducing risk was not universally accepted, but it is true that Markowitz'

[1] Expected returns can, for example, be calculated using historical return values adjusted by some market expectations.

Fig. 3.1 Risk and return

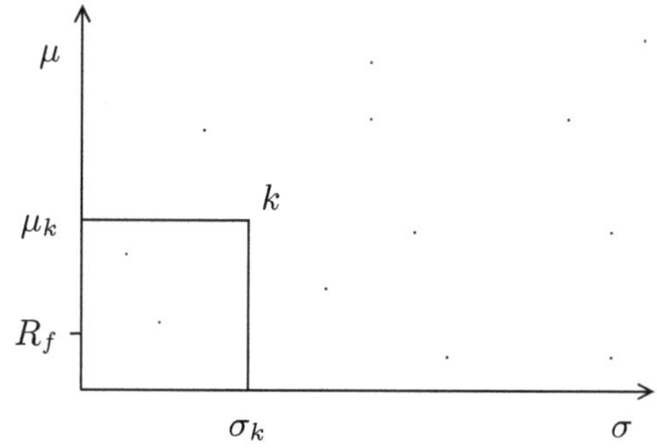

portfolio theory and their risk diversification, as we derive it in this section, was very controversial. To quote J. M. Keynes [Key88]:

> To suppose that safety-first consists of having a small gamble in a large number of different [companies] … strikes me as a travesty of investment policy.

Later the impact of the idea of diversification made such criticism look queer.[2]

Let us look back on the mean-variance model. What are the effects of diversification in this model mathematically?

If we combine two risky assets k and j we obtain an expected portfolio return of $\mu_\lambda := \lambda\mu_k + (1-\lambda)\mu_j$, where λ is the portion of wealth invested in asset k. The portfolio variance is

$$\sigma_\lambda^2 := \lambda^2\sigma_k^2 + (1-\lambda)^2\sigma_j^2 + 2\lambda(1-\lambda)\,\mathrm{cov}_{k,j}, \tag{3.1}$$

where $\mathrm{cov}_{k,j}$ is the covariance between asset k and j. How much one can gain by combining risky assets depends on this covariance. The smaller the covariance, the smaller is the portfolio risk, and the higher is the *diversification* potential of mixing these risky assets. Note however, that there is no diversification potential of mixing risky assets with the riskless security, since the covariance of the returns is equal to zero.

To see how portfolio risk changes with covariance it is convenient to standardize the covariance with the standard deviation of assets returns. The result is the *correlation* coefficient between returns of assets k and j defined as $\mathrm{corr}_{k,j} := \mathrm{cov}_{k,j}/\left(\sigma_k\sigma_j\right)$. The correlation takes values between -1 (perfectly negatively correlated) and $+1$ (perfectly positively correlated), see Sect. A.2. We consider the two extreme cases:

- If $\mathrm{corr}_{k,j} = +1$, we get $\sigma_\lambda^2 = \lambda^2\sigma_k^2 + (1-\lambda)^2\sigma_j^2 + 2\lambda(1-\lambda)\sigma_k\sigma_j$, thus: $\sigma_\lambda = \lambda\sigma_k + (1-\lambda)\sigma_j$.

[2]For information on the historical development of the mean-variance approach and the CAPM see [Var93a] from whom we have taken the above quote.

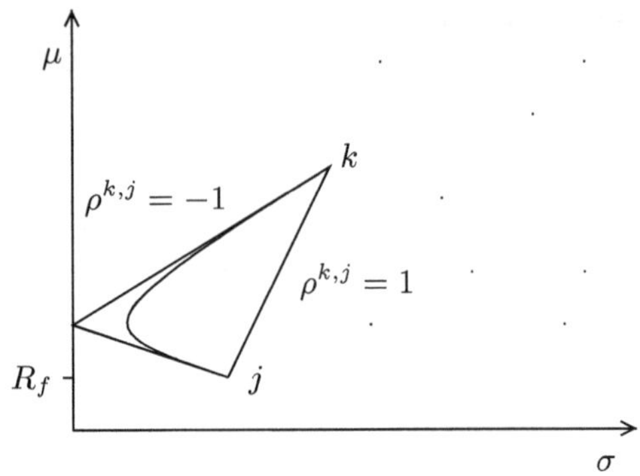

Fig. 3.2 Diversification

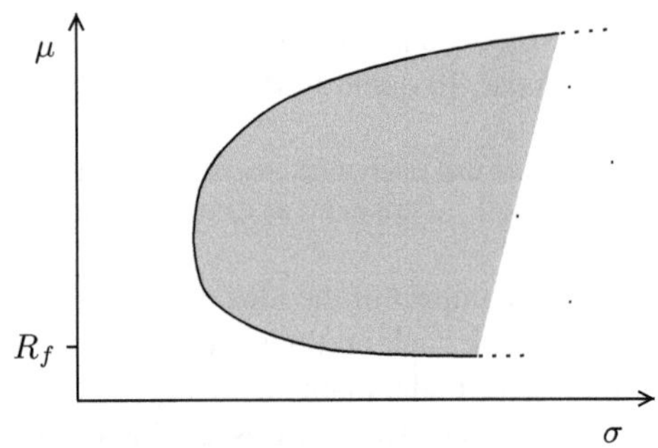

Fig. 3.3 Mean-variance opportunity set

- If $\mathrm{corr}_{k,j} = -1$, we get $\sigma_\lambda^2 = \lambda^2\sigma_k^2 + (1-\lambda)^2\sigma_j^2 - 2\lambda(1-\lambda)\sigma_k\sigma_j$, thus: $\sigma_\lambda = |\lambda\sigma_k - (1-\lambda)\sigma_j|$.

We see: the portfolio variance reaches its minimum, when the risky assets are perfectly negatively correlated, i.e., when $\mathrm{corr}_{k,j} = -1$. In this case, the portfolio may even achieve an expected return, which is higher than the risk-free rate without bearing additional risk. The portfolio consisting of risky assets does not contain risk because whenever the return of asset k increases, the return on asset j decreases, so if one invests positive amounts in both assets, the variability in portfolio returns cancels out (on average); see Fig. 3.2.

Investors can build portfolios from risky and riskfree assets but also portfolios from other portfolios etc. The set of possible μ-σ-combinations offered by portfolios of risky assets that yield minimum variance for a given rate of return is called *minimum-variance opportunity set* or *portfolio bound* (see Fig. 3.3). We assume that this set has approximately a shape as depicted in this figure, then the following arguments are valid. The skeptical reader can consult Sect. 3.1.4 for mathematically rigorous arguments.

The investor's problem when choosing an optimal portfolio is to pick a portfolio with the highest expected returns for a given level of risk. This is similar to the problem of minimizing portfolio variance for different levels of expected return,

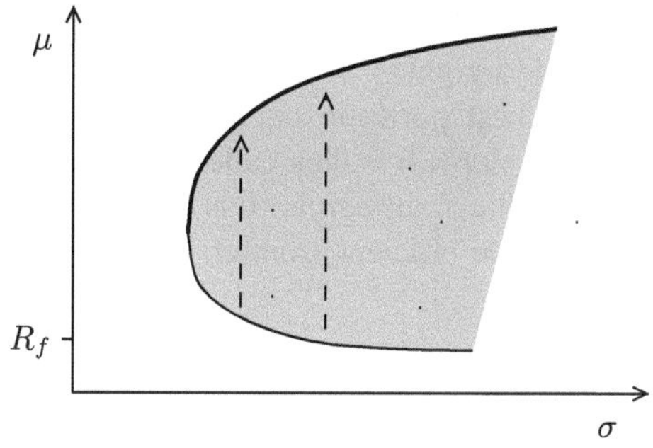

Fig. 3.4 Efficient frontier

i.e., to the following optimization problem:

$$\min_{\lambda_k,\lambda_j} \sum_k \sum_j \lambda_k \operatorname{cov}_{k,j} \lambda_j \quad \text{such that} \sum_k \lambda_k \mu_k = \text{const and} \sum_k \lambda_k = 1, \tag{3.2}$$

where λ_k denotes the proportion of money invested in asset k.

In this problem, however, we might end up suboptimal, if we choose the mean below the tip of the convex set in Fig. 3.3. In this case, increasing the desired mean allows us to reduce the variance further.

3.1.2 Efficient Frontier

The solution of problem (3.2) gives the mean-variance opportunity set or the portfolio bound. In order to identify the *efficient portfolios* in this set, one has to focus on that part of the mean-variance efficient set that is not dominated by lower risk and higher return. This is the upper part of the portfolio bound, since every portfolio on it has a higher expected return but the same standard deviation as portfolios on the lower part (see Fig. 3.4).

Thus, all the portfolios on the efficient frontier have the highest attainable rate of return given a particular level of standard deviation. The efficient portfolios are candidates for the investors optimal portfolio.

3.1.3 Optimal Portfolio of Risky Assets with a Riskless Security

The best combination of risky assets for μ-σ-investors lies on the efficient frontier. Every point on it is associated with the highest possible return for a given certain risk level.

If an investor desires to combine a risky asset (or a portfolio of risky assets) with a riskless security, he must choose a point on the line connecting both assets. This is a straight line, since the covariance between R_k and R_f (denoted by $\operatorname{cov}(R_k, R_f)$)

is zero and therefore the portfolio standard deviation σ_λ is a *linear* function of the portfolio weights.

The best portfolio combination is found when the line achieves its highest possible slope. It is then called Capital Market Line (CML). The slope of the CML is called the *Sharpe ratio*. It is equal to $(\mu_\lambda - R_f)/\sigma_\lambda$. The point at which the CML touches the efficient frontier gives the best portfolio of risky assets, the *tangent portfolio*.[3]

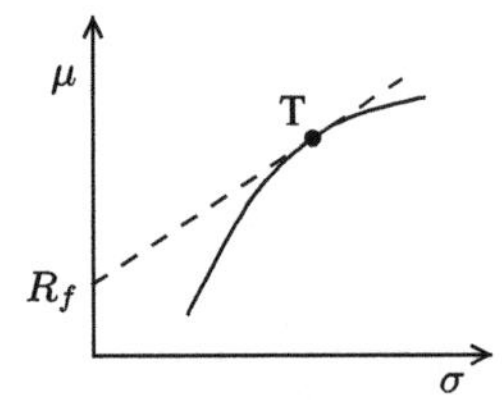

3.1.4 Mathematical Analysis of the Minimum-Variance Opportunity Set*

In the following we make the arguments rigorous that led to the definition of the tangent portfolio. The mathematically less inclined reader can skip this subsection.

It is sometimes said that the minimum-variance opportunity set is convex (as it is depicted in Fig. 3.3 and mathematically defined in Sect. A.1). This is, however, not always the case: the mean-variance opportunity set does not need to be convex, as we can already see in the case of two assets where the opportunity set is *only* convex if their correlation is $+1$ (compare Fig. 3.2). However, we don't need this convexity to prove the existence of a tangent portfolio, but before we can obtain any existence result, we need first to distinguish whether we allow for short-selling or not.

This decision has two sides: a modeling one and a mathematical one. First, it is not so clear whether allowing for short-sales is appropriate or not for our model. We could argue that in most developed markets short-selling is possible and hence our model should include it. On the other hand, there are markets where it is not possible (it might be banned or infeasible due to a lack of liquidity) and even on the most developed markets there are many market participants (small private investors) who do not have the chance to short-sell assets, at least not without steep costs. The mathematical side of the story is even more difficult: we will see that without short-selling we can find a rigorous proof for the existence of a tangent portfolio

[3] We will see that economically spoken, this portfolio is such that the marginal rate of substitution between the investor's preferences for risk and return equals the marginal rate of transformation offered by the minimum variance opportunity set.

under quite natural assumptions, but when we allow for short-selling then existence might fail if we do not impose rigid assumptions. Later, however, when we derive the capital asset pricing model, we will need to allow short-selling. This inherent problem of the geometric and "intuitive" approach presented in this chapter can only be fixed by studying the more rigorous no-arbitrage approach that we will follow in Chap. 4.

Let us now first consider the existence of the tangent portfolio when we exclude short-selling.

The main property of the opportunity set that we need is in this case that it is closed (compare Sect. A.3 for a definition). Moreover we need certain minor properties that we summarize later.

Lemma 3.1 *If we have finitely many assets, the minimum-variance opportunity set is closed and connected.*

Proof We give two proofs, the first based on the Bolzano-Weierstrass Theorem the second based on a property of continuous functions:

By construction it is clear that the opportunity set is connected. To see that the opportunity set is closed if we have finitely many assets is easy: let K denote the number of assets and let us consider a sequence of points $x_n = (\mu_n, \sigma_n)$ $(n = 1, 2, \ldots)$ in the opportunity set with $x_n \to x = (\mu, \sigma)$. Each of the x_n corresponds to a portfolio characterized by asset weights $\lambda_1^n, \ldots, \lambda_K^n$ with $\lambda_k^n \geq 0$ for all $k = 1, \ldots, K$ and $\sum_{k=1}^K \lambda_k^n = 1$. Therefore the vector $\boldsymbol{\lambda}^n := (\lambda_1^n, \ldots, \lambda_K^n)$ is for all $n \in \mathbb{N}$ in a compact set.[4] According to the Bolzano-Weierstrass Theorem A.3 we can select a converging subsequence of the $\boldsymbol{\lambda}^n$. Let us denote its limit by $\boldsymbol{\lambda}$, then $\boldsymbol{\lambda}$ defines a portfolio with mean μ and variance σ^2, since mean and variance depend continuously on the asset weights. Thus any limit of points in the opportunity set is again in the opportunity set, in other words we have proved that the opportunity set is closed.

The second proof uses the function f that assigns mean and variance to a portfolio $\boldsymbol{\lambda}$:

$$
\begin{aligned}
f: \quad S := &\left\{ \boldsymbol{\lambda} \in \mathbb{R}_+^{K+1} \,\middle|\, \sum_{k=0}^{K} \lambda^k = 1 \right\} \\
&\to \left\{ (\mu, \sigma) \,\middle|\, \mu = \sum_{k=0}^{K} \mu^k \lambda^k,\ \sigma^2 = \lambda^k \operatorname{cov}^{jk} \boldsymbol{\lambda} \lambda^j,\ \boldsymbol{\lambda} \in S \right\} .
\end{aligned}
$$

Now, since S is obviously closed and connected and since f is continuous, we can deduce that $f(S)$, i.e. the opportunity set, also is closed and connected, compare Sect. A.3. □

[4]We use in this chapter bold face characters for vectors to increase readability.

What about if we have infinitely many assets? In this case the opportunity set does not have to be closed. As a simple example think about perfectly correlated assets with $\mu_k = 1 - 1/k$ and $\sigma_k = 1$. The opportunity set is given by $\{(\mu, 1) | \mu \in [0, 1)\}$ and is obviously not closed. In this case we see also why we need closedness: the efficient frontier in the example does not exist, since any portfolio with mean μ and variance σ^2 in the opportunity set can be improved. (The potential "best" portfolio with $\mu = \sigma = 1$ is not contained in the opportunity set.) We see that we better stick to the case of finitely many assets.

Since the opportunity set is closed, we can in fact construct the efficient frontier. To construct a tangent portfolio, however, we need to know a little bit more about the geometric structure of the efficient frontier:

Lemma 3.2 *If we have finitely many assets, the efficient frontier can be described as the graph of a function* f: $[a, b]$, *where* $0 \leq a \leq b < \infty$. *Moreover there exists a point* $c \in [a, b]$ *such that* f *is concave and increasing on* $[a, c]$ *and decreasing on* $[c, b]$.

Proof By construction it is clear that the function f exists. It is also clear that $b < \infty$, since there are no points in the minimum-variance set with $\sigma > \max_{k=1,\ldots,K} \sigma_k$, compare (3.1). Suppose now that f is increasing and strictly convex on some interval $[s_1, s_2]$, where $s_2 > s_1$. Then we can combine the portfolios A, with mean $f(s_1)$ and variance s_1^2, and B, with mean $f(s_2)$ and variance s_2^2. Using again the formula (3.1) we can find a $\lambda \in (0, 1)$ such that the new portfolio $\lambda A + (1 - \lambda)B$ has the variance $(s_1 + s_2)^2/4$. The mean of this portfolio depends on the correlation between A and B, but can be estimated from below by $(f(s_1) + f(s_2))/2$, as a small computation shows. Given the strict convexity of f, however, $f((s_1 + s_2)/2) < (f(s_1) + f(s_2))/2$, thus we have found a portfolio that is "better" than the efficient frontier (i.e., its variance is the same, but its mean larger). This is a contradiction, thus f has to be concave when it is increasing.

With a similar construction we can prove that if f is decreasing at some point s then it cannot be increasing at any point larger than s. Putting everything together, we have proved the lemma. □

We mention that it is possible that the efficient frontier is a decreasing and strictly convex function and that the efficient frontier does not have to be continuous, see Fig. 3.5 for an example.

Using the above lemmas we can now prove the existence of a tangent portfolio:

Proposition 3.3 *If we have finitely many assets, and at least one asset has a mean which is not lower than the return* R_f *of the risk-free asset, then a tangent portfolio exists.*

Proof Given the above conditions, an efficient frontier exists according to Lemma 3.1. Using Lemma 3.2, we know that there are points a, b, such that the efficient frontier is the graph of a function f on $[a, b]$, which is concave and

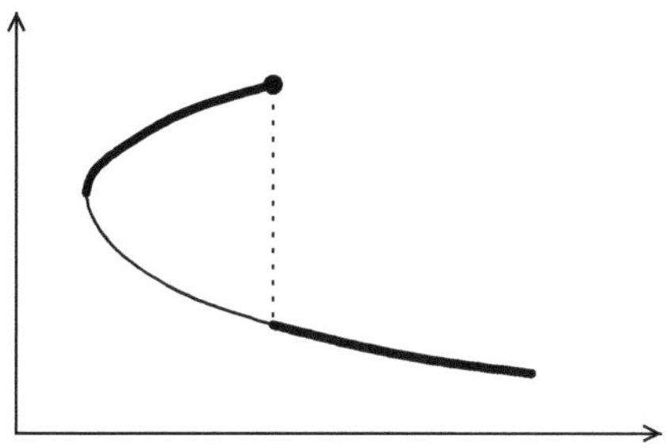

Fig. 3.5 An example for a discontinuous, partially decreasing and strictly convex efficient frontier

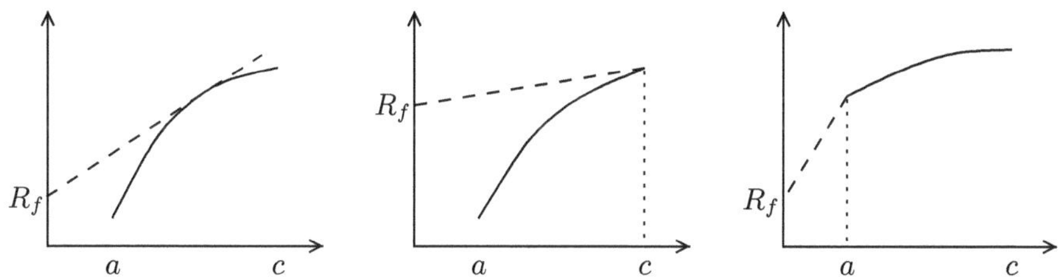

Fig. 3.6 The three cases for the construction of the tangent portfolio

increasing on $[a, c]$. We denote this part of the graph by F. Using the condition on the asset returns, we see that $f(c) \geq R_f$.

Now, we have to distinguish three cases: If there is a tangent on F through the point $(R_f, 0)$, i.e., the risk-free asset, we have found a tangent portfolio by taking a tangent point in F on this line. Otherwise, if a line from $(R_f, 0)$ to $(f(c), c)$ lies nowhere below F, the point $(f(c), c)$ is the tangent point. If both is not the case, then the tangent portfolio is given by the point $(f(a), a)$. Compare Fig. 3.6 for an illustration of the three cases.

In all three cases, the constructed line cannot lie below other points of the efficient frontier, since f is decreasing for values larger than c, but the tangent line is increasing (or at least horizontal), since $f(c) \geq R_f$. □

What changes in this argument if we allow for short-selling? In a nutshell: everything. – In fact, the existence is not guaranteed anymore! Take as simple example two assets $(\mu_1, \sigma_1) = (1, 1)$ and $(\mu_2, \sigma_2) = (1.1, 1)$ with a correlation of $+1$. Then we have for a portfolio of $\lambda \in \mathbb{R}$ units of asset 1 and $1 - \lambda$ units of asset 2 that $(\mu_\lambda, \sigma_\lambda) = (\lambda + 1.1(1-\lambda), 1) = (1.1 - 0.1\lambda, 1)$. Thus we can construct portfolios with arbitrarily large returns and a variance of 1 by choosing λ negative enough. It is now easy to see that we are not able to construct any tangent portfolio in this case.

The example can be modified such that $\sigma_1 \neq \sigma_2$: even then we will usually not be able to define a tangent portfolio. A similar construction is possible for correlation -1.

Assuming that no pair of assets has a correlation of -1 or $+1$ does not fix this problem either, since a combination of two such assets might have a correlation of $+1$ or -1 with another asset.

If we exclude this possibility as well, then we finally get an existence result:

Theorem 3.4 *Let X be a mean-variance opportunity set (with short-selling) and assume that for any two portfolios $X_j, X_k \in X$ with $j \neq k$ the correlation between their returns is in $(-1, +1)$, i.e. neither -1 nor $+1$, then there exists a tangent portfolio for X.*

Proof Let $\Delta = \{\lambda \in \mathbb{R}^K | \sum_{k=1}^K \lambda_k = 1\}$. Substituting $\lambda_k = 1 - \sum_{k=1}^{K-1} \lambda_k$, we can transform Δ to $\mathbb{R}^{K-1}$. Now we consider the compactification C^{K-1}. We define this in two steps: first transform $\mathbb{R}^{K-1}$ to $\mathring{D}^{K-1}$ via

$$(x_1, \ldots, x_{K-1}) \mapsto \left(\frac{2}{\pi} \arctan(x_1), \ldots, \frac{2}{\pi} \arctan(x_n) \right).$$

Now add $S^{K-1} = \partial \mathring{D}^{K-1}$ and use the standard topology of D^{K-1}. Now consider a sequence λ^n that maximizes $\frac{\mu(\lambda^n) - R_f}{\sigma(\lambda^n)}$. A subsequence of λ^n converges in C^{K-1}, since C^{K-1} is compact. Now consider the following two cases:

Case 1: The limit is in $\mathring{D}^{K-1}$, then also λ_k is finite. Now let $\lim_{n\to\infty} \frac{\mu(\lambda^n) - R_f}{\sigma(\lambda^n)} < \infty$, since otherwise we have a finite portfolio which is riskless and this could only happen if it is composed of two risky portfolios with correlation $+1$ or -1 – a contradiction.

Case 2: The limit is in S^{K-1}, then also λ_k is infinite. Define A_1 as asset K, A_2 as all other assets in the relative weights specified by the limit in S^{K-1}.

Then we can find a sequence $\tilde{\lambda}^n$ with the same limit, but being composed only of two portfolios A_1, A_2 with $\tilde{\lambda}_1^n A_1 + \tilde{\lambda}_2^n A_2$ where $\tilde{\lambda}_1^n \to +\infty$ and $\tilde{\lambda}_2^n \to -\infty$.

By assumption, $\text{corr}(A_1, A_2) \in (-1, +1)$, thus

$$\left(\mu(\tilde{\lambda}_1^n A_1 + \tilde{\lambda}_2^n A_2), \sigma(\tilde{\lambda}_1^n A_1 + \tilde{\lambda}_2^n A_2) \right)_{n \in \mathbb{N}}$$

is a curve with slope going to zero (computation in Chap. 3). Therefore $\lim_{n\to\infty} \frac{\mu(\tilde{\lambda}^n) - R_f}{\sigma(\tilde{\lambda}^n)} = 0$, but since μ and σ are contained in the portfolio weights, the original sequence λ^n could not have been maximizing which contradicts our initial assumption. □

This result, however, is not as useful to determine the existence of a tangent portfolio: we would have to check all (infinitely many) portfolios of our assets and their correlations with each other to verify the condition of the theorem. We will see in the next chapter how the no-arbitrage condition can help us to avoid this problem and secure the existence of a tangent portfolio under a more reasonable condition. Up to then, we will tacitly assume the existence of a tangent portfolio, although we know now that this is not a trivial matter.

By the way: whether we allow for short-selling or not, the tangent portfolio does not have to be unique. Non-uniqueness, however, occurs only in very specific situations and is not important for practical applications where we are usually happy with finding an optimal portfolio and do not care that much about whether there would have been another equally good portfolio...

3.1.5 Two-Fund Separation Theorem

The optimal asset allocation consisting of risky assets and a riskless security depends on the investor's preferences, which are for example[5] given by the utility function $U^i(\mu_\lambda, \sigma_\lambda^2) := \mu_\lambda - \frac{\rho^i}{2}\sigma_\lambda^2$, where ρ^i is a risk aversion parameter of investor i. We denote it by ρ and not by α or λ (as is standard) in order to avoid confusion with the portfolio weights (λ) and the excess return (α) of an investment, see Sect. 3.3. The higher this parameter, the higher is the slope of the utility function.[6] The higher the risk aversion, the higher is the required expected return for a unit risk (required risk premium).

Different investors have different risk-return preferences. Investors with higher (lower) level of risk aversion choose portfolios with a low (high) level of expected return and variance, i.e., their portfolios move down (up) the efficient frontier.

If there is a risk-free security, the *Separation Theorem* of Tobin (1958) states that agents should diversify between the risk free asset (e.g., money) and a single optimal portfolio of risky assets. Since the *Tangent Portfolio* gives the optimal mix of risky assets, a combination with the risk-free assets means that every investor has to make an investment decision *on* the *Capital Market Line*. Different attitudes toward risk result in different combinations of the risk-free asset and the optimal portfolio of risky assets. More conservative investors for example will choose to put a higher fraction of their wealth into the risk free asset; on the other hand, more aggressive investors may decide to borrow capital on the money market (go short in risk-free assets) and invest it in the Tangent Portfolio.

[5]For the purpose of deriving the Two-Fund Separation Theorem this single utility function is sufficient. Using a more general function like $V^i(\mu_\lambda, \sigma_\lambda)$ would result in expressions similar to those we derive here. In this case we get $\rho^i = -\frac{\partial_\sigma V^i(\mu_\lambda,\sigma_\lambda)}{\partial_\mu V^i(\mu_\lambda,\sigma_\lambda)}$. But as we see below, the point of the Two-Fund Separation Theorem is to show that ρ^i anyway cancels out from the portfolio of risky assets.

[6]The risk aversion concept is often discussed in the expected utility context. Recall, however, that there it is measured by the *curvature* of a utility function.

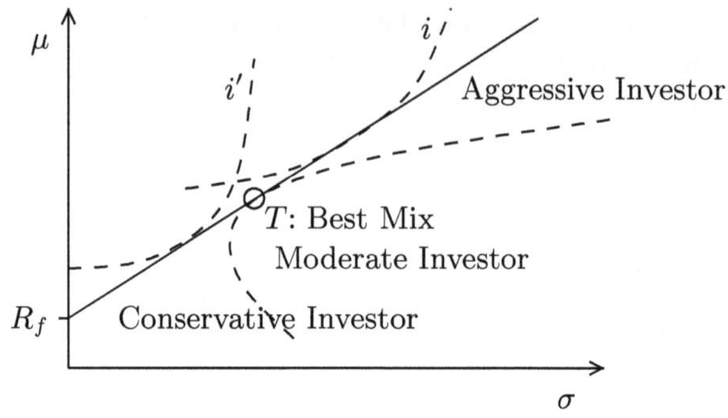

Fig. 3.7 Two-Fund Separation

Thus, the asset allocation decision of investor i is described by the vector of weights[7] $\boldsymbol{\lambda}^i = (\lambda_0^i, (1-\lambda_0^i)\boldsymbol{\lambda}^T)$, $i = 1, \ldots, I$, where $\boldsymbol{\lambda}^i \in \mathbb{R}^{K+1}$, $\lambda_0^i \in \mathbb{R}$, and $\boldsymbol{\lambda}^T \in \mathbb{R}^K$ (Fig. 3.7).

This property, known as Two-Fund Separation, has been summarized nicely by Campbell and Viceira [CV02]:

> The striking conclusion of his [Markowitz'] analysis is that all investors who care only about mean and standard deviation will hold the same portfolio of risky assets.

3.1.6 Computing the Tangent Portfolio

According to the Two-Fund Separation an investor with utility $U^i(\mu_{\boldsymbol{\lambda}}, \sigma_{\boldsymbol{\lambda}}^2) = \mu_{\boldsymbol{\lambda}} - \frac{\rho^i}{2}\sigma_{\boldsymbol{\lambda}}^2$ has to decide how to split his wealth between the optimal portfolio of risky assets with a certain variance-covariance structure (Tangent Portfolio) and the riskless asset. The structure of the Tangent Portfolio can be found either by maximizing the Sharpe Ratio subject to a budget constraint or by solving the simplest μ-σ maximization problem[8]:

$$\max_{\boldsymbol{\lambda} \in \mathbb{R}^{K+1}} U^i(\mu_{\boldsymbol{\lambda}}, \sigma_{\boldsymbol{\lambda}}^2) = \mu_{\boldsymbol{\lambda}} - \frac{\rho^i}{2}\sigma_{\boldsymbol{\lambda}}^2 \quad \text{such that} \sum_{k=0}^{K} \lambda_k = 1. \tag{3.3}$$

In this equation, λ_0 denotes the fraction of wealth invested in the riskless asset.[9] λ_0 can be eliminated from the optimization problem by substituting the budget

[7]Note: there is no index i on the Tangent Portfolio $\boldsymbol{\lambda}^T$ since this portfolio is the same for every investor.

[8]Note that solving the simplest (μ, σ)-problem is as good as any other (μ, σ)-problem, since by the Two-Fund Separation property all mean-variance utility functions deliver the same Tangent Portfolio.

[9]λ_0 is the first component of $\boldsymbol{\lambda}$.

constraint $\lambda_0 = 1 - \sum_{k=1}^{K} \lambda_k$ into the utility function. Using the definition of μ_{λ} and σ_{λ}^2 we get:

$$\max_{\boldsymbol{\lambda}} (\boldsymbol{\mu} - R_f \mathbf{1})' \boldsymbol{\lambda} - \frac{\rho^i}{2} \boldsymbol{\lambda}' COV \boldsymbol{\lambda}$$

where, from now on, $\boldsymbol{\lambda} \in \mathbb{R}^K$ is the vector of risky asset weights in the Tangent Portfolio, $\boldsymbol{\mu}$ is the vector of risky assets' mean returns, and COV is their covariance matrix. The first order condition of the problem is

$$COV \boldsymbol{\lambda} = \frac{1}{\rho^i} (\boldsymbol{\mu} - R_f \mathbf{1}).$$

If there are no constraints on $\boldsymbol{\lambda}$, then the solution is

$$\boldsymbol{\lambda} = COV^{-1} \frac{1}{\rho^i} (\boldsymbol{\mu} - R_f \mathbf{1}). \tag{3.4}$$

With short-sales constraints, $\boldsymbol{\lambda} \geq 0$, for example, one can apply standard algorithms for linear equation systems to solve the problem.

Say, the solution to the first order condition is $\boldsymbol{\lambda}^{\text{opt}}$, then the Tangent Portfolio can be found by a renormalization:

$$\lambda_k^T = \frac{\lambda_k^{\text{opt}}}{\sum_j \lambda_j^{\text{opt}}}.$$

Note that the risk aversion parameter ρ^i cancels after the renormalization, which is the Two-Fund Separation property.

Furthermore, the composition of the Tangent portfolio does not depend on the form of the utility function. Using more sophisticated functions than (3.3) will not change the result obtained in (3.4).

3.2 Market Equilibrium

We want to study market equilibria, therefore we make the following observation: if individual portfolios satisfy the Two-Fund Separation then by setting demand equal to supply the sum of the individual portfolios must be proportional to the vector of market capitalization[10] $\boldsymbol{\lambda}^M$, as we will prove in Sect. 4.3. Hence in equilibrium, the normalized *Tangent Portfolio* will be identical to the *Market Portfolio*.[11]

[10]The market capitalization of a company for example is the market value of total issued shares.

[11]Note that this equality is barely supported by empirical evidence, i.e., the Tangent Portfolio does not include all assets. The reason for this mismatch could for example be that not every investor

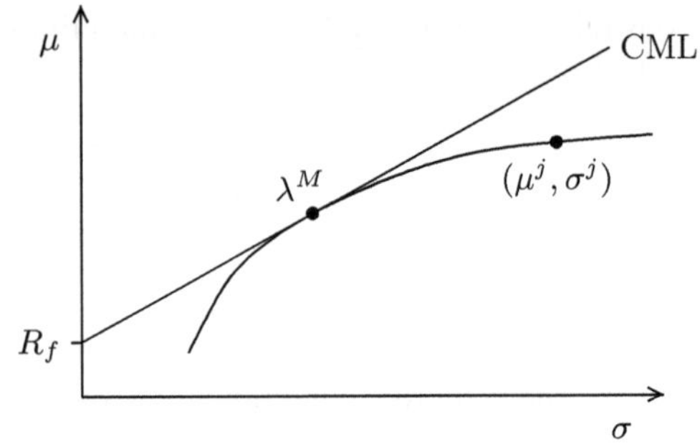

Fig. 3.8 Market Portfolio

3.2.1 Capital Asset Pricing Model

To understand the link between the individual optimization behavior and the market, compare the slopes of the Capital Market Line and a curve j that is obtained by mixing a portfolio of any asset j with the market portfolio. By the tangency property of $\boldsymbol{\lambda}^M$ these two slopes must be equal![12] (See Fig. 3.8.)

Curve j is obtained by a combination of some asset j with the market portfolio.

The slope of the Capital Market Line can be calculated as

$$\frac{\frac{\mathrm{d}}{\mathrm{d}\lambda}\mu\left(\lambda R_f + (1-\lambda)R^M\right)\big|_{\lambda=0}}{\frac{\mathrm{d}}{\mathrm{d}\lambda}\sigma\left(\lambda R_f + (1-\lambda)R^M\right)\big|_{\lambda=0}} = \frac{R_f - \mu^M}{-\sigma_M}.$$

The slope of the j-curve is

$$\frac{\frac{\mathrm{d}}{\mathrm{d}\lambda}\mu\left(\lambda R_j + (1-\lambda)R^M\right)\big|_{\lambda=0}}{\frac{\mathrm{d}}{\mathrm{d}\lambda}\sigma\left(\lambda R_j + (1-\lambda)R^M\right)\big|_{\lambda=0}} = \frac{\mu_j - \mu^M}{\left(\operatorname{cov}\left(R_j, R^M\right) - \sigma_M^2\right)/\sigma_M}.$$

From the slope's equality at point $\boldsymbol{\lambda}^M$ follows:

$$\frac{(\mu_j - \mu^M)\sigma_M}{\operatorname{cov}\left(R_j, R^M\right) - \sigma_M^2} = \frac{\mu^M - R_f}{\sigma_M}$$

optimizes over risk and return as suggested by Markowitz. For further ideas on this asset allocation puzzle see also [CMW97, BX00]

[12]If the j-curve would intersect with the CML then the Sharpe Ratio could still be increased, as can be verified graphically.

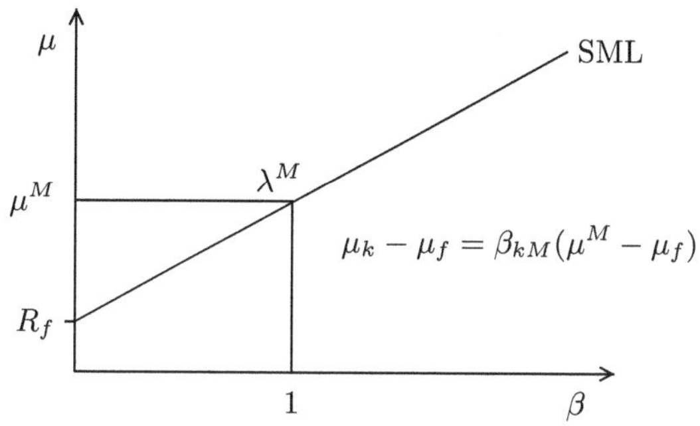

Fig. 3.9 Security Market Line

or equivalently

$$\mu_j - R_f = \beta_{j,M}(\mu^M - R_f) \quad \text{where } \beta_{j,M} := \frac{\text{cov}\left(R_j, R^M\right)}{\sigma_M^2}. \tag{3.5}$$

The result is the Security Market Line (SML, see Fig. 3.9).

The difference to the mean-variance analysis is the risk measure. In the CAPM the asset's risk is captured by the factor β instead of the standard deviation of asset's returns. It measures the sensitivity of asset j returns to changes in the returns of the market portfolio. This is the so-called "systematic risk".

3.2.2 Application: Market Neutral Strategies

The Capital Asset Pricing Model has many applications for investment managers and corporate finance. Even professionals dealing with alternative investments consider it while building portfolios. One example is a form of *Market Neutral Strategy* followed by some hedge funds. This strategy aims a zero exposure to market risk. To exclude the impact of market movements, it takes simultaneous long and short positions on risky assets. These assets have the same Beta (as measure for market risk) but different market prices. Under the assumption that market prices will eventually return to their fundamental value defined by the CAPM, hedge fund managers take long positions in underpriced assets and short positions in overpriced assets. In terms of expected returns, the long (short) positions are in assets with higher (lower) expected returns than in the CAPM.[13] We will discuss later the potential risks of this strategy.

[13]When prices revert and increase (decrease) in order to reach their fundamental value, the expected returns are decreasing (increasing).

3.2.3 Empirical Validity of the CAPM

As a portfolio model the mean-variance rule is nice and simple. However, claiming that all agents will hold the same portfolio of risky assets is certainly wrong since agents – in contrary to what we assumed above – do certainly have different expectations. A related but deeper critique on the two-fund separation was pointed out by Canner, Mankiw and Weil [CMW97] who studied the advice of one advisor given to differently risk averse agents. An advisor should apply the same expectations when giving recommendations to different clients and hence, following the two-fund separation property, he should recommend the same portfolio of risky assets scaled up and down with the risk-free asset in order to match the clients' risk aversion. Canner, Mankiw and Weil [CMW97] showed that this simple rule is, however, not followed by advisors. For example, the portfolio weight of S&P500 relative to government bonds changes from 15 % to 45 %, going from a conservative to an aggressive portfolio.

Even though the portfolio implications of mean-variance analysis are clearly not found in reality, one could still try to find the asset pricing implications, the validity of the SML. One of the nice properties of the SML is that it suggests a linear relation between the Beta and the excess returns. Hence simple linear regression studies can be used to test the SML and indeed there are very many of those studies. It is found that market risk, the Beta, indeed explains the excess returns of assets – at least to some extent. But more factors are needed to get a really good fit. The most famous additional factors are *value*, *size* and *momentum*. It turns out that investing in value[14] stocks gives significantly higher returns – even with lower Beta – than investing in glamour stocks. Also, investing in small cap stocks has this feature. Finally, investing in stocks that have gone up is increasing returns in the short run and the reverse is true in the long run. Famous empirical studies on the CAPM are Fama and French ([FF92] and [FF98]) and Lakonishok, Shleifer and Vishny [LSV94]. The size effect, i.e., the fact that small cap assets have higher risk adjusted returns than large cap assets, was first shown by Banz [Ban81].

3.3 Heterogeneous Beliefs and the Alpha

So far we have mentioned two motives for trade: smoothing intertemporal consumption and risk diversification. The first motive is served by fixed income markets, the second by reinsurance markets, stock markets and any other markets which allow diversifying risks, as for example markets for credit risk. As the two-fund separation principle showed, the diversification motive is best served by mutual funds that try to offer market exposure at minimal costs. That's why exchange traded funds, ETFs, are very popular. In ETFs the market portfolio is built without active management

[14]Value stocks are for example characterized by high multiples, i.e., book to price ratios, cash flow to price ratio, dividend yield etc.

of a fund manager. However, not all agents go for ETFs and there are many mutual funds claiming to stay close to ETFs and yet to outperform them. Last but not least there is a rapidly growing industry, called hedge funds, in which asset managers exert strategies that are totally different to those of mutual funds. Hedge funds claim to offer returns that are as high as those of stocks with a volatility as low as that of bonds, which is a clear violation of the CAPM – at least if we understand the market portfolio as the sum of all investments, not only as the stock market index. Hedge funds claim to generate the "Alpha", i.e., excess returns that cannot be explained by market risk. The *Alpha* has become a magic selling word. Banks offer Alpha funds,[15] hedge funds call themselves "*AlphaSwiss*", or "*Alpha Lake*", for example. Analysts write about the *future of the Alpha*, or the *pure Alpha* etc. Yet, "*the Alpha has no theory*", as Alexander Ineichen[16] from UBS states in his AIS-report [Ine05]. Do banks and hedge funds sell dreams like a perpetuum mobile that do not exist? In this section we show that the lack of theory can quite easily be removed by extending the standard CAPM towards heterogeneous beliefs. The assumption of homogeneous beliefs was always under scrutiny among finance theorists. Removing it we can model the Alpha *within* the CAPM, i.e., as a property of financial market equilibria! We show that in a CAPM with heterogeneous beliefs every investor who holds beliefs different to the average market belief, *sees some Alpha*. However, the sum of these Alphas is zero, i.e., the hunt for Alphas is a *zero-sum game*, in which one can only win at the expense of someone else, as it is nicely stated in [Ine05], page 31:

> The returns are achieved by the managers' ability to exploit inefficiencies left behind by other (less informed, less intelligent, less savvy, ignorant, or uneconomically motivated) investors in what is largely considered a zero or negative sum game.

The question then arises for how long the losers in the zero-sum game are willing to finance the gains of the winners. Since every market participant has the option to play a passive strategy by investing in the market portfolio, the less informed, less intelligent or less savvy investors will learn to stay passive so that it becomes more and more difficult for the active managers to outperform each other.[17] Hence, the market converges to a situation in which only the best informed determine market prices. This long-run outcome of the zero-sum game is consistent with the efficient market hypothesis and also with the CAPM based on homogeneous beliefs. The model described in this section predicts a departure from market efficiency in the

[15]To list some examples: Goldman Sachs offers "Global Alpha", Merrill Lynch "Absolute Alpha Fund" and UBS "Alpha hedge" and "Alpha select".

[16]Managing Director, Senior Investment Officer, Alternative Investment Solutions at UBS Global Asset Management.

[17]There are two tacit assumptions behind this argument that may or may not be true, namely first that the "bad" investors are capable and willing to learn from their mistakes and second that there are not sufficiently many new "bad" investors entering the market to compensate for their dropped out predecessors.

short run while the long run trend follows the efficient market hypothesis. This prediction finds good support in economic data.[18]

This section is a first step in understanding the Alpha. Later on other important aspects of the Alpha, as for example, generating returns that are of higher order than the mean (first order) and the variance (second order) will be addressed. This section is based on Gerber and Hens [GH06].[19] It is structured as follows. First we give a definition of the Alpha based on the security market line of the CAPM. Then we show that "hunting for Alpha opportunities", i.e., successively including investment opportunities with positive Alpha, leads to a mean-variance optimal portfolio. The main point of this section is then to model a CAPM with heterogeneous beliefs. We show that every investor will form a portfolio such that given his beliefs all Alpha opportunities are exhausted. In a sense we derive a personalized security line. The security market line of the CAPM with homogeneous beliefs finds its analogue in our model with heterogeneous beliefs by a linear relation between the *average* belief of the agents and the Beta of the market portfolio. Note that in the CAPM with heterogeneous beliefs the Security Market Line holds without the unrealistic two-fund separation property. In the model every investor has two options: being active, i.e., following his personal beliefs or being passive, i.e., following the average belief. While the former may incur some costs, the latter can easily be done by buying the market portfolio. Then we show that, as mentioned above, hunting for Alpha opportunities is a zero sum game and we draw some conclusions from this result for market efficiency.

3.3.1 Definition of the Alpha

The Alpha is a departure from the Security Market Line, SML. (It should not be mixed up with the risk aversion that usually is denoted by the same letter α – or alternatively by λ which we need to denote the portfolio weights! This is the reason why we denote this risk aversion by ρ.) Recall that according to the SML the excess return of any asset is proportional to the excess return of the market portfolio with the proportionality factor being the Beta, i.e., the assets' covariance to the market portfolio, standardized by the variance of the market portfolio, formally:

$$\mu(R_k) - R_f = \beta_{k,M}(\mu(R^M) - R_f), \quad \text{where} \quad \beta_{k,M} := \frac{\text{cov}\left(R_k, R^M\right)}{\text{var}\left(R^M\right)}.$$

[18]See, for example the long run data provided by Robert Shiller on his webpage: http://www.econ.yale.edu/~shiller/data.htm

[19]Compare also [Abe89].

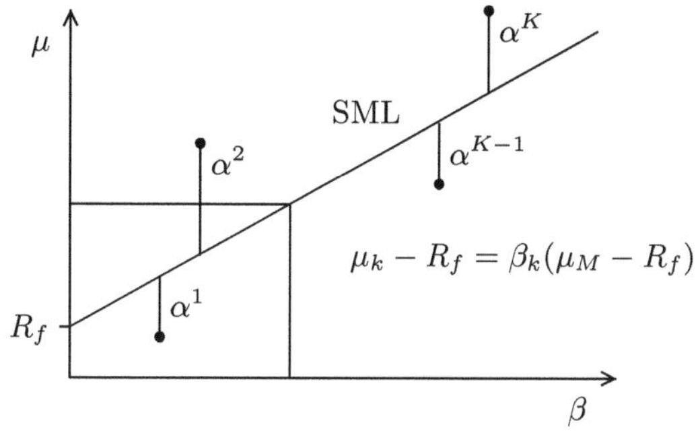

Fig. 3.10 The Alpha of an asset

Hence, the only way of getting higher excess returns is to take more market risk. Some asset managers claim to be able to depart from the straightjacket of the SML. They claim to deliver an excess return higher than that rewarded by market risk. To this end, define the Alpha of asset k as the gap between the claimed excess return and the theoretically justified return:

$$\alpha_{k,M} := \mu(R_k) - R_f - \beta_{k,M}(\mu(R^M) - R_f), \quad \text{where} \quad \beta_{k,M} := \frac{\operatorname{cov}\left(R_k, R^M\right)}{\operatorname{var}\left(R^M\right)}.$$

In principle the Alpha of an asset could be positive or negative. Figure 3.10 displays the Alpha graphically.

Is the standard selling argument correct, that a positive Alpha is a desirable property of an asset? To answer this question, recall that we assumed agents care about means and standard deviations and not about the Alpha itself. That is to say, we need to check the desirability of positive Alpha in the mean standard-deviation diagram and not in the mean-Beta diagram. Clearly, the SML in the mean-Beta diagram is the image of the CML in the mean-standard-deviation diagram and vice versa, i.e., on changing the portfolio weights in a portfolio consisting only of the risk-free and the market portfolio one moves along the SML as high as one moves along the CML. To see this, let λ be the portfolio weight of the market portfolio and accordingly let $(1-\lambda)$ be the weight on the risk-free asset. Then the SML and the CML are obtained by variations of λ and the resulting portfolio means coincide:

$$\begin{aligned}
\text{SML:} \quad & \mu(\lambda R^M + (1-\lambda)R_f) \\
&= R_f + \frac{\operatorname{cov}\left(\lambda R^M + (1-\lambda)R_f, R^M\right)}{\operatorname{var}\left(R^M\right)}(\mu(R^M) - R_f) \\
&= R_f + \lambda(\mu(R^M) - R_f). \\
\text{CML:} \quad & \mu(\lambda R^M + (1-\lambda)R_f) \\
&= R_f + \frac{\sigma\left(\lambda\right) R^M + (1-\lambda)R_f}{\sigma\left(R^M\right)}(\mu(R^M) - R_f) \\
&= R_f + \lambda(\mu(R^M) - R_f).
\end{aligned}$$

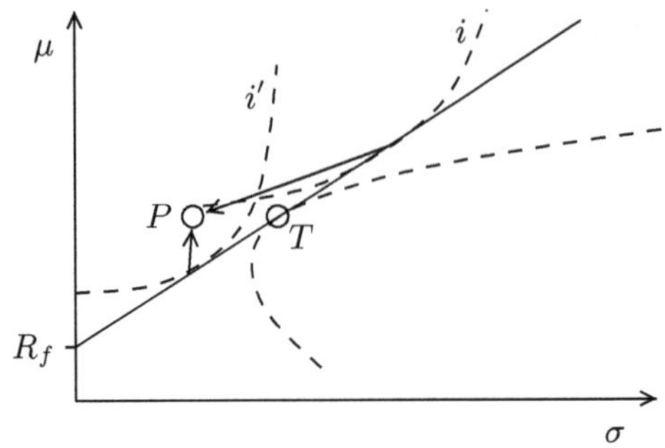

Fig. 3.11 Switching to portfolio P improves i' but not i. However, both can improve by investing *some* wealth in P

But is a point above the SML indeed also a point above the CML and if so, is any point above the CML also an improvement for the agent? Figure 3.11 suggests the following relation between points above the SML and improvements of the asset allocation: Not every point above the SML is an outright improvement of the agents' portfolio. However, adding *some* of it to the agent's portfolio makes the agent better off. We will show now that this is generally true. Therefore, a portfolio with a positive Alpha can be used to improve the agent. On the other hand a portfolio below the SML will always make the existing portfolio worse. Actually we show that the Alpha is the *direction* in which the mean-variance utility of the agent has its steepest increase!

Suppose an investor currently forms an optimal portfolio of the risk-free asset and $k = 1, \ldots, K$ risky assets. Recall his mean-variance utility function:

$$V(\mu_{\boldsymbol{\lambda}}, \sigma_{\boldsymbol{\lambda}}^2) = (\boldsymbol{\mu} - R_f \mathbf{1})'\boldsymbol{\lambda} - \frac{\rho}{2}\boldsymbol{\lambda}' COV \boldsymbol{\lambda},$$

where $\boldsymbol{\lambda} \in \mathbb{R}^K$. The gradient of a function is the vector of its first derivatives. Let $\alpha_{\boldsymbol{\lambda},k}$ be the k-th component of the gradient, i.e., the derivative of V in the direction of asset k. The gradient points into the direction of steepest ascent [S+05, Chap. 16, pp. 540f.]. The gradient of the mean-variance utility function with respect to $\boldsymbol{\lambda}$ is the vector with the entries:

$$\alpha_{\boldsymbol{\lambda},k} := (\mu_k - R_f) - \rho \operatorname{cov}\left(R_k, \sum_{k=1}^{K} R_k \lambda_k\right), \quad k = 1, \ldots, K.$$

If the investor has chosen an optimal portfolio then the derivative of his utility with respect to any asset weight of his portfolio is zero. (This is just the usual first order condition for optimality.) This implies in our notation that $\alpha_{\boldsymbol{\lambda},k} = 0$ where we take into account only the weights of the K assets over which we have already optimized.

Multiplying each equation by λ_k and adding over all assets, we can eliminate ρ and substitute it back into the first order condition. We obtain[20]:

$$\alpha_{\lambda,k} = (\mu_k - R_f) - \beta_{k,\lambda}(\mu_\lambda - R_f(1-\lambda_0)), \quad \text{where} \quad \beta_{k,\lambda} := \frac{\operatorname{cov}(R_k, R_\lambda)}{\operatorname{var}(R_\lambda)},$$

with $R_\lambda := \sum_{k=1}^K R_k \lambda_k$. This is the Alpha of any asset k. The Alpha of a portfolio of assets is accordingly $\sum_{k=1}^K \alpha_{\lambda,k}\lambda_k$. The Alpha of portfolio λ is the *directional derivative* of the mean-variance utility. The first order condition implies that in no direction we find portfolios composed of the K assets which can be an improvement for the investor. If the investor however considers investing in a portfolio that includes new assets, i.e., assets he did not consider before, then a positive Alpha of the new portfolio with respect to the existing optimal portfolio points at a direction of improvement. Hence, a simple rule like "Hunting for Alpha Opportunities" can indeed lead to an optimal asset allocation if the Alpha opportunities are included in small steps. Note that in any such a step the Alpha has to be computed with respect to the currently optimal portfolio. Thus, the reference at which we compute the utility gradient changes along the process.[21]

In the exercise book we also prove that adding any amount of Alpha opportunity to improve a benchmark portfolio may make a suboptimal portfolio worse. Hence, general selling initiatives that are typical in large banks, in which all clients are suggested to add the same Alpha opportunity computed on the basis of a benchmark portfolio, may be bad for many clients with suboptimal portfolios. It would be better to first move the suboptimal portfolios towards the benchmark portfolios. Figure 3.12 shows this effect graphically in the mean-variance diagram.

At this point we have to discuss a natural counter argument to this observation: Alpha opportunities improve the efficient frontier, therefore they should always improve the overall quality of portfolios, shouldn't they? In fact, this line of argument is right and wrong at the same time, and it is important to understand the different notions of "improvement" here. Let us explain this by a simple example: if you decided for a nice menu in a restaurant and now the set of available items is suddenly enlarged by a wonderful red wine for a reasonable price, this is obviously an improvement. However, your particular dinner, let's say fish and white wine, is probably not improved if you add a little bit of red wine to it: the red wine would fit neither to the fish nor to the white wine. The better approach is to choose a completely new menu, and then the red wine can really be part of an

[20]The factor $(1 - \lambda_0)$ in the security line appears since we did not normalize the asset allocation in the risky portfolio to sum up to one. Using the notation incorporating this normalization, i.e., the $\hat{\lambda}^i_k$, the term would not appear any more.

[21]To be more precise: since the optimization problem is concave, we can indeed find the optimal portfolio by optimizing iteratively over the assets, where we improve the portfolio every time we find a positive Alpha, as long as there is one. Concavity is needed in order that this method does not lead to a local optimum, but in fact to the globally optimal portfolio.

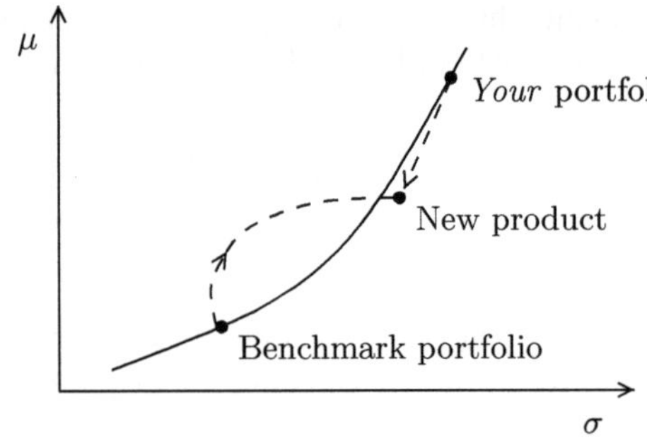

Fig. 3.12 Adding a new product that improves upon a benchmark portfolio to a portfolio different from the benchmark portfolio may make things worse. The *solid line* marks an indifference curve for an investor, the *dashed line* marks combinations with the new product. Clearly, there is a diversification advantage when starting with the benchmark portfolio, but only a disadvantage when starting with your portfolio

optimized menu. Unfortunately, Alpha opportunities are often added to existing portfolios without further checks and might (as Fig. 3.12 illustrates) make things simply worse, although they allow to construct a completely new portfolio with better performance.[22]

The same idea can also be displayed in the Mean-Beta diagram. Suppose, as displayed in Fig. 3.12, that agents have considered to invest in different sets of assets. Say $K^i \subset \{1, \ldots, K\}$ is the subset of assets investor i has so far considered to invest in. Let accordingly $\boldsymbol{\lambda}^i$ be the optimal portfolio he has built with the assets in K^i. His first order condition thus defines an individual security line by the condition that

$$\text{for all } k \in K^i: \quad \mu_k - R_f = \beta_{k,\boldsymbol{\lambda}^i}(\mu_{\boldsymbol{\lambda}^i} - R_f(1 - \lambda_0^i)),$$

where $\beta_{k,\boldsymbol{\lambda}^i} := \text{cov}\left(R_k, R_{\boldsymbol{\lambda}^i}\right) / \text{var}\left(R_{\boldsymbol{\lambda}^i}\right)$. Hence, even if investors shared the same beliefs their security lines differ if they a priori consider to invest in different assets. As Fig. 3.13 shows, it is then well possible that a new asset has a positive Alpha for one investor but not for another. A numerical example for this is given in the exercise book.

[22]If the fish and the red wine example doesn't convince you, you may finally look at a trivial example: imagine a person who only holds a fixed interest rate asset. Would you recommend this person to buy commodities in order to improve the performance through diversification, based on the argument that the efficient frontier will be improved by adding commodities? You probably won't, since a riskless portfolio can obviously not profit from any diversification effects: its covariance to any other asset is always zero.

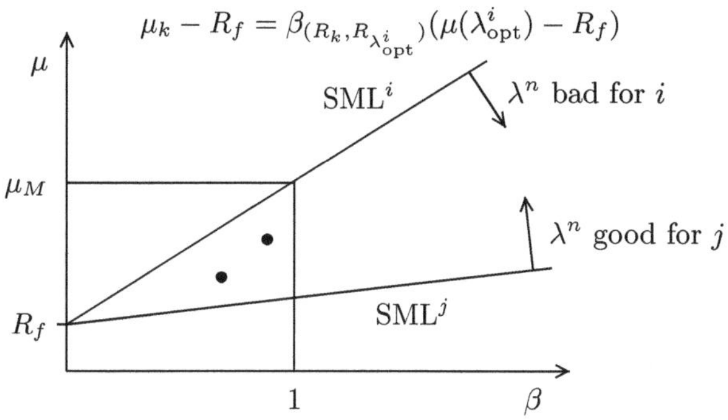

Fig. 3.13 The new product has a positive Alpha for investor j but it has a negative Alpha for investor i. The two investors differ by the set of assets they have so far invested in

3.3.2 CAPM with Heterogeneous Beliefs

While the previous section showed the relation between the Alpha, the SML and the CML for any given investor, this section extends the analysis towards heterogeneous investors. In the standard CAPM investors differ with respect to their initial endowments and their degree of risk aversion, but they share the same beliefs about the expected returns and covariance of returns. Now we allow the investors to also differ with respect to their beliefs on the assets' expected returns, i.e., in principle we could have that $\boldsymbol{\mu}^i \neq \boldsymbol{\mu}^j$ for any two investors i and j. However, we keep the assumption that investors agree on the covariances of the assets. This can be justified by two descriptive and one pragmatic argument.[23] First, errors in means are much more detrimental to the agents' utility than errors in covariances. To see this, let

$$\boldsymbol{\lambda}^{\text{opt}} := \frac{1}{\rho^i} COV^{-1}(\boldsymbol{\mu}^i - R_f\mathbf{1})$$

be the optimal portfolio of agent i, allowing for short sales. Then the optimal level of utility is

$$V^{i,\text{opt}} := \frac{1}{2\rho^i}(\boldsymbol{\mu}^i - R_f\mathbf{1})' COV^{-1}(\boldsymbol{\mu}^i - R_f\mathbf{1}).$$

Hence errors in covariances are of linear order to the utility while errors in means change the utility in a quadratic way.[24] Second, covariances tend to be better predictable, since they are less time-dependent. Take as an example bonds and

[23] See Gerber and Hens [GH06] for a generalization towards heterogeneous beliefs on covariances.

[24] Note that we have defined the utility on gross returns, i.e., expected returns are larger than one.

stocks: In the medium run (2–3 years) bond and stock returns are negatively correlated. In a boom, stocks shoot up but bonds do poorly, in an economic recession, bonds do fine but stocks do depreciate. However, whether on medium-run horizons stock returns are higher than bond returns is much more difficult to predict since this would include a prediction on the stage of the business cycle. Finally, there is a pragmatic reason to keep the assumption of homogeneous covariance expectations which is perhaps most compelling – at least from a didactical point of view: the assumption of heterogeneous expectations on means is already sufficient to explain all the phenomena we mentioned in the introduction of this section. – So why should we make things more complicated than necessary?

In the following we derive the SML for the case of heterogeneous beliefs. We state the result in a proposition and then give the proof of it.

Proposition 3.5 *In the CAPM with heterogeneous beliefs the Security Market Line holds for the average beliefs, i.e., for all assets* $k = 1, \ldots, K$,

$$\bar{\mu}_k - R_f = \beta_{k,M}(\bar{\mu}^M - R_f),$$

where as usual $\beta_{k,M} := \operatorname{cov}\left(R_k, R^M\right) / \operatorname{var}\left(R^M\right)$ *and* $\bar{\mu}^M := \sum_{i=1}^I a^i \mu^i$, *with* $a^i := \frac{r^i}{\rho^i} / \sum_{j=1}^I \frac{r^j}{\rho^j}$ *and* $r^i = w_f^i / \sum_{j=1}^I w_f^i$, $w_f^i = (1 - \lambda_0^i) W_0^i$, *where* W_0^i *denotes the first period income.*

Proof In the CAPM with heterogeneous beliefs an investor maximizes

$$(\boldsymbol{\mu}^i - R_f \mathbf{1})' \boldsymbol{\lambda}^i - \frac{\rho^i}{2} \boldsymbol{\lambda}^{i\prime} COV \boldsymbol{\lambda}^i.$$

The first-order condition is $COV\boldsymbol{\lambda}^i = \frac{1}{\rho^i}(\boldsymbol{\mu}^i - R_f\mathbf{1})$. Multiplying this equation with the relative financial wealth of investor i, which is given by $r^i := w_f^i / \sum_{j=1}^I w_f^i$, where w_f^i denotes the financial wealth of investor i, and summing up over all investors on the market, we get $COV\boldsymbol{\lambda}^M = \sum_{i=1}^I \frac{r^i}{\rho^i}(\boldsymbol{\mu}^i - R_f\mathbf{1})$, $\boldsymbol{\lambda}^M := \sum_{i=1}^I r^i \boldsymbol{\lambda}^i$, which by the definition of R^M is equivalent to

$$\operatorname{cov}\left(R_k, R^M\right) = \sum_{i=1}^I \frac{r^i}{\rho^i}(\mu_k^i - R_f), \quad k = 1, \ldots, K.$$

In these expressions λ_k^M denotes the relative market capitalization of asset k. We have $\lambda_k^M = \sum_{i=1}^I r^i \lambda_k^i$, i.e., the relative market capitalization of asset k is equal to the average percentage of wealth the investors put into asset k.

Multiplying the last expression with λ_k^M and summing up, we get

$$\operatorname{var}\left(R^M\right) = \sum_{i=1}^I \frac{r^i}{\rho^i}(\mu^{i,M} - R_f) \quad \text{where} \quad \mu^{i,M} := \sum_{k=1}^K \mu_k^i \lambda_k^M.$$

Dividing $\text{cov}\left(R_k, R^M\right)$ and $\text{var}\left(R^M\right)$ by $\sum_{i=1}^{I}\frac{r^i}{\rho^i}$ we obtain:

$$\frac{\text{cov}\left(R_k, R^M\right)}{\sum_{i=1}^{I}\frac{r^i}{\rho^i}} = (\bar{\mu}_k - R_f), \quad \text{where} \quad \bar{\mu}_k := \sum_{i=1}^{I}\underbrace{\frac{\frac{r^i}{\rho^i}}{\sum_{j=1}^{I}\frac{r^j}{\rho^j}}}_{=a^i}\mu_k^i,$$

and

$$\frac{\text{var}\left(R^M\right)}{\sum_{i=1}^{I}\frac{r^i}{\rho^i}} = (\bar{\mu}^M - R_f), \quad \text{where} \quad \bar{\mu}^M := \sum_{i=1}^{I} a^i\mu^{i,M}.$$

Eliminating $\sum_{i=1}^{I}\frac{r^i}{\rho^i}$ from the last equation and inserting into the previous one yields

$$\frac{\text{cov}\left(R_k, R^M\right)}{\text{var}\left(R^M\right)}(\bar{\mu}^M - R_f) = \beta_{k,M}(\bar{\mu}^M - R_f) = (\bar{\mu}_k - R_f),$$

which reads for asset k as $(\bar{\mu}_k - R_f) = \beta_{k,M}(\bar{\mu}^M - R_f)$. □

This is the *Security Market Line (SML)* with average expectations, as shown in Fig. 3.14. Note that the averaging is done taking into account both the relative wealth and also the risk aversion of the agents. The wealthier and the less risk averse agents determine the average more than the poor and more risk averse agents. Since agents have the same covariance expectations, the Beta factors are as in the model with homogeneous beliefs.

When we have heterogeneous beliefs, just like with heterogeneous investor sets, we get individual security lines along which all assets are lined up if they form an optimal portfolio. The derivation is done as above. We consider the first order condition for maximizing the mean-variance utility function, multiply each equation with the portfolio share and add these equations up to eliminate the risk aversion

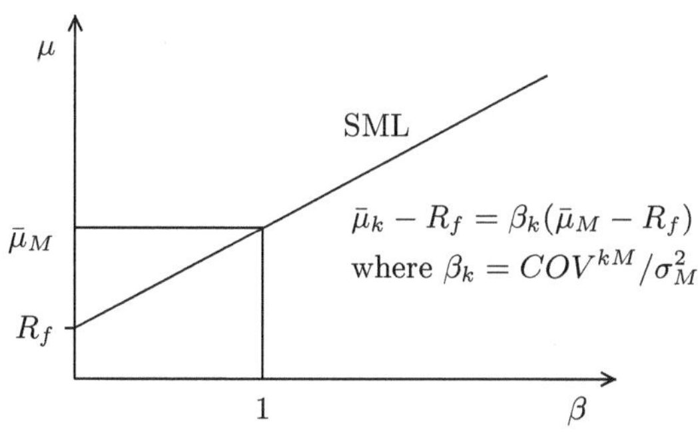

Fig. 3.14 The Security Market Line for average expectations

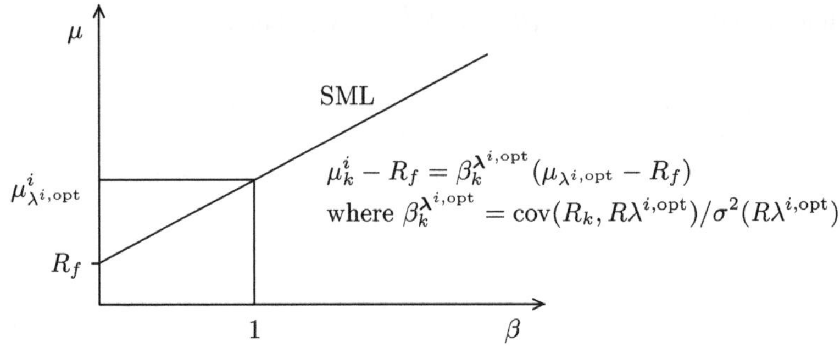

Fig. 3.15 Security Line of investor i

parameter ρ^i. As a result we obtain the individual security line: for all k we have

$$\mu^i_k - R_f = \beta_{k,\lambda^i}(\mu_{\lambda^i} - R_f(1 - \lambda^i_0)),$$

where $\beta_{k,\lambda^i} = \text{cov}\left(R_k, R_{\lambda^i}\right) / \text{var}\left(R_{\lambda^i}\right)$ (Fig. 3.15).

If an investor happens to have beliefs equal to the average belief, i.e., if $\boldsymbol{\mu}^i = \bar{\boldsymbol{\mu}}$ then he will hold a portfolio of risky assets that coincides with the market portfolio. In general this may not be the case and agents can have under-diversified portfolios,[25] two-fund separation fails and a new asset with positive Alpha vis-à-vis the market portfolio can have negative Alpha for some agents. The last point is exemplified in the exercise book. We come back to this point later, after we have analyzed the zero-sum game property.

3.3.3 Zero Sum Game

A zero sum game is a game in which one agent can benefit only at the expense of some other agent. A typical situation arises in allocation problems, i.e., in situations in which a given set of resources is allocated to various agents. Sharing a pie is a simple example of an allocation problem. One may argue that the CAPM is indeed a pie sharing model. It is one method to allocate the market returns among the investors. Indeed, as we showed more generally above, the market return is equal to the average return of the investors. Hence seen ex-post,[26] in this respect the CAPM (and any other equilibrium model) is a zero sum game. Any return given to some investor has to be taken away from some other investor. This suggests that

[25]Note that underdiversified portfolios do not need to be worse than well-diversified portfolios. Based on 78'000 households portfolios observed from 1991 to 1996 Ivković, Sialm and Weisbenner [ISW08] find that the more wealthy have more underdiversified portfolios achieving a positive Alpha to the market.

[26]*Ex-post* means after a state $s = 1, \ldots, S$ of the world has realized. *Ex-ante* means before the resolution is known.

any allocation of returns to investors is efficient in the sense that no agent can be improved without making some other agent worse off. However, from an ex-ante point of view it may happen that for all agents some allocations are better than others because individuals prefer different returns in different states, for example. One way of extending the zero sum game property in order to reflect the ex-ante point of view is to analyze whether the Alphas investors obtain at a CAPM equilibrium add up to zero. In doing so, we distinguish the Alphas by the reference portfolio, the portfolio based on the average market expectation and the one based on the correct expectation.[27] Before doing so, we remind the reader that in utility terms the CAPM is clearly not a zero sum game since it still involves trade to share risks which is beneficial to *all* investors.

We start our formal analysis with the equilibrium property on asset allocations known from the general notion of financial market equilibria, the average portfolio allocation of the investors, weighted by their relative wealth, coincides in equilibrium with the market capitalization:

$$\sum_{i=1}^{I} \lambda_k^i r^i = \lambda_k^M, \quad k = 0, \dots, K.$$

Multiplying each equation by the return of asset k and adding up over all assets, we obtain

$$\sum_{i=1}^{I} R_s^i r^i = R_s^M, \qquad s = f, \dots, S,$$

where $R_s^i := \sum_{k=0}^{K} R_{ks}\lambda_k^i$ and $R_s^M := \sum_{k=0}^{K} R_{ks}\lambda_k^M$. Hence, in each state the market return is the average return of the individual investors, where each investor has a weight equal to his relative wealth. Assuming $r^i > 0$, for all investors, this implies that indeed the return of any investor can only be increased if the return of some other investor is decreased, which is a first result concerning the zero sum property. However, this argument holds state by state and realizing that returns are risky and that investors may care differently about the size of returns in different states, one may conjecture that in terms of risk-adjusted returns the market is no zero sum game. One way of adjusting for risk is given by the Alpha. To make this point we first need a good definition of the Alpha.

Recall that each agent chooses his portfolio so that from his point of view no asset has an Alpha. Hence, defining the Alpha as a deviation of a portfolio from the individual security line, i.e., defining it as

$$\alpha^i_{k,\lambda^{i,\text{opt}}} := \mu_k^i - R_f - \beta_{k,\lambda^{i,\text{opt}}}(\mu_{\lambda^{i,\text{opt}}} - R_f),$$

[27]Note that in a model with rational expectations all investors are assumed to know the true market returns that can be expected. In particular they have then homogeneous expectations.

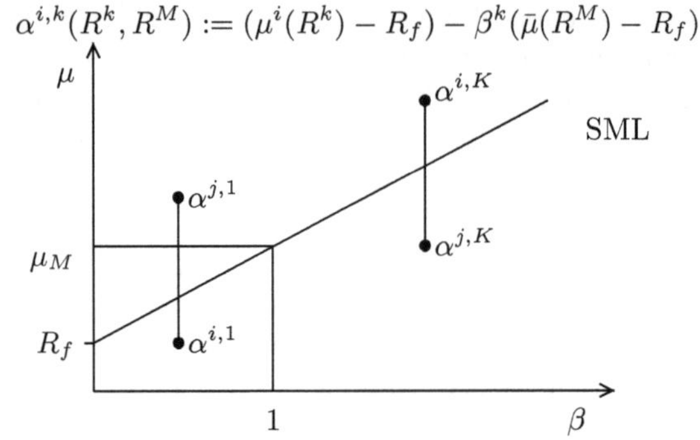

Fig. 3.16 Alphas of assets as compared to the excess return adjusted by the market risk. If one agent sees a positive Alpha some other agent needs to see a negative Alpha

where $\beta_{k,\lambda^{i,\mathrm{opt}}} := \operatorname{cov}\left(R_k, R_{\lambda^{i,\mathrm{opt}}}\right) / \operatorname{var}\left(R_{\lambda^{i,\mathrm{opt}}}\right)$, the CAPM clearly is a zero sum game, since each of these Alphas is zero, so that any weighted sum of those Alpha also needs to be zero. Thus for the zero sum property to be interesting one needs to take different portfolios or different beliefs as benchmarks. One candidate is the market portfolio respectively the average beliefs. Going this way, the Alpha of any asset k is the excess return that agent i sees in asset k over and above the return seen by the market, formally:

$$\alpha^i_{k,M} := (\mu^i_k - R_f) - \beta_{k,M}(\bar{\mu}^M - R_f).$$

As Fig. 3.16 illustrates, for any asset k some agent will see a positive Alpha while some other agent will see a negative Alpha.

Given this definition of the Alpha of asset k as seen by investor i, we define the Alpha of the portfolio of investor i as the market average of the Alphas he sees for the assets:

$$\alpha^i := \sum_{k=1}^{K} \lambda^M_k \alpha^i_{k,M}.$$

We call this way of defining the Alpha the beliefs point of view since in the definitions we used individuals' expectations of returns. We now get the zero sum property:

Proposition 3.6 *Defining the Alpha as the excess return that agent i sees in asset k over and above the return seen by the market, the weighted average of the individual investors' Alphas is zero. The weights are given as in the security market line.*

Proof The proposition claims that

$$\sum_{i=1}^{I} a^i \alpha^i = 0, \quad \text{where} \quad a^i := r^i \rho^i \Big/ \sum_{j=1}^{I} \frac{r^j}{\rho^j}.$$

Recalling the definition of the Alphas we get

$$\sum_{i=1}^{I} a^i \sum_{k=1}^{K} \lambda_k^M \alpha_{k,M}^i = \sum_{i=1}^{I} a^i \sum_{k=1}^{K} \lambda_k^M \left((\mu_k^i - R_f) - \beta_{k,M}(\bar{\mu}^M - R_f)\right)$$
$$= \sum_{i=1}^{I} a^i \sum_{k=1}^{K} \lambda_k^M (\mu_k^i - R_f) - \sum_{i=1}^{I} a^i \sum_{k=1}^{K} \lambda_k^M \beta_{k,M}(\bar{\mu}^M - R_f).$$

And hence, by the weighting factors and the market returns we get what we claimed:

$$\sum_{i=1}^{I} a^i \alpha^i = (\bar{\mu}^M - R_f) - \beta^M(\bar{\mu}^M - R_f) = 0. \qquad \square$$

One interpretation of Proposition 3.6 is that even if we do not know who is right we can still agree that not everybody can do better than average.

Yet, a different way of defining Alphas is to define them with respect to the true average returns. Suppose every agent forms his portfolio based on his beliefs about the average returns and then we let the model run for a while to compare the expected returns with the average of the realized returns. We can then ask who is best in guessing the true average returns and also whether benchmarked to those returns the zero sum property holds. To this end let $\hat{\mu}_k$, $k = 1, \dots, K$, denote the true average return of the assets and define $\hat{\mu}^M := \sum_{k=1}^{K} \lambda_k^M \hat{\mu}^k$ and the Alpha of asset k as the realized average return exceeding the expected average return based on market expectations:

$$\hat{\alpha}_{k,M} := \left(\hat{\mu}_k - R_f\right) - \beta_{k,M}\left(\hat{\mu}^M - R_f\right), \quad k = 1, \dots, K.$$

Figure 3.17 illustrates this notion of Alphas.

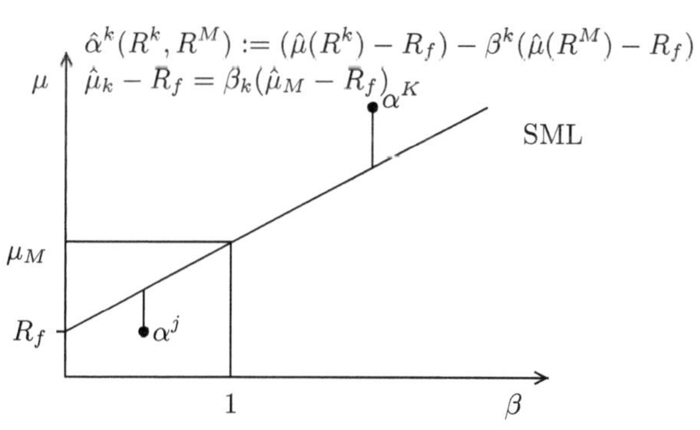

Fig. 3.17 Defining Alphas by the difference of the true returns and the risk adjusted true returns we see that not all assets can have a positive Alpha

For this notion of Alphas for any asset k we define the Alpha of the portfolio of investor i as the assets' Alphas weighted by his asset allocation:

$$\hat{\alpha}^i := \sum_{k=1}^{K} \lambda_k^i \hat{\alpha}_{k,M}.$$

Now we get the zero sum property when weighting the investors' Alpha with their relative wealth.

Proposition 3.7 *Defining the Alpha of asset k as the excess return that asset k realizes over and above the return justified by the security market line, the weighted average of the individual investors' Alphas is zero. The weights are given by the relative wealth of the investors.*

Proof The claim is $\sum_{i=1}^{I} r^i \hat{\alpha}^i = 0$, where $\hat{\alpha}^i := \sum_{k=1}^{K} \lambda_k^i \hat{\alpha}_{k,M}$ and

$$\hat{\alpha}_{k,M} := \left(\hat{\mu}_k - R_f\right) - \beta_{k,M}\left(\hat{\mu}^M - R_f\right).$$

And indeed:

$$\begin{aligned}\sum_{i=1}^{I} r^i \sum_{k=1}^{K} \lambda_k^i \hat{\alpha}_{k,M} &= \sum_{k=1}^{K} \lambda_k^M \left((\hat{\mu}_k - R_f) - \beta_{k,M}(\hat{\mu}^M - R_f)\right) \\ &= \left(\hat{\mu}^M - R_f\right) - \beta^M \left(\hat{\mu}^M - R_f\right) = 0.\end{aligned}$$

□

So far we have seen that the zero sum property can be obtained for three different definitions of the Alpha. We conclude this series of definitions of Alphas by one for which we do *not* get the zero sum property. If the Alpha that agent i gets for asset k is defined as his expected return over and above the realized return, i.e.

$$\left(\mu_k^i - R_f\right) - \beta_{k,M}\left(\hat{\mu}^M - R_f\right),$$

then the Alphas may not add up to zero because it could well be that all agents are too optimistic or too pessimistic for all assets.

We finish these ideas on the zero sum property of Alphas in the CAPM with heterogeneous beliefs by reminding once more that in utility terms all agents could benefit from trade due to diversification. The CAPM for homogeneous beliefs is still a special case of the model with heterogeneous beliefs. Hence, nobody shall come to the conclusion that financial markets as such are useless since they only offer zero alpha sum games.

3.3.4 Active or Passive?

An active investor in the CAPM optimizes his portfolio given his beliefs. The active investor thus invests in his Tangent Portfolio. A passive investor invests in the market portfolio as if he shared the average belief of the investors. While the CAPM with homogeneous beliefs states that there is no difference between active and passive investing,[28] the extension of the CAPM to heterogeneous beliefs can give a more realistic advice on the active/passive decision. In this section we analyze who shall be active and who shall be passive. We will assume that active asset management is costly while being passive is for free. Hence every investor has the choice to "passify" if he discovers himself to be a loser of the zero sum game. The remaining active investors chase the Alphas among themselves. Thus, if investors learn about their active market skills and more and more unskilled investors drop out, the remaining investors will have an ever harder task. Eventually, only the best active manager determines asset prices, which is a conclusion in the line of the efficient market hypothesis. However, since the active investors need to pay for their superior information while the passive investors get this information for free this is not a stable situation.

To begin with we first verify that an active investor forming his beliefs on the basis of the SML indeed chooses the market portfolio. Recall the first-order condition that determines the portfolio of risky assets of an active investor: $\boldsymbol{\lambda}^i = (1/\rho^i)COV^{-1}(\boldsymbol{\mu}^i - R_f\mathbf{1})$. Now suppose the active investor determines his belief from the SML, i.e., he sets $\mu_k^i := R_f + \beta_{k,M}(\bar{\mu}^M - R_f)$, where $\beta_{k,M} := \frac{COV^{k,M}}{\sigma_M^2}$. Then, on noting that $COV^{k,M} = (COV\boldsymbol{\lambda}^M)_k$ we get that his portfolio of risky assets is proportional to the market portfolio:

$$\boldsymbol{\lambda}^i = \frac{1}{\rho^i}COV^{-1}COV\boldsymbol{\lambda}^M\frac{(\bar{\mu}^M - R_f)}{\sigma_M^2} = \underbrace{\frac{1}{\rho^i}\left(\frac{\bar{\mu}^M - R_f}{\sigma_M^2}\right)}_{\text{scalar}}\boldsymbol{\lambda}^M.$$

In describing the active-passive decision we once more allude to the thought experiment made above: Suppose every agent forms his portfolio based on his beliefs about the average returns and then we let the model run for a while to compare the expected returns with the average of the realized returns. The agents then evaluate the result of their portfolio choice by what has happened on the market. If agents were only interested in the Alpha they would then evaluate their choice by the Alphas as defined for Proposition 3.7. The game they are playing would be zero sum with an outside option (being passive) by which every agent could guarantee to him the payoff zero. Hence, eventually none of the agents would be active. This seems like a compelling argument. However, the correct way of evaluating the

[28] This is a consequence of two-fund separation. Every active investor (holding his tangential portfolio) holds the market portfolio, i.e. turns out to be a passive investor in equilibrium.

situation is according to the agents *utility* derived from their investments. To this end let $U^i_{\hat{\mu}}(\mu^i)$ be the mean-variance utility that agent i gets in the course of his investment when he bases his decision on his belief μ^i while the true average returns are given by $\hat{\mu}$. We assume that if $\mu^i \neq \bar{\mu}$ then the agent pays a cost $C^i > 0$ for being active. Hence, optimizing with respect to his own beliefs an active investor achieves the utility

$$U^i_{\hat{\mu}}(\mu^i) = \lambda^i_0 R_f + (1 - \lambda^i_0)\hat{\mu}(R\boldsymbol{\lambda}^i) - \frac{\rho^i}{2}(1 - \lambda^i_0)^2 \operatorname{var}\left(R\boldsymbol{\lambda}^i\right) - C^i.$$

Accordingly, a passive investor who optimizes his portfolio given the markets beliefs achieves the utility

$$U^i_{\hat{\mu}}(\bar{\mu}) = \lambda^i_0 R_f + (1 - \lambda^i_0)\hat{\mu}(R\bar{\boldsymbol{\lambda}}^i) - \frac{\rho^i}{2}(1 - \lambda^i_0)^2 \sigma^2(R\bar{\boldsymbol{\lambda}}^i).$$

Which of the two utilities is larger depends on the market efficiency and also on the skill of the investor (how much his expectations deviate from reality).

One can show[29] that the agent should be active if and only if:

$$U^i_{\hat{\mu}}(\mu^i) - U^i_{\hat{\mu}}(\bar{\mu}) = \frac{1}{2\rho^i}\left(\|\hat{\mu} - \bar{\mu}\|^2 - \left\|\hat{\mu} - \mu^i\right\|^2\right) > C^i,$$

where $\|\boldsymbol{x}\|^2 := \boldsymbol{x}' COV^{-1}\boldsymbol{x}$. Here, $\|\hat{\mu} - \bar{\mu}\|^2$ is a market inefficiency term and $\left\|\hat{\mu} - \mu^i\right\|^2$ measures the deviation of expectations from reality.

This result shows that the investor should be active

- the less efficient the market,
- the more skilled the investor,
- the smaller his costs to be active and
- the less risk averse he is.[30]

Market efficiency is in turn depending on who is active! This implies for example that an active investor erodes his investment opportunities the more successful he becomes. This "winner's curse problem" is well-known for hedge funds. It may be one reason why the best funds are closed. The endogeneity of market opportunities also leads to the following pattern of market opportunities. All investors whose beliefs are farther away from the true belief than the average belief should rather be passive, which makes the market belief closer to the true beliefs and more investors will drop out of the group of activists. Eventually, only the most skilled investor

[29] See Gerber and Hens [GH06]

[30] That agents may trade actively because they are not at all risk averse but trade for entertainment has recently been observed in Dorn and Segmueller (2006).

will be active. However, at this point he can "pull his legs", i.e., he can – as every other passive investor – get the best belief for free by passively investing into the market portfolio. The consequence of this is that the market portfolio is no longer informative and the game starts all over again. That is to say, there are no stable market outcomes – a result which is known for other models as the Grossman-Stiglitz Information Paradox. Certainly, Alpha opportunities change when important unforeseen events that are difficult to value – like the commercialization of the Internet – occur. But our model suggests that even without those major events Alpha opportunities are not constant over time since the model generates cycles in Alpha opportunities within itself.

We close this section by noting that the Alpha is not itself a good criterion for the active–passive decision. It may well be that an agent has a positive Alpha but should rather be passive, as shown in the exercise book. Also the converse is true, an agent can have a negative Alpha but he should rather be active, as it is shown in the final example of [GH06].

3.4 Alternative Betas and Higher Moment Betas

The general topic of this chapter is to explain trade and valuation of risk and return in financial market equilibria. In the CAPM we analyzed trading for diversification purposes and, in the case of heterogeneous beliefs, also for trading motivated by different expectations, i.e., for "betting". The CAPM gives a first intuition about the valuation of risk and return in a financial market equilibrium. However, it is build around quite restrictive assumptions. It ignores intertemporal trade and cannot explain the excess return of the market portfolio, also called the equity premium, which has to be taken as exogenous in the CAPM.[31] Moreover, strictly spoken the CAPM is a model for the determination of *stock* market risk, the Beta. But in many applications the CAPM is claimed to hold also for non-stock market risk, like risk from alternative asset classes, e.g., commodities, private equity, real estate, art, gems and wine etc. Taking these risks may yield higher return than justified by its correlation to stock market risk. Yet, this excess return is no Alpha but should better be called *Alternative Beta*. Recall that we defined the Alpha as a return that is not justified by the risk factors of a model, but that arises from superior information or skill. Hence the fact that alternative asset classes yield returns that are higher than justified by their correlation to the stock market implies that one should rather extend the definition of the market portfolio to include alternative risk factors. Moreover, in contrast to the CAPM, excess returns may be received from holding skewed and fat tailed returns. Again these are no Alphas but rewards for other types of risk, which one should call *higher Moment Betas*.

[31] Applying the SML to the market portfolio itself results in the tautology $1 \equiv 1$, hence the CAPM is an asset pricing model valuing assets relative to the market portfolio, but it does not evaluate the market portfolio itself.

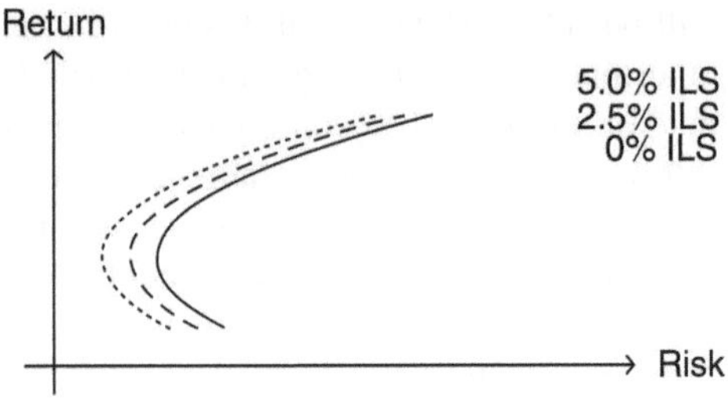

Fig. 3.18 Adding insurance-linked securities (ILS) can enlarge the efficient frontier, but does this imply that we are better off by adding them to our portfolios?

3.4.1 Alternative Betas

Some financial intermediaries (banks, hedge funds, asset managers) try to sell their products in the following way. They frame the asset allocation problem in terms of means and variances and then they show that including their product enlarges the efficient frontier. Figure 3.18 gives one such example for the case of investing in bonds that default when a catastrophe happens.

As we show in the exercise book, an enlargement of the efficient frontier is a better argument for investing in a product than the argument based on the Alpha would be. But is the enlargement argument sufficient for investing in the product? This is the case provided that investors do only care about the mean and the standard deviation of the payoffs of the product as related to other investments (stocks and bonds) that they have made. But if the investors are already quite loaded with the underlying risk of the product then they should take this into consideration and the attractiveness of the product may decrease. Consider for example the Goldman Sachs Commodity Index, GSCI, which to a large extent is related to the oil price. For the CEO of a car producer it may make sense to invest in the GSCI, since this may compensate him for a smaller bonus if the demand for cars drops due to a rise in oil prices. But for the CEO of a solar technical firm the opposite may be true. Besides this obvious correlation to the background risks, the product may also yield a high return due to factors like illiquidity which may be bad for some investors. Moreover, the product should be benchmarked to the correct notion of market portfolio so that the investors see whether it delivers the highest possible return given the risks it offers. Ultimately, one may use factor analysis to reveal all risk factors the product is based on. If the investor can determine his sensitivity to these factors one could see whether the product is really a good investment for him. Hence, the APT can be seen as the extension of the CAPM towards background risks which then allows including alternative betas. Formally we can derive the APT analogously to the derivation of the standard CAPM.

Section 3 has shown that all risks except that of the market portfolio can be diversified. If the market portfolio consists of more than stock market risk, then also the security market line will reflect these other sources of risk: Suppose

$$R_M = \sum_{j=1}^{J} x_j R_j,$$

where R_j are factors like stock market risk (traditional risk) and alternative risk. Supposing the factors are mutually exclusive,[32] the security market line is obtained as:

$$\mu_k - R_f = \sum_{j=1}^{J} \beta_{kj}(\mu_j - R_j)$$

where

$$\beta_{kj} := \frac{x_j \operatorname{cov}(R_k, R_j)}{\sigma_M^2}, \quad j = 1, \ldots, J.$$

This resembles the APT.

3.4.2 Higher Moment Betas

As the CAPM, the APT is based on correlations. Hence it ignores higher moments of the return distribution. Yet, if agents have realistic preferences like those coming from Expected Utility Theory or Prospect Theory then they may not only care for mean and variance. Some investments that look very attractive in the mean-variance framework may lose their attraction once higher moments are taken into account. For example, applying Prospect Theory to standard data on value and size portfolios, one can conclude that due to skewness and fat tails[33] the deep value and the small cap returns while being very attractive for a mean-variance agent, are not more attractive than the stock market index. This point has recently been made by De Giorgi, Hens and Post [DGP08]. Table 3.1 shows descriptive statistics of the standard size and value portfolio as they can be found on the webpage of Kenneth French. First we observe that the equity index has an excess return over bonds of about 6.6 %. Second we observe that small cap and value portfolios give higher return than the equity index and also than large cap and glamour portfolios.

[32] I.e., the covariance between each pair of factors is zero.

[33] Skewness is measured by the third moment of a distribution and fat tails are measured by the kurtosis, i.e., the fourth moment, of a distribution in excess of the kurtosis of a normal distribution, see Sect. A.2.

Table 3.1 Descriptive statistics (average, standard deviation, skewness, excess kurtosis, max and min) for the annual real returns of the value-weighted CRSP all-share market portfolio, the intermediate government bond index of Ibbotson and the size and value decile portfolios from Kenneth French' data library. The sample period is from January 1927 to December 2002 (76 yearly observations)

	Avg.	Stdev.	Skew.	Kurt.	Min	Max
Equity	8.59	21.05	0.19	0.36	−40.13	57.22
Bond	2.20	6.91	0.20	0.59	−17.16	22.19
Small	16.90	41.91	0.92	1.34	−58.63	155.29
2	13.99	37.12	0.98	3.10	−56.49	169.71
3	13.12	32.31	0.69	2.13	−57.13	139.54
4	12.53	30.56	0.46	0.83	−51.48	115.32
5	11.91	28.49	0.44	1.60	−49.57	119.40
6	11.65	27.46	0.31	0.61	−49.49	102.17
7	11.09	25.99	0.30	1.14	−47.19	102.06
8	10.15	23.76	0.29	1.19	−42.68	94.12
9	9.63	22.33	0.02	0.46	−41.68	78.15
Large	8.06	20.04	−0.22	−0.52	−40.13	48.74
Growth	7.84	23.60	0.02	−0.64	−44.92	60.35
2	8.77	20.41	−0.27	−0.27	−39.85	55.89
3	8.52	20.56	−0.10	−0.47	−38.00	51.90
4	8.25	22.49	0.49	2.39	−45.02	96.33
5	10.29	22.82	0.36	1.92	−51.55	93.77
6	10.06	23.04	0.19	0.63	−54.39	73.57
7	11.00	24.73	0.18	1.22	−51.13	97.91
8	12.82	27.01	0.67	1.95	−46.56	113.53
Value	13.32	33.05	0.43	1.40	−59.78	134.46

However, these higher returns come along with higher volatility, higher skewness and fatter tails of the return distribution. Finally, the range of observed returns is higher for the portfolios which have higher average returns.

For a mean-variance investor the small cap and the value portfolios are attractive since he does not care about the higher moments of the distribution, as it can be seen from Table 3.2.

We see that on a 5 % significance level[34] the Sharpe-ratios of small cap and of value portfolios are higher than those of the stock market and the bond market index. To make significant numbers more visible, in Table II they are in bold face letters and their cells are shaded in grey. How does a Prospect Theory investor care about the higher moments of the distribution? Well this depends on the functional form of his value function, as [DGP08] has shown. Evaluated with the piecewise power value

[34]The p-values displayed in Table 3.2 mirror the significance level. That is to say a utility value with a p-value not exceeding 0.05 indicates a significance on a 5 % level.

Table 3.2 For each portfolio: Sharpe ratio, CPT statistic and adjusted CPT statistic with the piecewise-exponential value function, compare (2.13); bootstrap p-values. Numbers in bold refer to portfolios that yield a significantly higher value than the market portfolio at a 5 % significance level

	MV		CPT		CPT (exp.)	
	Statistic	p-value	Statistic	p-value	Statistic	p-value
Equity	0.380		−1.590		−1.496	
Bond	**0.329**	**0.007**	**−0.788**	**0.008**	−1.105	0.240
Small	0.384	0.140	**2.290**	**0.030**	−2.172	0.933
2	0.357	0.317	1.053	0.085	−1.981	0.888
3	0.384	0.215	0.654	0.085	−1.749	0.749
4	0.387	0.212	0.278	0.066	−1.509	0.514
5	0.394	0.180	0.197	0.070	−1.411	0.377
6	0.400	0.153	**0.101**	**0.043**	−1.441	0.413
7	0.402	0.142	**0.076**	**0.033**	−1.416	0.347
8	0.403	0.140	**−0.006**	**0.020**	−1.342	0.233
9	0.404	0.116	**−0.552**	**0.035**	−1.322	0.224
Large	0.376	0.457	−1.767	0.741	−1.427	0.279
Growth	0.308	0.821	−2.673	0.863	−2.012	0.920
2	0.410	0.104	−1.352	0.410	−1.286	0.129
3	0.392	0.219	−1.299	0.251	−1.503	0.516
4	0.336	0.591	−0.695	0.158	−1.484	0.465
5	0.420	0.075	**0.502**	**0.039**	−0.985	0.059
6	0.403	0.137	0.176	0.147	−1.380	0.336
7	0.419	0.076	−0.018	0.101	−1.234	0.273
8	**0.447**	**0.027**	**2.083**	**0.003**	−1.163	0.233
9	**0.449**	**0.026**	**1.905**	**0.008**	−1.098	0.203
Value	0.383	0.174	−0.050	0.202	−1.422	0.436

function (2.7) from Sect. 2.4.3, the size and the value premium puzzle would even deepen. However, if one applies the piecewise exponential value function (2.13) from Sect. 2.4.4, the size and the value premium puzzle is gone. For the small payoffs, as they are usually considered in experiments, the two value functions look very similar, while for larger payoffs the piecewise exponential value function is more concave (compare Fig. 2.15). Applied to the size and value premium data this implies that the extremely high returns get less utility than in the mean-variance case or with the piecewise power value function.

In Table 3.2 the column CPT refers to the piecewise power value function while the column CPT (exp.) refers to the piecewise exponential value function. We see that for CPT (exp.) on a 5 % significance level none of the portfolios is any better than the stock market index.

While the argument of [DGP08] is only based on the shape of the value function, probability weighting in Prospect Theory may explain why investors are reluctant to invest in Insurance Linked Securities (ILS). The return distribution of ILS is very fat-tailed to the left, i.e., every now and then a real catastrophe happens and investors have to face huge losses. A Prospect Theory investor exaggerates these

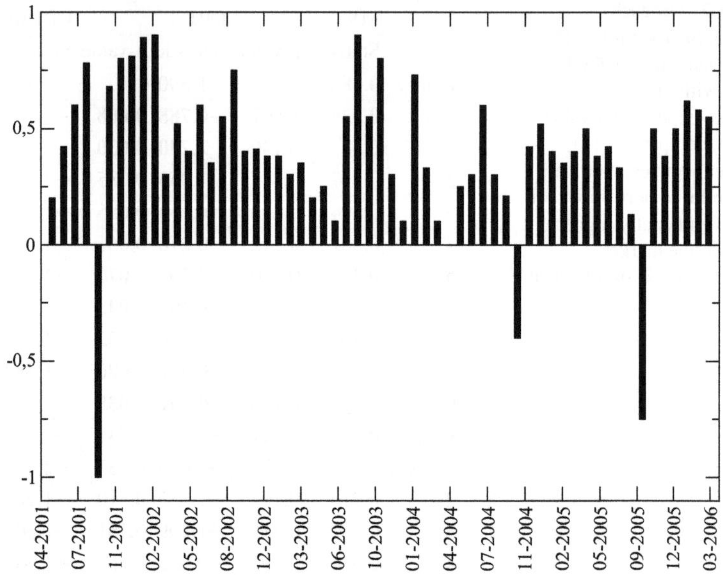

Fig. 3.19 Performance Track Record of ILS. The catastrophes imply a return distribution that is very skewed and fat-tailed on the left

small probability events and may hence not invest into ILS. Figure 3.19 shows the returns of ILS.

3.4.3 Deriving a Behavioral CAPM

Analogously to the derivation of the standard CAPM based on the mean-variance diagram, we can derive a SML that incorporates loss aversion and asymmetric risk aversion – two key properties of prospect theory. We ignore behavioral heterogeneity and use a "representative investor".[35] To begin we need to have a diagram for rewards and risks like mean and variance that however captures the main aspect of prospect theory: gains and losses. Since we want to generalize the basic CAPM to incorporate aspects of prospect theory, we will choose a piecewise quadratic value function for prospect theory:

$$u(\Delta x) = \begin{cases} \Delta x - \frac{\alpha^+}{2}(\Delta x)^2, & \text{if } \Delta x \geq 0 \\ \beta\left(\Delta x - \frac{\alpha^-}{2}(\Delta x)^2\right), & \text{if } \Delta x < 0 \end{cases}$$

[35]For the pros and cons of this approach see Sect. 4.6 of the next chapter.

where $\Delta x = x - RP$. The overall prospect utility then is

$$PT_u(\Delta x) = \sum_s p_s u(\Delta x_s),$$

where we ignored probability weighting to keep things simple.[36] Note that for $\beta = 1$ and $\alpha^+ = \alpha^- =: \alpha$ the prospect utility plus the RP is a function of mean and variances only since the second moment, i.e. the expectation of the square of a random variable, is a function of mean and variance. In particular, for $RP = \mu$, maximizing $PT_u(\Delta x) + RP$ is equivalent to maximizing the simple mean-variance utility $\mu - \frac{\alpha}{2}\sigma^2$. Hence mean-variance analysis is still a special case of our analysis here.

What would be an appropriate reward-risk diagram for prospect theory? PT divides outcomes into two aspects: gains and losses. Hence it is natural to use gains as the reward and losses as the risk dimension of a PT-diagram. To be precise, we write the value function slightly different:

$$v(c) = \begin{cases} u(c), & c \geq RP \\ -\frac{1}{\beta}u(c), & c < RP. \end{cases}$$

Then we define prospect gains,

$$pt^+(c) = \sum_{c_s > RP} p_s v(c_s),$$

and prospect losses,

$$pt^-(c) = \sum_{c_s < RP} p_s v(c_s).$$

We can express the overall prospect utility as the difference of prospect gains and beta times prospect losses:

$$PT_u(c) = pt^+(c) - \beta pt^-(c).$$

This suggest the following reward-risk diagram for prospect theory [DGH06] (Fig. 3.20).

Now we can derive a Behavioral CAPM, B-CAPM, based on this diagram in complete analogy to the geometric derivation of the simple CAPM which was based on the mean-variance diagram.

[36]You may include probability weighting by replacing p_s with $w_s = w(p_s)$. However, then some care has to be taken in the empirical analysis!

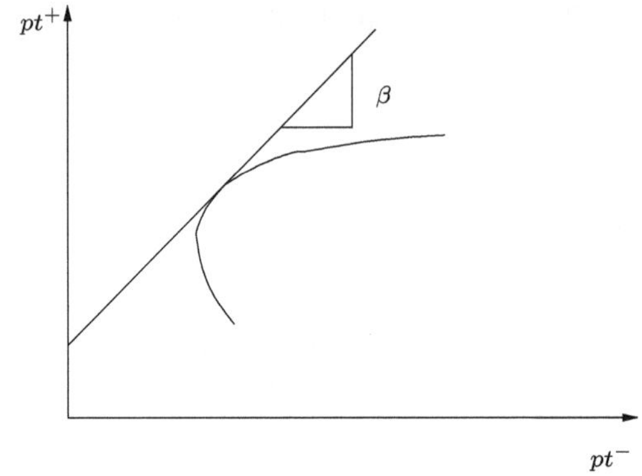

Fig. 3.20 The reward-risk diagram for prospect theory. The behavioral efficient frontier maximizes the prospect utility from gains given any level of the prospect utility from losses. It's upper left part is the analog to the efficient frontier of the mean-variance diagram. The optimal point is determined by the tangency of the highest line with slope equal to the loss aversion and the efficient frontier

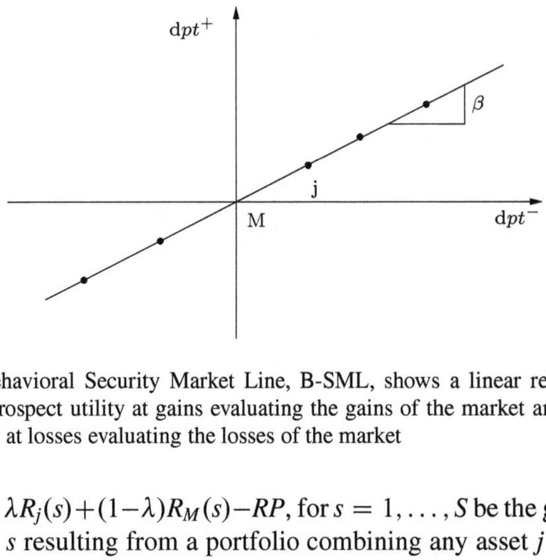

Fig. 3.21 The Behavioral Security Market Line, B-SML, shows a linear relation between the derivative of the prospect utility at gains evaluating the gains of the market and the derivative of the prospect utility at losses evaluating the losses of the market

Let $\Delta_\lambda(s) = \lambda R_j(s) + (1-\lambda)R_M(s) - RP$, for $s = 1, \ldots, S$ be the gain respectively the loss in state s resulting from a portfolio combining any asset j with the market portfolio M. Then we know that the curve resulting in the prospect theory diagram must be tangent to the line that determines the optimal portfolio. To understand the link between the individual optimization behavior and the market, compare the slopes of the Capital Market Line and the j-curve. By the tangency property of λ_M, they must be equal. Note that (Fig. 3.21)

$$pt^+(\lambda R_j + (1-\lambda)R_M) = \sum_{\Delta_\lambda(s)\geq 0} p_s \Big\{ \lambda(R_j(s) - RP) + (1-\lambda)(R_M(s) - RP) \\ - \frac{\alpha^+}{2} \left[\lambda(R_j(s) - RP) + (1-\lambda)(R_M(s) - RP)\right]^2 \Big\}$$

which, resolving the square, is equivalent to

$$pt^{+}(\lambda R_j + (1-\lambda)R_M) = \sum_{\Delta_\lambda(s)\geq 0} p_s \Big\{ \lambda(R_j(s) - RP) + (1-\lambda)(R_M(s) - RP) - \frac{\alpha^{+}}{2}[\lambda^2(R_j(s) - RP)^2 + (1-\lambda)^2(R_M(s) - RP)^2 + 2\lambda(1-\lambda)(R_j(s) - RP)(R_M(s) - RP)] \Big\}.$$

We want to derive a first order condition for the optimal portfolio weights. Taking the derivative with respect to λ, we obtain

$$\left.\frac{\mathrm{d}pt^{+}(\lambda R_j + (1-\lambda)R_M)}{\mathrm{d}\lambda}\right|_{\lambda=0} = \sum_{\Delta_\lambda(s)\geq 0} p_s \Big\{ (R_j(s) - R_M(s)) - \alpha^{+}(R_j(s) - R_M(s))(R_M(s) - RP) \Big\}.$$

We get a completely analogous expression for the loss term and then we can equate:

$$\left.\frac{\mathrm{d}pt^{+}(\lambda R_j + (1-\lambda)R_M)}{\mathrm{d}\lambda}\right|_{\lambda=0} = +\beta \left.\frac{\mathrm{d}pt^{-}(\lambda R_j + (1-\lambda)R_M)}{\mathrm{d}\lambda}\right|_{\lambda=0}.$$

This can be written as

$$\sum_{R_M(s)\geq RP} p_s \left\{(R_j(s) - R_M(s)) - \alpha^{+}(R_j(s) - R_M(s))(R_M(s) - RP)\right\} = +\beta \sum_{R_M(s)<RP} p_s \left\{(R_j(s) - R_M(s)) - \alpha^{-}(R_j(s) - R_M(s))(R_M(s) - RP)\right\}$$

We see that gains and losses are determined according to the market portfolio's return being higher or lower than the reference point. Moreover, these gains and losses are evaluated by the gradient of the value function.

Finally, note that for the asset j being the market portfolio considering the B-CAPM doesn't give us any information, since the identity $0 = 0$ is obtained. This resembles the fact that the valuation is a relative valuation – as the original SML was, too.

3.5 Summary

We developed in this chapter a simple model for asset pricing, the *Capital Asset Pricing Model (CAPM)*. Let us summarize the main ideas of this derivation: we consider only two time periods and assume that all investors have mean-variance preferences. Under this assumption we can represent assets in a mean-variance diagram. By combining two assets (diversification) we can find portfolios with different mean and variance. Let us now consider for a moment only the risky assets. The set of all possible portfolios of risky assets in the mean-variance set is called *opportunity set*. Its upper boundary is called the *efficient frontier*; only these portfolios and their combinations with the riskless asset are interesting for investors. The Two-Fund Separation Theorem states that all investors should hold as risky assets the same portfolio (the *Tangent Portfolio*). Depending on the risk attitudes, this portfolio can then be combined with the riskless asset. If we consider a market with several investors (Sect. 3.2), we can show that the Tangent Portfolio corresponds exactly to the market portfolio, i.e., the portfolio of *all* assets on the market. From this we can derive a formula for the price of assets, the CAPM. This price depends not only on mean and variance of the asset, but also on its covariance with the market portfolio.

In Sect. 3.3 we studied *the Alpha*, i.e., situations where some assets are seemingly under- or overpriced as compared to the price given by the CAPM. First, we have seen that adding Alpha opportunities does not improve every portfolio. However, in a financial market in equilibrium there wouldn't be any Alpha opportunities if we do not consider *heterogeneous beliefs*. This means that we generalize our simple model and take into account that investors have different expectations on the future development of assets. This leads to the effect that investors can perceive assets with positive Alpha, i.e., underpriced assets, whereas other investors do not consider them as underpriced. If expectations differ, only one of them can be right, the hunt for Alpha opportunities is therefore a zero-sum game and only the better-informed investors can profit from investing into such subjectively underpriced assets. On the long run, less-informed investors might return to a passive strategy, so that the market converges finally to the prices predicted by the CAPM. Some Hedge Fund strategies speculate that this will eventually be the case (e.g., using a *Market Neutral Strategy*).

Alternative investments often seem to outperform the market when considering mean and variance. In Sect. 3.4 we studied some reasons for this: Alternative or Higher Moment Betas. For instance, we noticed that the mean-variance approach as such is not sufficient to capture the nature of highly skewed or fat-tailed distributions. The attractiveness of such investments can often be better understood by considering Prospect Theory as underlying decision model.

In the next chapter we develop a more general model that includes a theory of the equity premium and that draws on alternative risk factors, higher order risk and background risk. Including all this the model should still give a simple risk-return interpretation that is compatible with the principle of no-arbitrage. This may sound like "squaring the circle", but on the contrary it is not only possible, but even not too difficult: Based on the no-arbitrage condition outlined in the next chapter and introducing intertemporal consumption we can develop the so-called consumption based CAPM of Breeden [Bre79], which is similar to the model of Lucas [LJ78]. First we will show that there is only one risk factor that can be used to price all risks by their covariance to that factor. Mathematically spoken this risk factor is a *likelihood ratio*, it is the ratio of the state price density and the physical probability.[37] Embedding the no-arbitrage idea in an economic model in which the ultimate goal of investing is to finance consumption, the likelihood ratio is given by the marginal rates of substitution evaluated at the stochastic consumption stream. For that reason it is also called the *stochastic discount factor*. The likelihood ratio will later turn out to be indeed proportional to the market portfolio if we work with the CAPM model, compare Sect. 4.4.2. This will lead to another proof of the SML-formula and allow for extensions to new models like APT and Behavioral CAPM (see Sect. 4.4).

Tests

1. What are the basic assumptions of this chapter?

   - ☐ We consider only two time periods.
   - ☐ The investors have all the same preferences.
   - ☐ The investors' preferences follow the Expected Utility Theory.
   - ☐ The investors' preferences follow the mean-variance approach.

2. Can combining two risky assets yield a portfolio with zero variance?

   - ☐ Yes, if the two assets are uncorrelated (correlation coefficient is zero).
   - ☐ Yes, if the two assets are negatively correlated (correlation coefficient is less than zero).
   - ☐ Yes, if the two assets are perfectly negatively correlated (correlation coefficient is -1).
   - ☐ No, this is not possible: we need the riskless asset.

3. What is the mean-variance opportunity set?

   - ☐ The set of all possible combinations of mean and variance that can be reached by a portfolio of the assets.
   - ☐ The set of all possible combinations of mean and variance that can be reached by a portfolio of the risky assets.
   - ☐ The set of all portfolios that can be optimal for an investor.

4. What is the efficient frontier?

   - ☐ The set of all portfolios that have maximal return for a given variance.
   - ☐ The set of all portfolios that have maximal return for a given standard deviation.

[37] In the continuous time version of the financial markets model the likelihood ratio is called the Radon-Nikodym derivative.

- ☐ The set of all portfolios that have minimal variance for a given return.
- ☐ The set of all portfolios in the mean-variance opportunity set for which there is no other point in the mean-variance opportunity set that improves mean and variance.
- ☐ The upper boundary of the mean-variance opportunity set in the mean-variance diagram.
- ☐ The left boundary of the mean-variance opportunity set in the mean-variance diagram.
- ☐ The set of optimal portfolios in the mean-variance diagram.

5. What is the tangent portfolio?

- ☐ The portfolio on the efficient frontier which admits a unique tangent.
- ☐ The portfolio on the efficient frontier such that a line from there to the risk-free asset has maximum slope.

6. What is the Sharpe ratio of the asset j?

- ☐ $(\mu_j - \mu_f)/\sigma_j$.
- ☐ μ_j/σ_j.
- ☐ The slope of the line in the mean-variance diagram that intersects with the risk-free asset and the asset j.

7. What does the Two-Fund Separation Theorem say for mean-variance investors?

- ☐ All investors should hold the same ratio of risky and riskless assets.
- ☐ All investors should hold the same portfolio of assets.
- ☐ All investors should hold the same portfolio of risky assets.
- ☐ The market portfolio consist only of two funds.

8. When does two-fund separation occur?

- ☐ In a market where everybody has mean-variance preferences.
- ☐ In a market where everybody has expected utility preferences.

9. What is the CAPM formula of an asset j with mean μ_j, variance σ_j^2 and returns R_j?

- ☐ $\mu_j - \mu_f = \text{cov}(R_j, R_M)(\mu_M - \mu_f).t$
- ☐ $\mu_j - \mu_f = \frac{\text{cov}(R_j,R_M)}{\sigma_M^2}(\mu_M - \mu_f)$.
- ☐ $\mu_j - \mu_f = \frac{\sigma_j^2}{\sigma_M^2}(\mu_M - \mu_f)$.

10. Given two assets with variance σ, which one has according to CAPM a smaller mean?

- ☐ Both have the same mean.
- ☐ The one with a smaller correlation with the market has a smaller mean.
- ☐ The one with a larger correlation with the market has a smaller mean.

11. What is "the Alpha"?

- ☐ The difference between the actual mean of an asset and its mean according to the CAPM.
- ☐ The difference between the mean of an asset and the risk-free return.
- ☐ An abbreviation for the Sharpe ratio.

12. What can be rational reasons for trading on financial markets?

- ☐ Intertemporal trade: I need the money later, you need it now, let's trade!
- ☐ Risk-trading: I want the risk, you don't want it, let's trade!
- ☐ Heterogenous beliefs: I think stocks go up, you think they go down, let's trade!

13. In which sense is the "hunt for Alpha" a zero-sum game?

- ☐ Whatever profit one person makes on the stock market, another person ought to pay for that. On average nobody makes any money.
- ☐ There are never any Alpha opportunities on the market, since the CAPM formula always holds, and therefore the Alphas are zero.
- ☐ Given heterogeneous beliefs, better informed investors can detect and exploit Alpha opportunities, but only on the expense of worse informed investors.

14. How can one explain that Insurance Linked Securities (ILS) are not as popular as they should be according to CAPM?

- ☐ They improve the performance of a benchmark portfolio, but not the performance of the portfolio of a given investor.
- ☐ The preferences of the investors are not mean-variance preferences, hence they do not only take mean and variance into account, and the return distribution of ILS is highly skewed, i.e., not at all normally distributed.
- ☐ One needs to consider a multi-period model to evaluate them correctly.
- ☐ There is not enough data on these investments yet, therefore the computation of mean and variance is still too imprecise.

15. What are the limitations of the CAPM?

- ☐ People are often not acting according to Mean-Variance preferences.
- ☐ People are often not acting according to Expected Utility Theory.
- ☐ We cannot model complicated derivatives in the CAPM, since we need more time-steps to do so.
- ☐ It is not possible to study the effect of differences in the investors' beliefs.
- ☐ Mean and variance are not sufficient to describe highly skewed distributions, as they are typical for hedge funds or other alternative investments.

16. What are the merits of CAPM?

- ☐ CAPM proves that it never makes sense to "hunt for Alpha opportunities", but instead passive investment into the market and a riskless asset is optimal.
- ☐ CAPM is a simple and useful tool to evaluate traditional investments.
- ☐ CAPM is the standard method to price options and derivatives.
- ☐ CAPM shows that, although investors might have different preferences, they all should invest into the same assets.
- ☐ CAPM is a model that helps to get an intuition into portfolio analysis that is useful when studying more complicated models.

Two-Period Model: State-Preference Approach 4

Toutes les généralisations sont dangereuses, y compris celle-ci.
(All generalizations are dangerous, even this one.)
ALEXANDRE DUMAS

In the last chapter we have assumed that investors base their decisions on the mean-variance approach. This helped us to develop a model for pricing assets on a financial market, the CAPM. In this chapter we want to generalize this model in that we relax the assumptions on the preferences of the investors.

The fundamental idea which allows this generalization is the Principle of No-arbitrage: in a well-functioning financial market it is not possible to get something for nothing. This principle is equivalent to a pricing rule in which all assets are priced with respect to a single abstract portfolio – similar to the security market line. To get some understanding of the abstract pricing portfolio – also called the likelihood ratio process or the stochastic discount factor – it is useful to analyze how it varies with the returns of the market portfolio that played a crucial role in the CAPM presented in the previous chapter. As we will see the likelihood ratio process is a decreasing function of the market portfolio since this property reflects the decreasing marginal utility of wealth – a standard assumption in finance. To get more content for the abstract pricing portfolio, one can then introduce assumptions on agents' preferences – some of them leading to the CAPM. In the case of the CAPM the likelihood ratio process turns out to be a linear function of the market portfolio. Finally we show that under certain conditions the heterogeneity of agents can be replaced with a single representative agent – supposing one does not want to do out-of-sample predictions.

T. Hens, M.O. Rieger, *Financial Economics*, Springer Texts in Business and Economics, DOI 10.1007/978-3-662-49688-6_4

4.1 Basic Two-Period Model

The basic model consists of a finite set of investors trading a finite set of assets at time-period zero that deliver payoffs at period one in a finite set of states of the world. In contrast to the previous chapter we are taking all of these payoffs into account and do not simplify the problem by only studying their mean and variance. Nevertheless, the mathematical tools needed are still very simple: finite dimensional linear algebra (vectors, matrices, scalar products etc.) in a finite dimensional Euclidean space is sufficient.[1] We will first describe the assets and then the agents trading the assets.

4.1.1 Asset Classes

Traditional asset classes are money market investments (e.g., certificates of deposit), bonds and stocks. The markets for trading these assets have quite a long history and are by now well established all over the world. Recently most investors have gained access to other markets like funds of real estate, commodities, private equity, and hedge funds. These asset classes are called alternative investments since they are an alternative to the traditional asset classes.

One important difference between assets is the way they deliver payoffs. *Bonds* deliver payoffs that are known when a bond is traded. These payoffs are called coupons because before financial markets became electronic the owner would deposit his bonds in a safe and every month he would cut off a piece, the coupon, which he presented to the issuer in order to receive the fixed payoff. Markets for bonds are also called fixed income markets.

Stocks are shares of firms. They entitle the owner to receive some dividends. Since dividends depend on the profit (after having paid the interest on bonds) the payoffs of stocks are not certain upon the purchase of stocks.

Some alternative assets do not pay off any coupon or dividends. Commodities for example can only be sold to get a payoff from them. Finally the classification of *hedge funds* within the class of alternative assets can be questioned because hedge funds are strategies and not assets. We will come back to the issue of hedge funds later.

Figure 4.1 displays the cumulative returns of asset classes in which a typical pension fund would invest. We see that stocks perform best but are also the most volatile. On the other extreme are bonds with a low average performance that, however, is more reliable. A counterexample to the rule "higher average return implies higher volatility" are hedge funds which in that period have delivered quite high average returns with very low volatility.

[1]For the unlikely case that the reader is not familiar with these topics, or in the more likely case that he wants to refresh his memory, the Appendix A.1 gives a quick review on basic linear algebra.

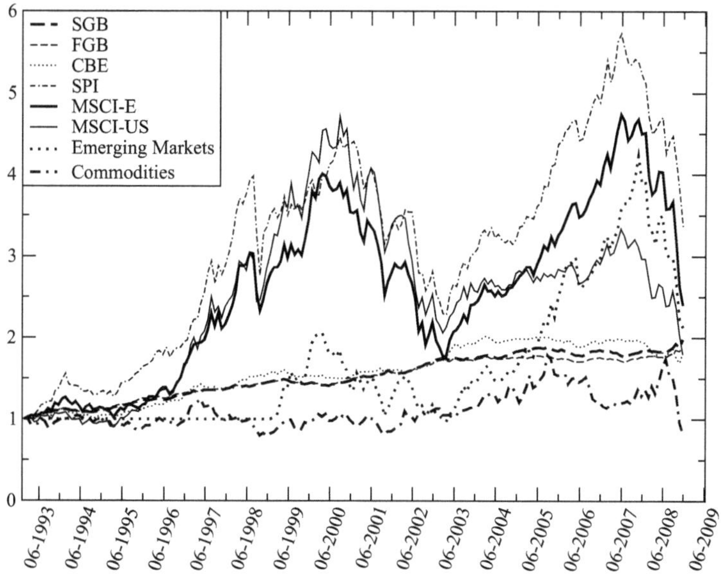

Fig. 4.1 Cumulative returns of several asset classes

How can these quite different assets be valued? A standard approach for assets with payoffs (stocks and bonds) is based on the representative agent asset pricing idea: the price of the asset is equal to the discounted sum of all future payoffs where the discount factors are the representative agent's marginal rates of substitution between future consumption (contingent on states of the world) and current consumption. These discount factors are also called the *stochastic discount factors* since they are not constant over time. Applying this valuation technique to assets without payoffs (commodities and hedge funds, for example) would obviously result in a zero price for these assets. Payoffs based on this asset class can only be realized through speculation, i.e., buying low and selling high. But then some other investor must have taken the complementary position so that, seen from the bird's eye view of the representative agent, on average the gains are zero. To understand such asset classes we clearly need to give up the aggregate perspective and look into the trades that are done.

4.1.2 Returns

How can we model the payoffs, prices and returns of assets? Certainly we need to model some uncertainty because payoffs and resale prices are unknown at the time the assets are purchased. The simplest such model has two periods, $t = 0, 1$. In period $t = 0$ we are in state $s = 0$. In period $t = 1$ a finite number of states of the world, $s = 1, 2, \ldots, S$ can occur. The time-uncertainty structure is thus described by a tree as in Fig. 4.2.

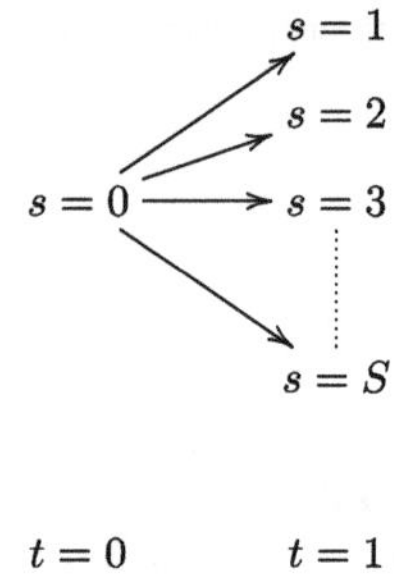

Fig. 4.2 Illustration of an event tree

We denote the assets by $k = 0, 1, 2, \ldots, K$. The first asset, $k = 0$, is the risk-free asset delivering the certain payoff 1 in all second period states. The assets' payoffs are denoted by A_s^k. The time 0 price is denoted by q^k, so that the gross return of asset k in state s is given by $R_s^k := \frac{A_s^k}{q^k}$. The net return is accordingly $r_s^k := R_s^k - 1$. We gather the structure of all asset returns in the so called states-asset-returns-matrix, the SAR-matrix:

$$\boldsymbol{R} := (R_s^k) = \begin{pmatrix} R_1^0 \cdots R_1^K \\ \vdots \quad\quad \vdots \\ R_S^0 \cdots R_S^K \end{pmatrix} = (\boldsymbol{R}^0 \cdots \boldsymbol{R}^K) = \begin{pmatrix} \boldsymbol{R}_1 \\ \vdots \\ \boldsymbol{R}_S \end{pmatrix},$$

with generic entry R_s^k and with columns denoted by $\boldsymbol{R}^k$ and rows denoted by $\boldsymbol{R}_s$. One simple way of filling the SAR-matrix with data is to take a sample of realized asset returns in some time periods $t = 1, 2, \ldots, S$ and then to identify each state s with one time period t, i.e. $s = n$ is $t = n$, for $n = 1, \ldots, T$ or S.

In this book we will mainly use the SAR-matrix representation of returns. However, to see the link to other textbooks we will now briefly show how other representations of returns can be derived from the SAR-matrix. One simple and commonly used description of returns is based on means and covariances. How do we compute mean returns and covariances of returns from the SAR-matrix? Given some probability measure on the set of states, prob_s, $s = 1, \ldots, S$, we compute the mean return of asset k as

$$\mu(R^k) = \sum_{s=1}^{S} \text{prob}_s R_s^k = \textbf{prob}' \boldsymbol{R}^k.$$

The covariance matrix

$$COV(\boldsymbol{R}) = \begin{pmatrix} \text{cov}(R^1, R^1) & \cdots & \text{cov}(R^1, R^K) \\ \vdots & & \vdots \\ \text{cov}(R^K, R^1) & \cdots & \text{cov}(R^K, R^K) \end{pmatrix}$$

is accordingly computed as

$$COV(\boldsymbol{R}) = \boldsymbol{R}' \begin{pmatrix} \text{prob}_1 & & \\ & \ddots & \\ & & \text{prob}_S \end{pmatrix} \boldsymbol{R} - (\boldsymbol{R}'\textbf{prob})(\textbf{prob}'\boldsymbol{R}).$$

Of course one can also go the other way round and compute the SAR-matrix for given means and covariances. A very simple model showing the direct link between mean and covariance is given in the exercise book, as well.

Yet another way of thinking about returns is to consider them being generated by some factors. Many such factors have been identified for stock returns. Those factors include inflation, interest rates, growth, oil prices, terrorism etc. Table 4.1 gives an overview of factors for stock market returns analyzed in various studies since 1986. To show the link between factors and asset returns, suppose you can identify $f = 1, \ldots, F$ factors in the $s = 1, \ldots, S$ states of the world. Let R_s^f be the value of factor f in state s. They can again be collected in a matrix, the factor value matrix (FV-matrix):

$$(R_s^f) = \begin{pmatrix} R_1^1 & \cdots & R_1^F \\ \vdots & & \vdots \\ R_S^1 & \cdots & R_S^F \end{pmatrix}.$$

The sensitivity of asset k's returns to factor f is typically denoted by β_k^f. Then the return of asset k can be thought of as being generated by the F factor values:

$$R_s^k = \sum_{f=1}^{F} R_s^f \beta_k^f,$$

which is in matrix notation: $(R_s^k) = (R_s^f) \cdot (\beta_k^f)$. In the exercise book we give examples for returns being generated by factors.

Finally, let us take a closer look at the assets' payoffs. In principle they are derived from two sources, dividends or coupons and resale values. The price difference $q_s^k - q^k$, if positive, is called a capital gain, otherwise a capital loss. Hence,

$$R_s^k = \frac{D_s^k + q_s^k}{q^k} := \frac{A_s^k}{q^k}$$

where D_s^k are the dividends or coupons paid by asset k in state s.

Table 4.1 Factors for stock market returns analyzed in various studies since 1986

Study		Identified factors driving stock market return
Chen, Roll, Ross	1986	Growth rate of industrial production
		Inflation rate
		Spread between short-term and long-term interest rates
		Default risk premia of bonds
Berry, Burmeister, McElroy	1988	Inflation rate
		Spread between short-term and long-term interest rates
		Default risk premia of bonds
		Growth rate of aggregated sells
		Return of the S&P 500
Salomon Brothers	1990	Inflation rate
		Growth rate of GDP
		Interest rate
		Rate of change of oil price
		Growth rate of defense budgets
Mei	1993	January dummy variable
		Return for a value-weighted portfolio
		One-month treasure bill rate
		Difference between 1-month treasury bill rate and long-term AAA corporate bond
		Dividend-yield on the value-weighted portfolio
Fama, French	1993	Premium of a diversified market portfolio
		Difference between returns of small cap and large cap portfolios
		Difference between returns of growth and value portfolios
Elton, Gruber, Mei	1994	Difference between returns of long-term government bonds and short-term treasury bills
		Change of returns of treasury bills
		Change of exchange rates between USD and foreign currency
		Change of GDP forecast
		Change of inflation forecast
		Portion of market return that cannot be explained by the above five factors
Davis	1994	Book to market equity
		Cash-flow/price ratio
		Earnings/price ratio
Lakonishok, Shleiter, Vishny	1994	Earnings/price ratio
		Cash-flow/price ratio
		Sales-growth variable
Gallati	1994	European market 1-month SFR interest rate
		Swiss obligation index EFFAS
		European market 3-month DM interest rate
		FTSE 100 index
Kothari, Shanken, Sloan	1995	Beta
		Firm size variable

4.1.3 Investors

So far we have described the objects of trade: bonds, stocks and alternative investments. Now we ask who is trading those assets – and why. Many agents trade assets for secondary reasons, but ultimately they are doing it to derive the highest utility for the investors, i.e., the principal owners of the assets. This is the individualistic paradigm on which market economies are built. A financial market may however have several layers of agents that help the ultimate investors benefit from the market.

Modern financial markets are populated by various investors with different wealth and objectives and quite heterogeneous beliefs. There are private investors with pension, housing and insurance concerns, firms implementing investment and risk management strategies, investment advisors providing financial services, investment funds managing pension or private capital and the government financing the public deficit. The investment decisions are implemented by brokers, traders, and market makers. Many financial markets are dominated by large investors. On the Swiss equity market, for example, more than 75 % of the wealth is managed by institutional asset managers providing services to private investors, insurance funds, and pension funds. Since the asset managers' investment abilities and efforts are not observable by their clients, the contract between the principal and the agent must be based on measurable variables such as relative performance. However, such contracts may generate herding behavior particularly among institutional investors. In the words of Lakonishok et al. [LSV92, page 25]:

> Institutions are herding animals. We watch the same indicators and listen to the same prognostications. Like lemmings we tend to move in the same direction at the same time. And that naturally exacerbates price movements.

Additionally, asymmetric information may create "window dressing" effects, i.e., agents change their behavior at the end of the reporting period.

To study such effects, it is first necessary to understand a general model for investors. Let us assume there are I investors on a market. We denote them by $i = 1, \ldots, I$. Each investor is described by his exogenous wealth in all states of the world $\boldsymbol{w}^i = (w_0^i, w_1^i, \ldots, w_S^i)'$. Given these exogenous entities and given the asset prices $\boldsymbol{q} = (q^0, q^1, \ldots, q^K)'$, the investors can finance consumption $\boldsymbol{c}^i = (c_0^i, c_1^i, \ldots, c_S^i)'$ by trading the assets. We denote by $\boldsymbol{\theta}^i = (\theta^{i,0}, \theta^{i,1}, \ldots, \theta^{i,K})'$ the vector of asset trades of agent i. Note that $\theta^{i,k}$ can be positive or negative, i.e., agents can buy or sell assets. The only restriction on asset trade is that the budget restrictions need to be satisfied. In the first period agents can use their exogenous wealth w_0^i for consumption c_0^i or to buy assets. The value of a portfolio of assets is $\sum_{k=0}^K q^k \theta^{i,k}$. If this value is non-positive we say the portfolio is *self-financing*, since it does not need extra wealth to be carried out. The first period budget constraint is thus:

$$c_0^i + \sum_{k=0}^{K} q^k \theta^{i,k} = w_0^i.$$

In every state of the world in the second period, the assets deliver payoffs A_s^k which can in principle be positive or negative. The second period budget constraints are given by:

$$c_s^i = \sum_{k=0}^{K} A_s^k \theta^{i,k} + w_s^i, \quad s = 1, \ldots, S.$$

Now we know what an agent can do, given the asset payoffs and the asset prices. The final point in the description of the agent is then to describe what the agent wants to achieve. As we said above, ultimately the agents are interested in consumption. But objectives like "I want the highest possible consumption in all states of the world" are not useful here because markets will not offer such fairy tale outcomes. In other words: markets will not offer "free lunches", i.e., arbitrage opportunities (compare Sect. 4.2 or [DR92] for a precise definition). What they offer instead are trade-offs, e.g., higher consumption today at the expense of lower consumption tomorrow or more evenly distributed consumption in all states at the expense of a really high payoff in one of the states. Hence, the agent needs to find a stand on those trade-offs.

The intertemporal trade-off is described by the agent's time preference. Suppose the agent discounts future utility back to current utility by some discount rate $\delta^i \in (0, 1)$, compare Sect. 2.7. If moreover the agent forms some beliefs over the occurrence of the states, then we can describe his (rational) preferences by a von Neumann-Morgenstern utility function (compare Sect. 2.2) and we obtain

$$U^i(c_0^i, c_1^i, \ldots, c_S^i) = u^i(c_0^i) + \delta^i \sum_{s=1}^{S} \text{prob}_s^i u^i(c_s^i).$$

If we increase one of the c_s^i, the utility U^i should also increase: higher consumption is always preferred. (Remember the Woody Allen quote: "More money is better, if only for financial reasons.") We might also assume that U is quasi-concave such that a more evenly distributed consumption is preferred over extreme distributions.

This form of a utility function is called "expected discounted utility" and we have already seen it applied in Sect. 2.7. It is the most convenient form to do calculations such as finding the optimal asset allocation and it is also the rational way of doing it.[2] It is, however, questionable whether this is a realistic utility function that describes the preferences of real investors: as we have seen in Sect. 2.4, there are many experiments that show strong deviations from rational behavior in decision problems. In other words: the model we study here is adequate for analyzing optimal investments, but only on markets with purely rational investors – a strong assumption. We will see later how to model markets with non-rational investors.

Before passing to the formulation of the decision problem we shall mention some general qualitative properties of utility functions that are often referred to in the remainder of this book:

1. Continuity: the utility function U is continuous on its domain $\mathbb{R}_+^{S+1}$.
2. Quasi-concavity: the upper contour sets are convex, i.e., $\{\boldsymbol{c} \in R_+^{S+1} \mid U(\boldsymbol{c}) \geq \text{const}\}$ is convex.[3]
3. Monotonicity: "More is better", or more precisely:
   (a) Strict monotonicity[4]: $\boldsymbol{c} > \boldsymbol{c}'$ implies $U(\boldsymbol{c}) > U(\boldsymbol{c}')$.
   (b) Weak monotonicity: $\boldsymbol{c} \gg \boldsymbol{c}'$ implies $U(\boldsymbol{c}) > U(\boldsymbol{c}')$.

Expected utility and Prospect Theory utility functions are typically strictly monotonic while mean-variance utility functions are not monotonic at all. The latter has been shown by the mean-variance paradox (Theorem 2.30).

Throughout this book we will assume that utility functions are strictly monotonic with respect to first period consumption c_0. Thus the three notions of monotonicity relate to the uncertain consumption in the second period.

We can now summarize the agent's decision problem as:

$$\boldsymbol{\theta}^i \in \arg\max_{\boldsymbol{\theta}^i \in R^{K+1}} U^i(\boldsymbol{c}^i) \quad \text{such that} \quad c_0^i + \sum_{k=0}^{K} q^k \theta^{i,k} = w_0^i$$

$$\text{and} \quad c_s^i = \sum_{k=0}^{K} A_s^k \theta^{i,k} + w_s^i \geq 0, \quad s = 1, \ldots, S.$$

[2] Compare, however, the remarks on time-discounting in Sect. 2.7.

[3] In the case of two commodities the upper contour set is the set that is included by the indifference curve. Hence, the utility function is quasi-concave if the indifference curves are convex.

[4] For two vectors $\boldsymbol{c}, \boldsymbol{c}'$ we use the notation $\boldsymbol{c} > \boldsymbol{c}'$ to mean that in each component the vector is at least as large as the vector $\boldsymbol{c}'$ and in at least one $\boldsymbol{c}$ is strictly greater than $\boldsymbol{c}'$. We use $\boldsymbol{c} \gg \boldsymbol{c}'$ if the vector $\boldsymbol{c}$ is strictly greater than $\boldsymbol{c}'$ in all components, see Appendix A.1.

To study this decision problem with the framework of an Edgeworth Box we need to reduce it and make three specializing assumptions:

1. There is no first period consumption,[5]
2. there are only two states, denoted s and z, and
3. there are two Arrow securities for the contingent delivery of wealth in each state, i.e.

$$A := \begin{pmatrix} 1 & 0 \\ 0 & 1 \end{pmatrix}.$$

Although the Edgeworth Box (compare Fig. 4.8) is a nice pedagogical toolbox, the above constraints limit the dimensionality substantially. Thus we will give up this approach and switch from geometrical tools like the Edgeworth Box to analytical tools like calculus and linear algebra.

There are alternative ways of writing this decision problem. One may for example extract from the exogenous wealth the part which consists of assets. Any vector $w^i \in \mathbb{R}^S$ can be decomposed into two vectors, w^i_A and $w^i_\perp$, where the first component w^i_A can be generated by a portfolio of assets, i.e., $w^i_A = \sum_{k=0}^K A^k \theta_A^{i,k}$, and $w^i_\perp$ is somehow orthogonal to the first component, thus the index $\perp$. Substituting $\hat{\theta}^{i,k} := \theta_A^{i,k} + \theta^{i,k}$ we can write the budget restriction as follows:

$$c_0^i + \sum_{k=0}^K q^k \hat{\theta}^{i,k} = \sum_{k=0}^K q^k \theta_A^{i,k} + w_0^i$$

$$\text{and} \quad c_s^i = \sum_{k=0}^K A_s^k \hat{\theta}^{i,k} + w^i_{\perp s}, \quad s = 1, \ldots, S.$$

So far you may wonder how we could include the notion of asset returns, from which we started this chapter. To do so, independently of the previous decomposition of the endowment vector, we will now transform the decision problem given above in economic terms like quantities and prices, into finance terms like asset allocations and returns. To this end we first define an agent's first period wealth, i.e., his total wealth in terms of assets and exogenous wealth, by $\mathrm{w}_0^i := \sum_{k=0}^K q^k \theta_A^{i,k} + w_0^i$. The agent splits this wealth among the various assets and first period consumption. Denote by $\lambda^{i,k} := q^k \hat{\theta}^{i,k} / \mathrm{w}_0^i$ the percentage of wealth the agent invests in asset k. Similarly let $\lambda^{i,\mathrm{con}} := c_0^i / \mathrm{w}_0^i$ be the percentage of wealth spent on consumption. We call $(\lambda^{i,0}, \lambda^{i,1}, \ldots, \lambda^{i,K})$ the asset allocation of agent i.

[5]The Edgeworth Box can alternatively be used to display intertemporal trade. In that case one would need to assume that there is only one state of the world tomorrow, i.e., $S = 1$, so all assets are identical to the risk-free asset. In this case, we would of course consider first-period consumption.

His first period budget constraint is then written as $\lambda^{i,\text{con}} + \sum_{k=0}^{K} \lambda^{i,k} = 1$. In finance, asset allocations are typically kept separate from consumption, i.e., they are normalized so that they themselves sum up to 1. To this end define

$$\hat{\lambda}^{i,k} := \frac{\lambda^{i,k}}{(1-\lambda^{i,\text{con}})}, \quad k = 0, 1, \dots, K.$$

Hence, the vector of budget shares is now written as

$$\left(\lambda^{i,\text{con}}, (1-\lambda^{i,\text{con}})\left(\hat{\lambda}^{i,0}, \hat{\lambda}^{i,1}, \dots, \hat{\lambda}^{i,K}\right)\right)$$

with $\sum_{k=0}^{K} \hat{\lambda}^{i,k} = 1$. By defining $w_0^{i,\text{fin}} := (1-\lambda^{i,\text{con}})w_0^i$, the agent's wealth he spends on financial assets, $\hat{\lambda}^{i,k}$, becomes the share of financial wealth that agent i invests in asset k. The budget restriction in the first period is then written as

$$\lambda^{\text{i,con}} + (1-\lambda^{\text{i,con}}) \sum_{k=0}^{K} \hat{\lambda}^{i,k} = 1.$$

In the second period the budget restriction is

$$c_s^i = \sum_{k=0}^{K} R_s^k \hat{\lambda}^{i,k} w_0^{i,\text{fin}} + w_s^i, \quad s = 1, \dots, S.$$

To be sure:

$$R_s^k \hat{\lambda}^{i,k} w_0^{i,\text{fin}} = \frac{A_s^k}{q^k} \frac{q^k \theta^{i,k}}{w_0^{i,\text{fin}}} w_0^{i,\text{fin}} = A_s^k \theta^{i,k}.$$

Summarizing, the finance way of presenting the decision problem is:

$$(\lambda^{i,\text{con}}, \boldsymbol{\lambda}^i) = \arg\max_{\boldsymbol{\lambda}^i \in \Delta^{K+2}} U^i(\boldsymbol{c}^i)$$

$$\text{such that} \quad c_0^i = w_0^i - (1-\lambda^{i,\text{con}}) \sum_{k=0}^{K} \hat{\lambda}^{i,k} w_0^i$$

$$= w_0^i - w_0^{i,\text{fin}} \sum_{k=0}^{K} \hat{\lambda}^{i,k}$$

$$\text{and} \quad c_s^i = \left(\sum_{k=1}^{K} R_s^k \hat{\lambda}^{i,k}\right) w_0^{i,\text{fin}} + w_s^i, \quad s = 1, \dots, S.$$

4.1.4 Complete and Incomplete Markets

A financial market is said to be complete if all second period consumption streams $\boldsymbol{c} \in \mathbb{R}^S$ can be achieved by asset trade, i.e., for all $\boldsymbol{c} \in \mathbb{R}^S$ there exists some $\boldsymbol{\theta} \in \mathbb{R}^{K+1}$ such that $\boldsymbol{c} = \sum_{k=0}^{K} \boldsymbol{A}^k \theta^k$. If some second period consumption streams are not attainable the market is said to be incomplete. A complete market is very useful because it allows insuring all future consumption plans. Also, it allows pricing all future consumption plans in a unique way. Whether financial markets are complete or incomplete depends on the states of the world one is modeling. If for example the states of the world are defined by the assets returns themselves, then the market is complete if the variation of the returns is not more frequent than the number of assets. A famous case of this sort is the binomial model[6] in which states of the world are defined by whether the price of an asset goes up or down. Together with the risk-free asset the market is then complete. If on the other hand the states of the world are given by the exogenous income $\boldsymbol{w}$ then it may be that there are insufficient assets to hedge all risks in this exogenous income. An example of this incompleteness is that students cannot buy securities to insure their future labor income.

The mathematical condition for completeness of a market is that the rank of the return matrix $\boldsymbol{R}$ needs to be equal to the number of states S. Since the return matrix is the payoff matrix post-multiplied by a diagonal matrix,[7] $\boldsymbol{R} = \boldsymbol{A}\boldsymbol{\Lambda}(\boldsymbol{q})^{-1}$, the return matrix is complete if and only if the payoff matrix is complete.

Example 4.1 Consider

$$\boldsymbol{A}_1 := \begin{pmatrix} 1 & 0 \\ 1 & 2 \end{pmatrix}, \quad \boldsymbol{A}_2 := \begin{pmatrix} 1 & 1 \\ 1 & 2 \\ 1 & 3 \end{pmatrix}, \quad \boldsymbol{A}_3 := \begin{pmatrix} 1 & 1 & 0 \\ 1 & 2 & 1 \\ 1 & 3 & 2 \end{pmatrix}.$$

$\boldsymbol{A}_1$ is complete, but $\boldsymbol{A}_2$ and $\boldsymbol{A}_3$ are incomplete!

4.1.5 What Do Agents Trade?

Obviously, in a financial market agents trade financial assets. However they are doing this in order to obtain the best possible consumption patterns. Hence, we may also say that agents trade consumption, i.e., they trade intertemporal consumption by buying and selling the risk-free asset and they insure consumption risk by trading the risky assets. If agents hold heterogeneous beliefs then we may say that they trade "opinions", i.e., they are betting on their beliefs. An alternative answer would be that

[6]We will use the binomial model in Chap. 5 and in Chap. 8 when we show how to price derivatives.

[7]The operator $\boldsymbol{\Lambda}$ transforms an n-dimensional vector into a $n \times n$ diagonal matrix with the vector being the main diagonal. The operator $^{-1}$ computes the inverse of a matrix. Compare Appendix A.1.

agents trade risk factors. They trade assets, but asset returns are determined by risk factors.

For each of these formulations one can define an objective function with an appropriate choice variable: to model asset trade we could define a utility on assets by $U_A(\boldsymbol{\theta}) := U(\boldsymbol{w} + \boldsymbol{A}\boldsymbol{\theta})$ or in finance terms by $U_R(\boldsymbol{\lambda}) := U(\boldsymbol{w} + \boldsymbol{R}\boldsymbol{\lambda} w_0^{i,\text{fin}})$. To model trade in risk factors one could define a utility function on risk factors by

$$U_F(z) := U(\boldsymbol{w} + \boldsymbol{F} \underbrace{\boldsymbol{B}\boldsymbol{\lambda}}_{=z} w_0^{i,\text{fin}}).$$

Note that the fundamental properties (1)–(3) stated for a consumption utility function are inherited by the asset and the risk factors utility function, provided asset returns are non-negative. Hence, whether a financial market model is written in terms of consumption (e.g., the general equilibrium model with incomplete markets[8]), asset trade (e.g., the CAPM) or factors (e.g. Ross' APT) is more a matter of convenience than a matter of substance.

4.2 No-Arbitrage Condition

4.2.1 Introduction

Suppose the shares of Daimler Chrysler are traded at the NYSE for 90 Dollars while the same shares are traded in Frankfurt for 70 Euros. If the Dollar/Euro exchange rate were 1:1 what would you do? Clearly you would buy Daimler Chrysler in Frankfurt and sell it in New York while covering the exchange rate risk by a forward on the Dollar. This arbitrage opportunity is so obvious that it can hardly ever be found. Indeed studies show that for double listings even very small differences of less than 1 % are erased within 30 s. How come? Prop traders at banks and hedge funds have written computer programs to spot and immediately exploit this arbitrage opportunity.

In general an arbitrage opportunity is a trading strategy that gives you positive returns without requiring any payments. Researchers and practitioners agree that arbitrage strategies are so rare that without making a big mistake one can assume they do not exist. This simple idea has far reaching conclusions, for example for the valuation of derivatives. Derivatives are assets whose payoffs depend on the payoff of other assets, the underlyings. In the simple case where the payoff of the derivative can be duplicated by a portfolio of the underlying and a risk-free asset, the price of the derivative must be the same as the value of the duplicating portfolio.[9] Why?

[8] See Magill and Quinzii [MQ02] for details.

[9] We neglect trading costs!

Suppose the derivative's price is higher than the value of the duplicating portfolio. Then one can build an arbitrage strategy by shorting the asset and hedging the payoff by holding the duplicating portfolio. If the price of the derivative were smaller than that of the duplicating portfolio one would trade the other way round. Hence, the same payoffs, even if they are generated by different combinations of assets, need to have the same price. This is the so called "Law of One Price". Any agent whose utility is increasing in current period consumption would like to exploit a departure of asset prices from the Law of One Price. Hence, he would try to exercise the arbitrage opportunity more and more so that he will not find an optimal strategy which conflicts with the idea of equilibrium.

The absence of arbitrage is however somewhat deeper than the Law of One Price. Formulated more generally an arbitrage opportunity is a trading strategy that carries an investor to Nirvana, i.e., to infinite utility. Note that in this formulation of an arbitrage the qualitative properties of the investors' utility come into play. In particular whether some trading strategy is indeed an arbitrage depends on the type of monotonicity of the investor's utility function. A mean-variance investor clearly benefits if he gets the risk-free asset for free but he may not want to scale up any asset with positive variance, as we have seen in Sect. 2.3.2. If the utility is weakly monotonic the investor will only benefit if he gets a positive payoff in all future states without requiring a payment today. If on the other hand the utility is strictly monotonic the investor will benefit if he gets a non-negative payoff in all future states of the world, which is not itself zero, without requiring a payment today.

As we will see in this chapter, the absence of arbitrage implies some restrictions on asset prices. Let us sketch the main ideas that lead to these restrictions: the Law of One Price requires that asset prices are linear, i.e., doubling all payoffs means doubling the price and the price of an asset that delivers the sum of two assets' payoffs has to be the sum of the two assets' prices. In mathematical terms, the asset pricing functional is *linear*. Therefore by the Riesz representation theorem (see Appendix A.1, Theorem A.1) there exist weights, called state prices, such that the price of any asset is equal to the weighted sum of its payoffs. Absence of arbitrage for mean-variance utilities then implies that the sum of the state prices are positive while absence of arbitrage under weak monotonicity implies that all state prices are non-negative and finally the absence of arbitrage for strictly monotonic utility functions is equivalent to the existence of strictly positive state prices that express asset prices as the weighted sum of the assets' payoffs.

In many textbooks on financial economics only this last version of the absence of arbitrage is considered. Since in this book we want to build a bridge between the economist's look at financial markets and the finance practitioner's point of view, it is important to include the case of mean-variance no-arbitrage. This gives two main cases. Having understood these two cases you will be able to do the other two cases (Law of One Price and weakly monotonic utilities) easily yourself.

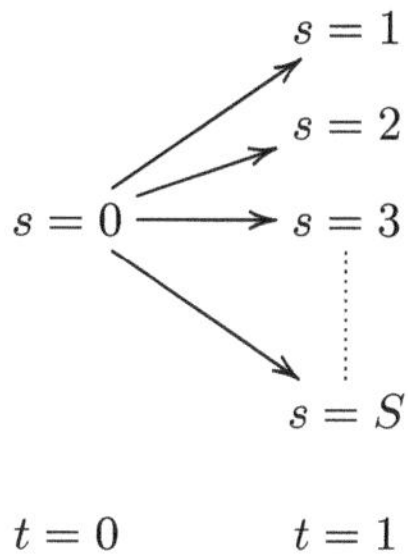

Fig. 4.3 Event tree

4.2.2 Fundamental Theorem of Asset Prices

We use the same model as outlined in the previous chapter. There are two periods, $t = 0, 1$. In the second period a finite number of states of the world, $s = 1, 2, \ldots, S$ can occur. The time-uncertainty structure is thus described by a tree as in Fig. 4.3:

There are $k = 0, 1, 2, \ldots, K$ assets with payoffs denoted by A_s^k. We gather the structure of all assets' payoffs in the states-asset-payoff matrix,

$$\boldsymbol{A} = \begin{pmatrix} A_1^0 \cdots A_1^K \\ \vdots \quad\quad \vdots \\ A_S^0 \cdots A_S^K \end{pmatrix} = (\boldsymbol{A}^0 \cdots \boldsymbol{A}^K) = \begin{pmatrix} \boldsymbol{A}_1 \\ \vdots \\ \boldsymbol{A}_S \end{pmatrix}$$

An *arbitrage* is a trading strategy that an investor would definitely like to exercise. Note that, as we mentioned above, this definition of arbitrage depends on the qualitative properties of the investor's utility function. For strictly monotonic utility functions an arbitrage is a trading strategy that leads to positive payoffs without requiring any payments. For mean-variance utility functions an arbitrage is a trading strategy that offers the risk-free payoffs without requiring any payments.

We first formalize an arbitrage opportunity for strictly monotonic utility functions. Under this assumption, an arbitrage is a trading strategy $\boldsymbol{\theta} \in \mathbb{R}^{K+1}$ such that

$$\begin{pmatrix} -\boldsymbol{q}' \\ \boldsymbol{A} \end{pmatrix} \boldsymbol{\theta} > \boldsymbol{0}.$$

Hence, the trading strategy never requires any payments and it delivers a non-negative and non-zero payoff. To give an example, let there be just two assets and two states. Say, the payoff matrix is

$$\boldsymbol{A} := \begin{pmatrix} 1 & 2 \\ 1 & 3 \end{pmatrix}$$

while the asset prices are $\boldsymbol{q} = (1,4)'$. Maybe you want to stop a moment and try to find an arbitrage opportunity, before reading on? In case you have not found it: by selling one unit of the second asset and using the receipts (four units of wealth) to buy three units of the first asset, you are left with one unit of wealth today, and tomorrow you will be hedged if the second state occurs while you have an extra unit of wealth if the first state occurs. How can we erase arbitrage opportunities in this example? Obviously asset 2 is too expensive relative to asset 1. Suppose now the asset prices are $\boldsymbol{q} = (1, 2.5)'$. Can you still find an arbitrage opportunity? We see that trying will not always be successful and is not helpful at all if there is no arbitrage opportunity. Instead we need a general result that tells us whether arbitrage opportunities exist. This is the content of the following theorem:

Theorem 4.2 (Fundamental Theorem of Asset Prices, FTAP) *The following two statements are equivalent:*

1. *There exists no $\boldsymbol{\theta} \in \mathbb{R}^{K+1}$ such that*

$$\begin{pmatrix} -\boldsymbol{q}' \\ \boldsymbol{A} \end{pmatrix} \boldsymbol{\theta} > \mathbf{0}.$$

2. *There exists a $\boldsymbol{\pi} = (\pi_1, \ldots, \pi_s, \ldots, \pi_S)' \in \mathbb{R}^S_{++}$ such that*

$$q_k = \sum_{s=1}^{S} A_s^k \pi_s, \quad k = 0, \ldots, K.$$

In the example above we see that for the state prices $\boldsymbol{\pi} := (0.5, 0.5)'$ the two asset prices can be displayed as the weighted sum of their payoffs and therefore, applying the FTAP we know that there are no arbitrage opportunities at the asset prices $\boldsymbol{q} = (1, 2.5)'$. Hence, any effort to find an arbitrage must fail!

The proof of the FTAP has an easy and a tough part. It is straightforward to show that (2.) implies (1.): Suppose (2.) holds and consider a portfolio such that $\boldsymbol{A\theta} \geq 0$. Then by the strict positivity of state prices $\boldsymbol{\pi}'\boldsymbol{A\theta} \geq 0$. But this implies $\boldsymbol{q}'\theta \geq 0$ ruling out arbitrage opportunities.

In the following we first give a geometric proof of the more difficult part of the FTAP for the case of two assets and two states of the world. This will provide us with some intuitive understanding on the FTAP. Afterwards we give a proof of the general result which will be based on the Riesz representation theorem (Theorem A.1).

Proof of Theorem 4.2 (simple case) In the case of two assets and two states the payoffs of the assets in the two states $s = 1, 2$ can be represented by the two dimensional vectors $\boldsymbol{A}_1$ and $\boldsymbol{A}_2$. To find the set of non-negative portfolio payoffs in a particular state, we first determine the set of assets where the asset payoff, $\boldsymbol{A}_s\boldsymbol{\theta}$, is

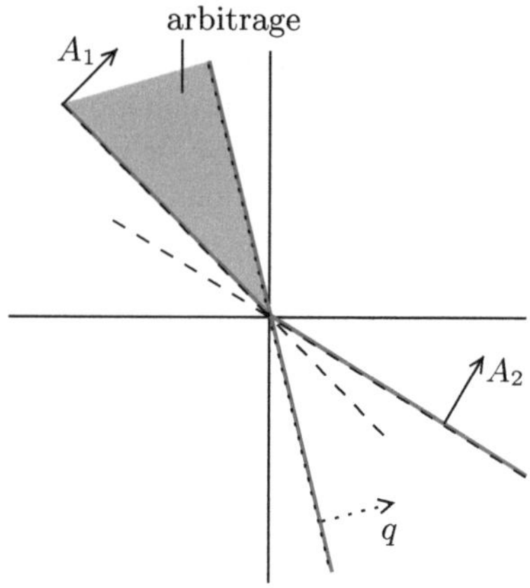

Fig. 4.4 Finding arbitrage opportunities

equal to 0. This is a line orthogonal to the payoff vector.[10] Plotting these orthogonal lines for the vectors A_1 and A_2, we determine the set of non-negative payoffs in both states as the area of the intersection of two half planes as shown in Fig. 4.4 below.

To determine the set of arbitrage opportunities, we have then to find a strategy requiring no investments, i.e. $-\boldsymbol{q}'\boldsymbol{\theta} \geq 0$ or $\boldsymbol{q}'\boldsymbol{\theta} \leq 0$ with a positive payoff in at least one of the states. To find the set of arbitrage portfolios we then plot the price vector $\boldsymbol{q}$ so that conditions $\boldsymbol{q}'\boldsymbol{\theta} \leq 0$ and $\boldsymbol{A}_s\boldsymbol{\theta} \geq 0$ are satisfied. This is possible if and only if $\boldsymbol{q}$ does not belong to the cone of $\boldsymbol{A}_1$ and $\boldsymbol{A}_2$, i.e. if there are no constants $\pi_1, \pi_2 > 0$ such that $\boldsymbol{q}' - \pi_1\boldsymbol{A}_1 + \pi_2\boldsymbol{A}_2$. □

Proof of Theorem 4.2 (general case) The general argument is easy if markets are complete, in which case it follows from the Law of One Price. For any given payoff asset matrix $\boldsymbol{A}$, consider the set of all payoffs that you can generate with alternative portfolios:

$$\text{span}\,\{\boldsymbol{A}\} = \{\boldsymbol{y} \in \mathbb{R}^{S+1} : \boldsymbol{y} = \boldsymbol{A}\boldsymbol{\theta} \text{ for some } \boldsymbol{\theta} \in \mathbb{R}^{K+1}\}$$

Let $q(\boldsymbol{y})$ be the price associated with the payoff $\boldsymbol{y}$ in span $\{\boldsymbol{A}\}$. The function q : span $\{\boldsymbol{A}\} \rightarrow \mathbb{R}$ is called the pricing functional on the set of attainable payoffs span $\{\boldsymbol{A}\}$. By the Law of One Price q is linear, i.e., for all $\boldsymbol{y}, \boldsymbol{y}' \in$ span $\{\boldsymbol{A}\}$ and $\alpha \in \mathbb{R}$ we have

(i) $q(\boldsymbol{y} + \boldsymbol{y}') = q(\boldsymbol{y}) + q(\boldsymbol{y}')$,
(ii) $q(\alpha \boldsymbol{y}) = \alpha q(\boldsymbol{y})$.

[10]The scalar product is positive (negative) if the angle is smaller (greater) than 90°. The scalar product of orthogonal vectors is equal to 0 (see Appendix A.1).

Why is this true? – Since otherwise, one could find an arbitrage opportunity with hedged payoffs tomorrow and positive payoff today.

By the Riesz representation theorem (Theorem A.1) linear functionals can be represented as $q(y) = \boldsymbol{\pi}'y$ for some vector of state prices $\boldsymbol{\pi} \in \mathbb{R}^{S+1}$.[11] From the various assumptions on the utility functions we get additional restrictions on the state prices. Suppose, for example, the utility function is increasing in the risk-free asset. Then the sum of the state prices must be positive because otherwise one could get the risk-free asset for free. If the utility function were strictly monotonic then each state price must be positive because otherwise the portfolio delivering a positive payoff in the state with zero or negative price would be an arbitrage opportunity for this type of investors. Note that this argument assumes that all portfolios can be built, i.e., that markets are complete. A proof for the general case can be found in the book by Magill and Quinzii [MQ02]. □

Let us now formulate the variant of the FTAP for mean-variance utilities, its proof is analogous to Theorem 4.2:

Theorem 4.3 (FTAP for mean-variance utility functions) *The following two conditions are equivalent:*

1. *There exists no $\boldsymbol{\theta} \in \mathbb{R}^{K+1}$ such that*

$$\boldsymbol{q}'\boldsymbol{\theta} \leq 0 \quad \text{and} \quad \boldsymbol{A\theta} = v\mathbf{1}, \quad \text{for some } v > 0.$$

   Note $\mathbf{1} = (1, \ldots, 1)'$.
2. *There exists a $\boldsymbol{\pi} \in \mathbb{R}^S$ with $\sum_{s=1}^{S} \pi_s > 0$ such that*

$$q^k = \sum_{s=1}^{S} A_s^k \pi_s, \quad k = 0, \ldots, K.$$

To conclude this section we will give different equivalent formulations of the No-arbitrage Principle. These formulations are very useful to deepen the understanding of the main idea. Also it is important to study them because different fields of finance use different formulations of it.

The first reformulation of the No-arbitrage Principle displays asset prices as their discounted expected payoffs. This formulation follows from a normalization of state prices: applying the linear pricing rule,

$$q^k = \sum_{s=1}^{S} A_s^k \pi_s, \quad k = 0, \ldots, K,$$

[11] It is obvious that a representation by state prices satisfies linearity. The converse is a bit harder to see (compare Appendix A.1).

to the risk-free asset, $k = 0$, we see that the risk-free rate is the reciprocal of the sum of the state prices:

$$q^0 = \frac{1}{R_f} = \sum_{s=1}^{S} \pi_s.$$

On defining the normalized state prices as $\pi_s^* := \pi_s/(\sum \pi_z)$ we get the discounted expected payoff formulation of asset prices[12]:

$$q^k = \frac{1}{R_f} \sum_{s=1}^{S} A_s^k \pi_s^* = \frac{1}{R_f} \mathbb{E}_{\pi^*}(A^k).$$

There are similar results for situations where the no-arbitrage condition is disturbed by transaction costs or by short sale constraints [JK95b, JK95a].

The normalized state prices are called the *martingale measure* in financial mathematics and they are called the *risk neutral probabilities* in finance. The latter is a bit confusing since π^* is actually accounting for the risk preferences of the agents, as we will see in Sect. 4.3. "Risk neutral probabilities" therefore means: probabilities that take the risk preferences already into account and can therefore be used *as if* they were physical probabilities and the investor were risk-neutral. Calling them *risk adjusted probabilities* would be less confusing.

From this way of writing the pricing formula we also get an immediate formulation in terms of returns:

$$R_f = \sum_{s=1}^{S} \frac{A_s^k}{q^k} \pi_s^* = \mathbb{E}_{\pi^*}(R^k).$$

Hence, in the light of the normalized state prices all assets deliver the risk-free rate. Indeed, a reformulation of the FTAP for returns does read like this:

Corollary 4.4 (FTAP for returns) *The following two statements are equivalent:*

1. *There exists no* $\boldsymbol{\lambda} \in \mathbb{R}^{K+1}$ *with*

$$\begin{pmatrix} -1 & \cdots & -1 \\ & \boldsymbol{R} & \end{pmatrix} \boldsymbol{\lambda} > \mathbf{0}. \tag{4.1}$$

[12]Note that we did not assume that all state prices are positive. For this formulation we only need that the sum of the state prices is positive, which holds, for example, with mean-variance utilities.

2. *There exists a* $\boldsymbol{\pi}^* \in \mathbb{R}^S_{++}$ *with*

$$R_f = \sum_{s=1}^{S} R_s^k \pi_s^*, \quad k = 0, \ldots, K.$$

Finally, we would like to mention how the FTAP looks like in the case of the Ross APT. Recall that according to the arbitrage pricing theory of Ross, asset returns are thought of as being determined by several factors. The asset returns matrix is the product of the factor matrix and the matrix of factor loadings: $\boldsymbol{R} = \boldsymbol{F}\boldsymbol{B}$. Accordingly asset payoffs can be written in terms of factors by noting that

$$\boldsymbol{A} = \boldsymbol{A}\boldsymbol{\Lambda}(\boldsymbol{q})^{-1}\boldsymbol{\Lambda}(\boldsymbol{q}) = \boldsymbol{R}\boldsymbol{\Lambda}(\boldsymbol{q}) = \boldsymbol{F}\boldsymbol{B}\boldsymbol{\Lambda}(\boldsymbol{q}).$$

By defining $\boldsymbol{z}$ as the allocation of factor risks, i.e., $\boldsymbol{z} := \boldsymbol{B}\boldsymbol{\Lambda}(\boldsymbol{q})\boldsymbol{\theta}$, we can write $\boldsymbol{A}\boldsymbol{\theta} = \boldsymbol{F}\boldsymbol{z}$ and the value of asset portfolios is

$$\boldsymbol{q}'\boldsymbol{\theta} = \boldsymbol{q}'\boldsymbol{A}^{-1}\boldsymbol{F}\boldsymbol{z} =: \hat{\boldsymbol{q}}'\boldsymbol{F}\boldsymbol{z}.$$

Hence, Corollary 4.4 can be rewritten as follows:

Corollary 4.5 (FTAP for Ross APT) *The following two statements are equivalent:*

1. *There exists no* $\boldsymbol{z} \in \mathbb{R}^{K+1}$ *with*

$$\begin{pmatrix} -\hat{\boldsymbol{q}}' \\ \boldsymbol{F} \end{pmatrix} \boldsymbol{z} > \boldsymbol{0}.$$

2. *There exists a* $\check{\boldsymbol{\pi}} \in \mathbb{R}^S_{++}$ *with*

$$\hat{q}_f = \sum_{s=1}^{S} F_s^f \check{\pi}_s, \quad f = 1, \ldots, F.$$

In other words: factors have a price that can be expressed as the weighted sum of their payoffs, weighted with some state prices. This concludes our remarks on the Fundamental Theorem of Asset Pricing. In the next section we introduce an important application of this theory: the pricing of derivatives.

4.2.3 Pricing of Derivatives

The Fundamental Theorem of Asset Pricing is essential for the valuation of assets such as derivatives. We first look at the case of redundant assets, i.e., those derivatives that can be duplicated with already priced assets. In general, there are two possible ways to determine the value of a derivative. The first approach is based on determining the value of a *hedge portfolio*. This is a portfolio of assets that delivers the same payoff as the derivative. The second approach uses the *risk-neutral probabilities* in order to determine the current value of the derivative's payoff.

Consider an example of the one-period binomial model. In this simplified setting, we are looking for the current price of a call option on a stock S. Assume that $S := 100$ and there are two possible prices in the next period: $Su := 200$ if $u = 2$ and $Sd := 50$ if $d = 0.5$. The riskless interest rate is 10 %. The value of an option with strike price X is given by $\max(Su-X, 0)$ if u and $\max(Sd-X, 0)$ if d is realized.

To determine the value of the call, we replicate its payoff using the payoffs of the underlying stock and the bond. If arbitrage is excluded, the value of the call is equal to the value of the *hedge portfolio*, which is the sum of the values of its constituents. The idea is that a portfolio that has the same cash flow as the option must have the same price as the call which we are looking for.

Calculating the call values for each of the states, we obtain $\max(Su - X, 0) = 200 - 100 = 100$ in the "up" state and $\max(Sd - X, 0) = \max(50 - 100, 0) = 0$ in the "down" state. The *hedge portfolio* then requires to borrow $1/3$ of the risk-free asset and to buy $2/3$ risky assets in order to replicate the call's payoff in each of the states:

$$\begin{aligned} \text{"up"}: \quad & \tfrac{2}{3}200 - \tfrac{1}{3}100 = 100 \\ \text{"down"}: \quad & \tfrac{2}{3}50 - \tfrac{1}{3}100 = 0 \end{aligned}$$

In general, we need to solve:

$$\begin{aligned} C_u &:= \max(Su - X, 0) = nSu + mBR_f \\ C_d &:= \max(Sd - X, 0) = nSd + mBR_f \end{aligned}$$

where n is the number of stocks and m is the number of bonds needed to replicate the call payoff. We get

$$n = \frac{C_u - C_d}{Su - Sd}, \qquad m = \frac{SuC_d - SdC_u}{BR_f(Su - Sd)}.$$

n is also called the *delta* of the option.

The value of the option is therefore:

$$\begin{aligned} C = nS + mB &= \frac{C_u - C_d}{u - d} + \frac{uC_d - dC_u}{R_f(u - d)} \\ &= \frac{1}{R_f} \frac{C_u(R_f - d) + C_d(u - R_f)}{u - d}. \end{aligned}$$

In the binomial model, we need two equations to match ("up" and "down") with two securities (stock and bond). In the trinomial model (if there is a state "middle"), we need a third security in order to replicate the call payoff etc.

The second approach to value derivatives is based on the FTAP result that in the absence of arbitrage we do not consider the "objective" probabilities associated with "up" and "down" movements, which are already considered in the equilibrium prices, instead we can value all securities "as if" we were in a risk-neutral world with no premium for risk. In this case, we can consider the probability of an "up" ("down") movement as being equal to the risk-neutral probability π^*, $(1-\pi^*)$. Thus, the expected value of the stock with respect to these risk neutral probabilities is

$$S_0 = \pi^* Su + (1-\pi^*)Sd.$$

In a riskless world, this must be the same as investing S today and receiving SR_f after one period. Then, $\pi^* Su + (1-\pi^*)Sd = SR_f$ or $\pi^* u + (1-\pi^*)d = R_f$.

The risk-neutral probabilities are then defined over the size of the up and down movements of the stock price and the risk-free rate

$$\pi^* = \frac{R_f - d}{u - d}, \quad 0 \le \pi^* \le 1.$$

Using the risk-neutral measure we can calculate the current value of the stock and the call:

$$S = \frac{\pi^* Su + (1-\pi^*)Sd}{R_f}, \qquad C = \frac{\pi^* C_u + (1-\pi^*)C_d}{R_f}.$$

Plugging in π^*, we get the price

$$\begin{aligned} C &= \frac{1}{R_f}\left(\frac{R_f - d}{u-d}C_u + \left(1 - \frac{R_f - d}{u-d}\right)C_d\right) \\ &= \frac{1}{R_f}\frac{C_u(R_f - d) + C_d(u - R_f)}{u-d}, \end{aligned}$$

i.e., the same as above. Similarly, put options and any other redundant derivatives can be priced.

But what about non-redundant derivatives? Well those can only exist in incomplete markets and applying the Principle of No-arbitrage will only give valuation bounds. This is easiest to see from an example. Let there be three states and two assets, a risk-free asset and the first Arrow security, i.e.,

$$A := \begin{pmatrix} 1 & 1 \\ 1 & 0 \\ 1 & 0 \end{pmatrix}.$$

The risk-free asset's price is 0.9 while the price of the Arrow security is 0.25. Suppose you need to find the arbitrage-free value of a third asset with payoffs, say $\boldsymbol{A}^3 := (2, 1, 0)'$. Obviously it cannot be worth less than two times the second existing asset, since it has payoffs dominating the payoffs of this asset. Also it cannot be worth more than the price of the risk-free asset plus the second asset, because this portfolio would dominate the payoff of the third asset. Hence the general principle to find valuation bounds is to look at all portfolios that dominate the payoff of the third asset and select the one with the least price in order to get an upper bound of the third asset's price while looking at all portfolios that are dominated by the payoff of the third asset and selecting the one with the highest price gives a lower bound. Formally, for any payoff $\boldsymbol{y}$ we get

$$\overline{q}(\boldsymbol{y}) = \min_{\boldsymbol{\theta}} \boldsymbol{q}'\boldsymbol{\theta} \quad \text{such that} \quad \boldsymbol{A}\boldsymbol{\theta} \geq \boldsymbol{y}, \quad \text{and}$$

$$\underline{q}(\boldsymbol{y}) = \max_{\boldsymbol{\theta}} \boldsymbol{q}'\boldsymbol{\theta} \quad \text{such that} \quad \boldsymbol{A}\boldsymbol{\theta} \leq \boldsymbol{y}.$$

In our example the least expensive dominating portfolio consists of one unit of asset 1 and one unit of asset 2, hence the upper bound is $\overline{q}(\boldsymbol{y}) = 1.15$. The most expensive dominated portfolio consists of two units of the Arrow security, hence the lower bound is $\underline{q}(\boldsymbol{y}) = 0.5$. And as with redundant assets these bounds could also be found using the state prices. The no-arbitrage restrictions on state prices are $0.9 = \pi_1 + \pi_2 + \pi_3$ for the risk-free asset and $0.25 = \pi_1$ for the first Arrow security. Hence, state prices have one degree of freedom expressed as $0.65 = \pi_2 + \pi_3$. Going to the one extreme we can choose $\pi_2 = 0$ and $\pi_3 = 0.65$ while the other extreme would be $\pi_2 = 0.65$ and $\pi_3 = 0$. Hence, again the third asset's price is bounded above by 1.15 and it is bounded below by 0.5. The general formulation of the no-arbitrage bounds in terms of state prices is:

$$\overline{q}(\boldsymbol{y}) = \min_{\pi} \boldsymbol{\pi}'\boldsymbol{y}, \quad \text{such that} \quad \boldsymbol{\pi}'\boldsymbol{A} = \boldsymbol{q}', \quad \text{and}$$

$$\underline{q}(\boldsymbol{y}), = \max_{\pi} \boldsymbol{\pi}'\boldsymbol{y}, \quad \text{such that} \quad \boldsymbol{\pi}'\boldsymbol{A} = \boldsymbol{q}'.$$

4.2.4 Limits to Arbitrage

The above considerations assumed that the investors were totally free in choosing any portfolios. One may however argue that in reality investors face *short-sales constraints* and some limits in horizon along which an arbitrage strategy can be carried out. Though, in the presence of limits of arbitrage like short-sales constraints, the arbitrage is limited and even the Law of One Price may fail in equilibrium. Let us consider first some examples.

4.2.4.1 3Com and Palm

On March 2, 2000, the company 3Com made an IPO of one of its most profitable units. They decided to sell 5 % of its Palm stocks and retain 95 % thereof. At the IPO day, the Palm stock price opened at \$38, achieved its high at \$165 and closed at \$95.06. This price movement was puzzling because the price of the mother-company 3Com closed that day on \$81.81. If we calculate the value of Palm shares per 3Com share, which is \$142.59,[13] and subtract it from the end price of 3Com, we get $\$81.81 - \$142.58 = -\$60.77$. If we additionally consider the available cash per 3Com share, we would come to a "stub" value for 3Com shares of −\$70.77! Clearly, this result is a contradiction of the Law of One Price since the portfolio value (the value of Palm shares, the rest of 3Com shares and the cash amount), which is negative, differs from the sum of its constituents, which is positive.

However, the relative valuation of Palm shares did not open an arbitrage strategy, since it was not possible to short Palm shares. Also it was not easy to buy sufficiently many 3Com stocks and then to break 3Com apart to sell the embedded Palm stocks. The mismatch persisted for a long time (see Fig. 4.5).

4.2.4.2 Volkswagen and Porsche

In October 2008, just when the stock market was in turmoil due to the financial crisis and in particular the bankruptcy of Lehmann Brothers, a larger investment bank, had just been announced, at a time when most stocks lost heavily, one stock was excelling them all: Volkswagen had been steadily increasing from 154.48 Euro per share on January 1, 2008 to 210.85 Euro on October 24. Then it started to rise like a rocket: only two trading days later, on October 28, it reached 1005.01 Euro, thus increasing by 377 % within a few days – quite an exciting performance for a producer of solid, but not that exciting cars. In fact the market value of Volkswagen at that point was higher than the market value of *all* other European car producers *together*! What had happened? Did Volkswagen invent a car driving with water

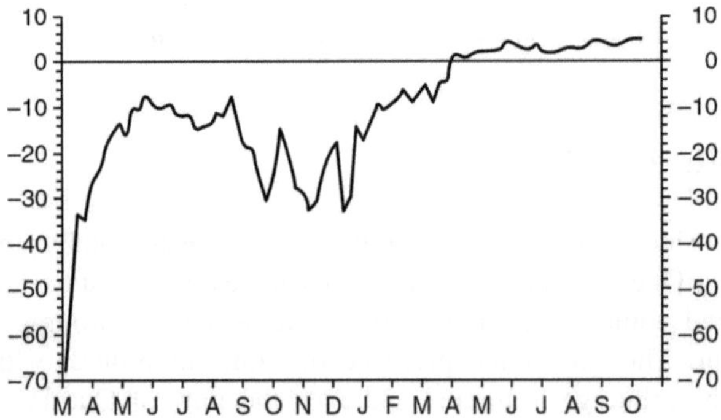

Fig. 4.5 Negative stub value of 3Com after the IPO of Palm

[13] $0.95 \cdot 95.06/$number of outstanding 3Com shares.

instead of gasoline? What effect could justify the sudden increase in the price of Volkswagen stocks? Or was this another example of mispricing?

Indeed, the case of Volkswagen is not dissimilar to the case of Palm, albeit much more extreme in its consequences as we will discuss later.

What happened is that Porsche, another, but much smaller car producer, had started a slow takeover of Volkswagen by buying stocks, but also options on stocks. This caused a steady increase of the Volkswagen stock prices in times where most stocks went down. The increase lured many investors (particularly hedge funds) into speculating on an eventual decline of the price of Volkswagen stocks – after all they seemed by all economic measures to be overpriced. Therefore, these investors went short in Volkswagen stocks. Porsche, however, still increased its position. On October 23, Porsche announced that it had already bought 42 % of the Volkswagen shares and held options for another large portion of stocks. Given that the state held 25 % of Volkswagen stocks, this implied that there were actually very few stocks freely available on the market. In fact, the amount of stocks that has been sold short must have exceeded this amount. Consequently, the stock price was rising once more and many investors who were short in Volkswagen were suddenly forced to liquidate their positions – by buying Volkswagen shares which increased the price even more. The increase also meant that the weight of Volkswagen on the DAX, the German stock market index, automatically increased. This pushed many fund managers to buy Volkswagen shares in order to hedge their DAX funds which increased the price even more and started a vicious cycle that caused a crisis on the stock market.

The next day things changed: Porsche was forced by public pressure to sell some of its options and the German stock exchange was pushed to reduce the weight of Volkswagen in the DAX by changing its composition rules. Moreover, the excessive mispricing probably triggered new investors to enter the short market and to buy put options on Volkswagen.[14] But even here the market efficiency was limited, since there were just no put options available with high strikes: the highest strike price was just around 240 Euro and thus by a large amount out-of-the-money. Nevertheless, the price of these puts more than doubled within a week after these events and the price of Volkswagen went down already the next day to 517 Euro and until the end of the year to 250 Euro.

As in the case of 3Com we can compute the stub value of Porsche when subtracting the current market value of Volkswagen from the market value of Porsche. Figure 4.6 shows this (highly negative) value over time. The computation is rough, since some important parameters are not known, but the order of magnitude should be correct. Again, it would be difficult to find a strategy that eliminates the mispricing quickly. Going short in Volkswagen was initially *not* a good strategy, as we have seen: several hedge funds must have lost billions of Euro in this way. One particularly tragic event was that a German entrepreneur committed suicide a few months after he had lost a fortune and his reputation due to a failed speculation on Volkswagen stocks.

---

[14]One of the authors was among them.

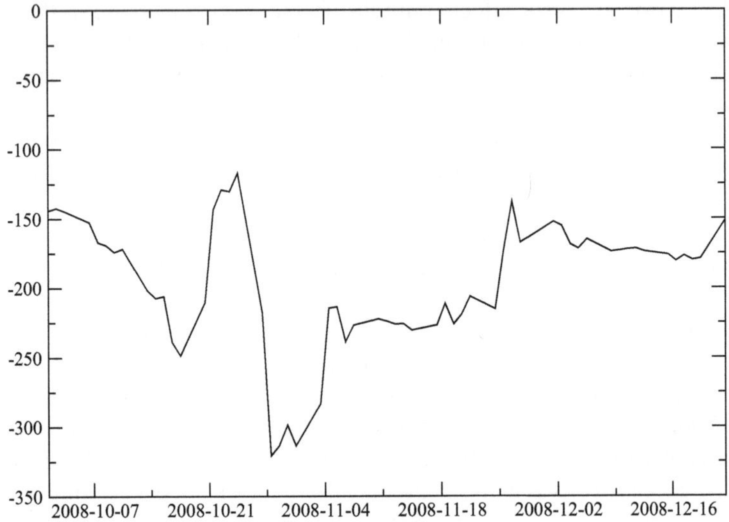

Fig. 4.6 Negative stub value of Porsche during the attempted takeover of Volkswagen

Finally, Porsche, heavily leveraged by the planned Volkswagen deal, was hit severely by the financial crisis. It went close to bankruptcy and was taken over by Volkswagen. The moral of the story: a shark can eat a herring, but a herring should not try to eat a shark!

4.2.4.3 Closed-End Funds

The case of closed-end funds[15] is more puzzling since the portfolio ingredients are not only known but also tradeable. Though, on average, the prices of fund shares are still not equal to the sum of the prices of its components as Fig. 4.7 shows.

The reason for this mismatch is the fact that no investor can unbundle the closed-end funds and trade their components on market prices. Additionally, buying a share of an undervalued closed-end fund and selling the corresponding portfolio until maturity does not work because closed-end funds typically do not pay out the dividends of their assets before maturity.

As in the 3Com-Palm case the violation of the Law of One Price does not constitute an arbitrage strategy because the discount/premium of the closed-end funds can deepen until maturity.

4.2.4.4 LTCM

The prominent LTCM case is an excellent example of the risks associated with seemingly arbitrage strategies. The LTCM managers discovered that the share price of Royal Dutch Petroleum at the London exchange and the share price of Shell

[15]A closed-end fund is a mutual fund with a fixed asset composition.

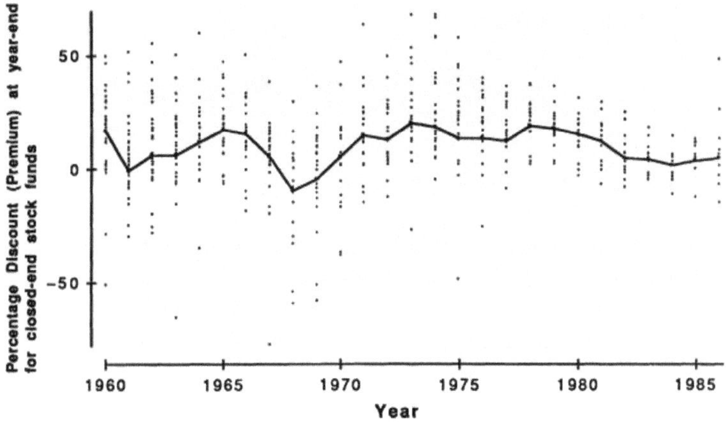

Fig. 4.7 Percentage discount (premium) at year-end for closed end stock funds (Lee, Shleifer and Thaler, *Journal of Finance*, 1991)

Transport and Trading at the New York exchange do not reflect the parity in earnings and dividends stated in the splitting contract between these two units of the Royal Dutch/Shell holding. According to this splitting contract, earnings and dividends are paid in relation 3 (Royal Dutch) to 2 (Shell), i.e., the dividends of Royal Dutch are 1.5 times higher than the dividends paid by Shell. However, the market prices of these shares did not follow this parity for long time but they followed the local markets' sentiment.

This example is most puzzling because a deviation of prices from the 3 : 2 parity invites investors to either buy or sell a portfolio with shares in the proportion 3 : 2 and then to hold this portfolio forever: doing this one can cash in a gain today while all future obligations in terms of dividends are hedged. There is however the risk that the company decides to change the parity.

4.2.4.5 No-Arbitrage with Short-Sales Constraints

To illustrate how limits to arbitrage enlarge the set of arbitrage-free asset prices, consider the case of non-negative payoffs and short-sales constraints, i.e., $A_s^k \geq 0$ and $\lambda_k^i \geq 0$. The short-sales restriction may apply to one or more securities. Then, the Fundamental Theorem of Asset Pricing reduces to:

Theorem 4.6 (FTAP with Short-Sales Constraints) *There is no long-only portfolio* $\boldsymbol{\theta} \geq \mathbf{0}$ *such that* $\boldsymbol{q}'\boldsymbol{\theta} \leq \mathbf{0}$ *and* $A\boldsymbol{\theta} > \mathbf{0}$ *is equivalent to* $\boldsymbol{q} \gg \mathbf{0}$.

Proof Suppose $\boldsymbol{q} \gg \mathbf{0}$ and $\boldsymbol{\theta} \geq \mathbf{0}$. For strategy $\boldsymbol{\theta}$ with $A\boldsymbol{\theta} > \mathbf{0}$ must be true that $\boldsymbol{q}'\boldsymbol{\theta} > 0$. In other words, every long-only portfolio must cost something. Conversely, suppose $q^k \leq 0$, then for some k,

$$\theta^* = (0, \ldots, \underset{\substack{\uparrow \\ k}}{1}, \ldots, 0)'$$

is an arbitrage, i.e., $A\theta^* > \mathbf{0}$ and $q'\theta^* \leq 0$. □

Hence, *all* positive prices are arbitrage-free because sales restrictions deter rational managers to exploit eventual arbitrage opportunities. Consequently, the no-arbitrage condition does not tell us anything and we need to look at specific assumptions to determine asset prices. This is done in the following section.

4.3 Financial Markets Equilibria

The Principle of No-arbitrage that we analyzed in the previous section gives a first idea about asset prices. The main strength of the principle is that it shows how the prices of redundant assets should be related to the prices of a set of fundamental assets. However, the Principle of No-arbitrage tells us nothing about how the prices of the fundamental assets should be related to each other. The Fundamental Theorem of Asset Prices shows that asset prices are determined by *some* state prices, but the value of the state prices is not determined by the No-arbitrage Principle! Hence, we must dig a bit deeper into financial market theory and find a theory that explains the state prices.

This brings us back to the idea expressed in the introduction that prices are determined by trade – but trades are in turn depending on prices, which looks like a "hen and egg" problem. The notion of a *competitive equilibrium* captures exactly this interdependence of decisions and prices. A competitive equilibrium is a price system where all agents have optimized their positions and all markets clear, i.e., we obtain the equality of supply and demand on every market. The asset prices then reflect on the one hand the agents' decision criteria (their utility functions) and on the other hand the agents' resources. Hence, the notion of equilibrium explains state prices by agents' time preferences, their risk preferences and the risk embodied in their resources. As a general rule we obtain that state prices are larger for those states the agents believe to be more likely to occur and that they are higher for those states in which there are less resources. For special cases like the CAPM, we can get more specific pricing rules. In the CAPM, asset prices are determined by the expected payoff of the assets adjusted by the scarcity of resources. This adjustment is measured by the covariance of the payoffs and the aggregate availability of resources. The latter is called the market portfolio.

We structure this section along the various motives for trade. If an asset has a positive covariance to the market portfolio it pays off a lot when resources are not scarce. Hence, other things being equal, it has a lower price than an asset with a negative covariance to the market portfolio. The next subsection generalizes the idea of a tradeoff between risk and return, as we have already seen it in the CAPM. Using this generalization, we can study competitive equilibria in financial markets – for short: *financial markets equilibria* – in terms of quantities and prices (the economics way), and also in terms of asset allocations and returns (the finance way). Then we look at intertemporal trade (interest rates), risk diversification (the Beta) and betting (the Alpha). Thereafter, we show how these three motives can be embedded in a general risk-return model which gives the foundation for the consumption based asset pricing model. Then we point at a great simplification technique to explain asset prices: *aggregation*. If markets are complete, asset allocations are Pareto-efficient[16] and hence asset prices can be described by a *single* decision problem, the optimization problem of a *representative investor*. We conclude this chapter with some warnings: the representative agent technique for asset prices may fail for predictions and it may give a wrong impression of market dynamics.

4.3.1 General Risk-Return Tradeoff

In this subsection we derive a general risk-return formula from the Principle of No-arbitrage. CAPM, APT and consumption-based asset pricing model will simply be special cases of this general result.

Recall that the absence of arbitrage is equivalent to the existence of state prices π^* such that $R_f = \mathbb{E}_{\pi^*}(R^k)$, for all $k = 1, \dots, K$.

Hence, evaluated with the normalized state prices, all risky assets are equivalent to the risk free asset. However actual return data are driven by the physical measure. Can we change the expectation under the state prices so that we can obtain a risk measure based on the observable return data? As the following calculation shows, this is easily done by defining the ratio of the state price measure and the physical probabilities, the so called likelihood ratio process[17] $\ell_s := \pi_s^*/p_s$:

$$R_f = \mathbb{E}_{\pi^*}(R^k) = \sum_s \pi_s^* R_s^k = \sum_s p_s \left(\frac{\pi_s^*}{p_s}\right) R_s^k = \sum_s p_s \ell_s R_s^k = \mathbb{E}_p(\ell R^k).$$

Furthermore, recall that by the definition of the covariance we can rewrite this expression to obtain $\mathbb{E}_p(R^k) = R_f - \text{cov}_p(R^k, \ell)$, where the covariance of the strategy returns to the likelihood ratio represents the unique risk measure. Hence, we found a simple risk-return formula which is based on the covariance to a unique factor.

---

[16]In other words, there exists no asset allocation where nobody is worse off and at least somebody is better off.

[17]We will always refer to the normalized state prices as the state price measure. However, as can be seen from the calculations, we do not actually need that all state prices are non-negative. Only the sum of the state prices needs to be positive. Hence, we can accommodate without any special considerations the case of mean-variance preferences for which the positivity of state prices was not guaranteed.

Thus we have found the ultimate formula for asset-pricing and can stop here, can't we? Not really: in a sense we only exchanged one unknown, the state price measure, with another unknown, the likelihood ratio process. Seen this way the remaining task is to identify the likelihood ratio process based on reasonable economic assumptions.

4.3.2 Consumption Based CAPM

A well known way to identify the likelihood ratio process is the consumption based CAPM. In the C-CAPM one assumes that agents maximize expected utility functions and that markets are complete. Then the likelihood ratio process coincides with the marginal rates of substitution of the investors.

To derive the C-CAPM recall the general decision problem of an agent with expected utility:

$$\max_{\hat{\theta}^i \in \mathbb{R}^{K+1}} U^i(c_0^i, \boldsymbol{c}_1^i) = u^i(c_0^i) + \delta^i \sum_{s=1}^{S} p_s u^i(c_s^i)$$

$$\text{such that} \quad c_0^i + \sum_{k=0}^{K} q^k \hat{\theta}^{i,k} = w_0^i + \sum_{k=0}^{K} q^k \theta_A^{i,k}, \quad c_s^i \geq 0,$$

where $\boldsymbol{c}_1^i = \sum_{k=0}^{K} A^k \hat{\theta}^{i,k} + w_{\perp 1}^i$.

Note that writing this we assumed homogeneous beliefs. Using the no-arbitrage relation to express asset prices in terms of state-price discounted asset payoffs, the budget restriction can be written as:

$$c_0^i + \sum_{s=1}^{S} \pi_s c_s^i = w_0^i + \sum_{s=1}^{S} \pi_s w_s^i \quad \text{and} \quad (\boldsymbol{c}_1^i - \boldsymbol{w}_1^i) \in \text{span}\{A\},$$

where the latter restriction can be skipped in the case of complete markets. Note, that the first order condition for this maximization problem is:

$$p_s \frac{\delta^i \partial_{c_s^i} u^i(c_s^0)}{\partial_{c_0^i} u^i(c_0^i)} = \pi_s, \quad s = 1, \ldots, S.$$

Hence, the likelihood ratio process is equal to the marginal rates of substitution and we can compute for $s = 1, \ldots, S$:

$$\ell_s = \frac{\pi_s^*}{p_s} = \frac{\frac{p_s \delta^i u'(c_s)}{u'(c_0)}}{\delta^i \frac{\sum_t p_t u'(c_t)}{u'(c_0)}} / p_s$$

$$= \frac{u'(c_s)}{\sum_t p_t u'(c_t)}$$
$$= \frac{u'(c_s)}{\mathbb{E}u'(c_t)}.$$

In principle it would thus suffice to know *any* utility function u^i and any consumption process c^i to determine the likelihood ratio process. But one may argue that individual decisions are subject to mistakes so that determining the likelihood ratio process from an arbitrarily chosen agent may be quite misleading. That is the reason why for empirical purposes the likelihood ratio process is determined from aggregate consumption assuming some simple parametric form of the utility function, like CRRA (see Sect. 2.2.3). How this aggregation will be justified is shown in Sect. 4.6. In any case we see that ℓ should be a decreasing function of aggregate consumption because u is typically concave. More specifically, for $u'(c_s) = a - bc_s$ ℓ is linear in c_s and for $u'(c_s) = c_s^{-\alpha}$ ℓ is convex in c_s etc.

Later, in Sect. 4.4, we give four examples for this identification. First we confirm that the CAPM is still a special case of our model, then we derive the APT by introducing background risk, we derive the C-CAPM by identifying the likelihood ratio process with the marginal rates of substitution of the investors and finally we derive a behavioral CAPM based on Prospect Theory.

4.3.3 Definition of Financial Markets Equilibria

We use the two-period model as outlined in Chap. 3 and first give the definition of financial markets equilibria in economic terms, i.e., in terms of asset prices and quantities of assets bought and sold. As before, the periods are enumerated $t = 0, 1$. In the second period $t = 1$ a finite number of states of the world, $s = 1, 2, \ldots, S$ can occur (compare Fig. 4.3).

As before, we denote the assets by $k = 0, 1, 2, \ldots, K$. The first asset, $k = 0$, is the risk-free asset delivering the certain payoff 1 in all second period states. The assets' payoffs are denoted by A_s^k. The time 0 price of asset k is denoted by q^k. Recall the states-asset-payoff matrix,

$$\boldsymbol{A} = (A_s^k) = \begin{pmatrix} A_1^0 \cdots A_1^K \\ \vdots \quad\quad \vdots \\ A_S^0 \cdots A_S^K \end{pmatrix} = (\boldsymbol{A}^0 \cdots \boldsymbol{A}^K) = \begin{pmatrix} \boldsymbol{A}_1 \\ \vdots \\ \boldsymbol{A}_S \end{pmatrix},$$

which gathers the essence of the asset structure.

Each investor $i = 1, \ldots, I$ is described by his exogenous wealth in all states of the world $\boldsymbol{w}^i = (w_0^i, \ldots, w_S^i)'$. Given these exogenous entities and given the asset prices $\boldsymbol{q} = (q^0, \ldots, q^K)'$ he can finance his consumption $\boldsymbol{c}^i = (c_0^i, \ldots, c_S^i)'$ by trading the assets. We denote by $\boldsymbol{\theta}^i = (\theta^{i,0}, \ldots, \theta^{i,K})'$ the vector of asset trade of agent i. Note that $\theta^{i,k}$ can be positive or negative, i.e., agents can buy or sell assets. In these terms, the agent's decision problem is:

$$\max_{\theta^i \in \mathbb{R}^{K+1}} U^i(c^i) \quad \text{such that} \quad c_0^i + \sum_{k=0}^{K} q^k \theta^{i,k} = w_0^i$$

$$\text{and} \quad c_s^i = \sum_{k=0}^{K} A_s^k \theta^{i,k} + w_s^i \geq 0, \quad s = 1, \ldots, S,$$

which, considering that some parts of the wealth may be given in terms of assets,[18] can be written as:

$$\max_{\hat{\theta}^i \in \mathbb{R}^{K+1}} U^i(c^i) \quad \text{such that} \quad c_0^i + \sum_{k=0}^{K} q^k \hat{\theta}^{i,k} = \sum_{k=0}^{K} q^k \theta_A^{i,k} + w_0^i$$

$$\text{and} \quad c_s^i = \sum_{k=0}^{K} A_s^k \hat{\theta}^{i,k} + w_{\perp s}^i, \quad s = 1, \ldots, S.$$

A *financial markets equilibrium* is a system of asset prices and an allocation of assets such that every agent optimizes his decision problem and markets clear, formally:

Definition 4.7 A financial markets equilibrium is a list of portfolio strategies $\hat{\boldsymbol{\theta}}^{\text{opt},i}$, $i = 1, \ldots, I$, and a price system q^k, $k = 0, \ldots, K$, such that for all $i = 1, \ldots, I$,

$$\hat{\boldsymbol{\theta}}^{\text{opt},i} = \arg\max_{\hat{\theta}^i \in \mathbb{R}^{K+1}} U^i(\boldsymbol{c}^i) \quad \text{such that} \quad c_0^i + \sum_{k=0}^{K} q^k \hat{\theta}^{i,k} = \sum_{k=0}^{K} q^k \theta_A^{i,k} + w_0^i$$

$$\text{and} \quad c_s^i = \sum_{k=0}^{K} A_s^k \hat{\theta}^{i,k} + w_{\perp s}^i, \quad s = 1, \ldots, S,$$

and markets clear:

$$\sum_{i=1}^{I} \hat{\theta}^{\text{opt},i,k} = \sum_{i=1}^{I} \theta_A^{i,k}, \quad k = 0, \ldots, K.$$

Note that we only required asset markets to clear. What about markets for consumption? Are we sure that they are also in equilibrium? Formally, can we show that also the sum of the consumption is equal to the sum of the available resources,

[18] See Chap. 4.1.3 for this transformation of the decision problem.

i.e.,

$$\sum_{i=1}^{I} c_0^i = \sum_{i=1}^{I} w_0^i \quad \text{and} \quad \sum_{i=1}^{I} c_s^i = \sum_{i=1}^{I} w_s^i, \quad s = 1, \ldots, S?$$

Noting that $w_s^i = \sum_{k=0}^{K} A_s^k \theta_A^{i,k} + w_{\perp s}^i$, this follows from the agents' budget restrictions:

$$\sum_{i=1}^{I} \left(c_0^i + \sum_{k=0}^{K} q^k \hat{\theta}^{\text{opt},i,k} \right) = \sum_{i=1}^{I} \left(w_0^i + \sum_{k=0}^{K} q^k \theta_A^{i,k} \right)$$

and

$$\sum_{i=1}^{I} c_s^i = \sum_{i=1}^{I} \left(\sum_{k=0}^{K} A_s^k \hat{\theta}^{\text{opt},i,k} + w_{\perp s}^i \right), \quad s = 1, \ldots, S,$$

because asset markets clear: $\sum_{i=1}^{I} \hat{\theta}^{\text{opt},i,k} = \sum_{i=1}^{I} \theta_A^{i,k}$, $k = 0, \ldots, K$. Hence, nothing is missing in the Definition 4.7.

It is immediate to see that in a financial market equilibrium there cannot be arbitrage opportunities. This is true, because otherwise the agents would not be able to solve their maximization problem since any portfolio they consider could still be improved by adding the arbitrage portfolio. Hence, deriving asset prices from an equilibrium model automatically leads to arbitrage-free prices.

As mentioned before, a financial markets equilibrium can be illustrated by an Edgeworth Box (Fig. 4.8). At the equilibrium allocation both agents have optimized their consumption by means of asset trade given their budget constraint and markets clear.

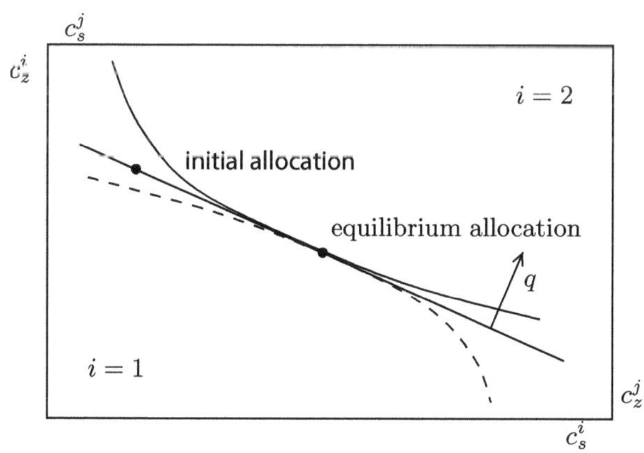

Fig. 4.8 A financial markets equilibrium in an Edgeworth Box

The geometry of the Edgeworth Box suggests that asset prices should be related to the agents' marginal rates of substitution. And indeed, on investigating the first order conditions for solving their optimization problems we see that the marginal rates of substitution are one candidate for state prices. The first order condition for any agent is:

$$q^k = \sum_{s=1}^{S} \underbrace{\frac{\partial c_s U^i(c_0^i, \ldots, c_S^i)}{\partial c_0 U^i(c_0^i, \ldots, c_S^i)}}_{\pi_s^i} A_s^k, \quad k = 0, \ldots, K.$$

In particular, for the case of expected utility

$$U^i(c_0^i, \ldots, c_S^i) = u^i(c_0^i) + \delta^i \sum_{s=1}^{S} \text{prob}_s^i u^i(c_s^i)$$

we get:

$$q^k = \sum_{s=1}^{S} \underbrace{\frac{\text{prob}_s^i \delta^i \partial c_s u^i(c_s^i)}{\partial c_0 u^i(c_0^i)}}_{\pi_s^i} A_s^k, \quad k = 0, \ldots, K.$$

Hence, we get a nice theory of state prices that links them to the agents' time preferences, their beliefs, their risk aversion and their consumption. The consumption is hereby dependent on the aggregate availability of resources.

We recall how to express a financial markets equilibrium in finance terms:

$$\max_{\lambda^i \in \Delta^{K+2}} U^i(\mathbf{c}^i) \quad \text{such that} \quad c_0^i = w_0^i - (1 - \lambda^c) \sum_{k=0}^{K} \hat{\lambda}^{i,k} w_0^i$$

$$\text{and} \quad c_s^i = \left(\sum_{k=1}^{K} R_s^k \hat{\lambda}^{i,k} \right) w_0^{i,\text{fin}} + w_{\perp s}^i, \quad s = 1, \ldots, S.$$

This puts us in a position to define a financial markets equilibrium in finance terms:

Definition 4.8 A financial markets equilibrium is a list of portfolio strategies $\boldsymbol{\lambda}^i$, $i = 1, \ldots, I$, and a system of returns R^k, $k = 0, \ldots, K$, such that for all $i = 1, \ldots, I$,

$$\boldsymbol{\lambda}^{\text{opt},i} = \arg\max_{\lambda^i \in \Delta^{K+2}} U^i(\boldsymbol{c}^i) \quad \text{such that} \quad c_0^i = w_0^i - \sum_{k=0}^{K} \hat{\lambda}^{i,k} w_0^{i,\text{fin}}$$

$$\text{and} \quad c_s^i = \left(\sum_{k=1}^{K} R_s^k \hat{\lambda}^{i,k} \right) w_0^{i,\text{fin}} + w_{\perp s}^i, \quad s = 1, \ldots, S,$$

and markets clear:

$$\sum_{i=1}^{I} \lambda^{\text{opt},i,k} r^i = \lambda^{M,k}, \quad k = 0, \ldots, K,$$

where $r^i := w_0^{i,\text{fin}} / (\sum_i w_0^{i,\text{fin}})$ and $\lambda^{M,k}$ is the relative market capitalization of asset k.

The market clearing condition in Definition 4.8 may look a bit unusual because it is not often stated explicitly in finance models.[19] So let us make sure it is indeed equivalent to the equality of demand and supply of assets:

Multiplying each market clearing condition for assets,

$$\sum_{i=1}^{I} \hat{\theta}^{\text{opt},i,k} = \sum_{i=1}^{I} \theta_A^{i,k}, \quad k = 0, \ldots, K,$$

by the price of that asset and extending the expressions by the financial wealth of the agents, $w^{i,\text{fin}} = \sum_{k=0}^{K} q^k \theta_A^{i,k}$, yields the equivalence:

$$\sum_{i=1}^{I} \lambda^{i,k,\text{opt}} r^i = \sum_{i=1}^{I} \frac{q^k \hat{\theta}^{\text{opt},i,k}}{w_0^{i,\text{fin}}} \frac{w_0^{i,\text{fin}}}{\sum_i w_0^{i,\text{fin}}} = \frac{q^k \sum_i \theta_A^{i,k}}{\sum_{k=0}^{K} (q^k \sum_i \theta_A^{i,k})} = \lambda^{M,k}.$$

Before passing on to the next section we should once more mention that everything can also be expressed in terms of factors. A financial markets equilibrium is then a system of factor returns such that all agents take the factor risk that suits best their consumption plans and markets clear. This is further discussed in the exercise book.

[19] Most finance models work right away with a representative investor being in equilibrium with himself. Hence, the market clearing condition is not stated explicitly.

4.3.4 Intertemporal Trade

One great service that a financial market offers for our society is to provide means for intertemporal trade, i.e., for savings and loans. Agents have different wealth along their life cycle, since their income is typically hump-shaped: they are quite poor when young, have the highest income when middle aged and have no income when old – unless they traded on the financial market, i.e., unless they saved before getting old. The motive for intertemporal trade explains interest rates by demand and supply on the savings and loans market. In general one would expect that interest rates are positive since agents should have a positive time preference, i.e., they discount future consumption, e.g., because the chances to survive till the money is returned are not 100 %. On the other hand, agents trade intertemporally to smooth their consumption path. Finally, one would expect that the aggregate resources relative to aggregate needs also determine interest rates. If, for example, too many want to retire at the same time, it may well be that the savings of that generation are worth less than at the time they were saving it. This phenomenon is called the "asset melt down".[20] In this subsection we want to shed some light onto all these puzzling aspects of intertemporal trade by exploring their fundamental economic ideas.

Consider an agent contemplating how much to save for the future. As before, there are two time periods, but to make things simple, we ignore uncertainty. The agent has an intertemporal utility with discount factor δ: $u(c_0) + \delta u(c_1)$. Without saving he would have to consume his exogenous wealth, which is, as usual, denoted by $\boldsymbol{w} = (w_0, w_1)$. If the wealth is quite different in the two periods, the agent can improve upon consuming his exogenous wealth by *consumption smoothing*, i.e., he may want to sacrifice some consumption when he is quite wealthy and transfer this to the other time period, because his utility function u will most certainly have a decreasing marginal utility of wealth. Figure 4.9 displays this idea in terms of the period utility from wealth $u(\boldsymbol{w})$.

Figure 4.9 is nice and simple but it does not show us easily what the optimal degree of consumption smoothing is. For this, one also needs to compare the time preference and the interest rate. Hence, one needs to formalize the intertemporal decision problem. Denoting the savings amount by s and the interest rate by r, the decision problem is given by:

$$\max_s u(c_0) + \delta u(c_1) \quad \text{such that} \quad c_0 + s = w_0$$
$$\text{and} \quad c_1 = w_1 + (1 + r)s.$$

[20] In politics, this is sometimes used as an argument against a private pension fund system. The asset melt down, however, is mitigated by the fact that demographic developments are very diverse between different countries and financial markets are global.

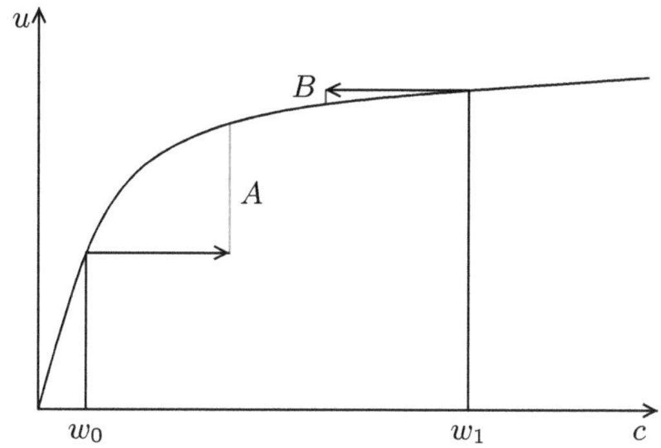

Fig. 4.9 Consumption smoothing. Transferring wealth from a time when one is rich to a time when one is poor increases the utility when being poor (A) by a larger amount than it reduces it when being rich (B)

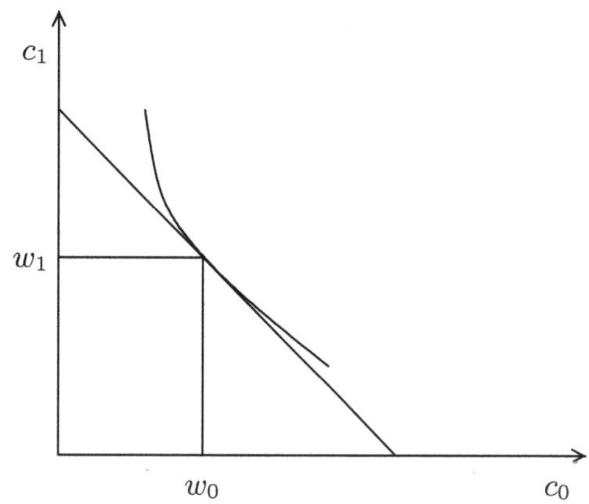

Fig. 4.10 The intertemporal consumption problem

Eliminating s, the two budget constraints can be combined into a single one written in terms of present values:

$$c_0 + \frac{1}{1+r}c_1 = w_0 + \frac{1}{1+r}w_1.$$

Hence, the decision problem can be displayed in a diagram showing the amount of consumption in both periods, as Fig. 4.10 does.

The first order condition to this problem is:

$$\frac{u'(c_0)}{\delta u'(c_1)} = (1+r).$$

Thus differences in the time preference and the interest rate are compensated by the marginal utilities. If you discount future consumption by more than the interest rate r, then you go for a higher consumption today than tomorrow. For the logarithmic utility this leads to a simple theory of interest rates:

$$1 + r = (1+g)/\delta, \quad \text{where} \quad c_1 = (1+g)c_0.$$

Hence, g is the growth rate of consumption. That is to say, interest rates increase if people become less patient and if consumption growth increases. The latter depends on the business cycle. In general interest rates increase when the growth of the GDP is strong and falling interest rates may be a signal for a recession. Note that r is the real rate of interest. Nominal interest rates do of course depend on inflation rates, as well.

If we compare this model with real data, we will see one prominent effect which is not captured: the interest rate for long-term investments is nearly always larger than for short-term investments. The interest rate as function of investment horizon is called the *yield curve*. The yield curve is usually increasing. One explanation for this effect is that short-term bonds are preferred, since their value fluctuates less, e.g. when interest rates change. Expected interest rates are typically higher than real interest rates.

We will come back to this topic in the next chapter and apply a multi-periods model to discuss the shape of the yield curve in more detail.

4.4 Special Cases: CAPM, APT and Behavioral CAPM

The general model that we have derived above can be used to find simple derivations for the CAPM, APT and a behavioral version of the CAPM, the B-CAPM. In all of these cases, diversification is the central motive for trading on financial markets.

In the following, we assume that the consumption in the first period is already decided. Moreover, we assume that all agents agree on the probabilities of occurrence of the states, $\text{prob}_s, s = 1, \ldots, S$. This assumption then separates diversification also from betting (compare Chap. 3). We will show that the CAPM can be embedded into our model as a special case. Afterwards, we derive its main conclusion, the Security Market Line (SML), from the financial markets equilibrium in economics terms (prices and quantities). As a result we recover the CAPM formula (3.5).

In the general two-period model outlined above, the CAPM is given by the following five assumptions:

Assumption 4.9

(i) There exists a risk-free asset, i.e., $(1, \ldots, 1)' \in \text{span}\,\{A\}$.
(ii) There is no first period consumption nor first period endowment.
(iii) Endowments are spanned, i.e., $(w_1^i, \ldots, w_S^i)' \in \text{span}\,\{A\}$, $i = 1, \ldots, I$.
(iv) Expectations are homogeneous, i.e., $\text{prob}_s^i = \text{prob}_s$, $i = 1, \ldots, I$ *and* $s = 1, \ldots, S$.
(v) Preferences are mean-variance, i.e.,

$$U^i(c_1^i, \ldots, c_S^i) = V^i(\mu(c_1^i, \ldots, c_S^i), \sigma(c_1^i, \ldots, c_S^i)),$$

where

$$\mu(c_1^i, \ldots, c_S^i) = \sum_{s=1}^{S} \text{prob}_s c_s^i$$

$$\text{and} \quad \sigma^2(c_1^i, \ldots, c_S^i) = \sum_{s=1}^{S} \text{prob}_s (c_s^i - \mu(c_1^i, \ldots, c_S^i))^2.$$

4.4.1 Deriving the CAPM by 'Brutal Force of Computations'

Note that we have up to now always made the first of these assumptions. For the sake of completeness we state it explicitly since the risk-free asset plays a special role in the CAPM. To make use of this special role we need to separate the risk-free asset from the risky assets. To this end we introduce the following notation. For vectors and matrices we define $\boldsymbol{A} = (\boldsymbol{1}, \hat{\boldsymbol{A}})$ where $\hat{\boldsymbol{A}}$ is the $S \times K$ matrix of *risky* assets. By $\mu(\hat{\boldsymbol{A}}) = (\mu(\hat{\boldsymbol{A}}^0), \ldots, \mu(\hat{\boldsymbol{A}}^K))$ we denote the vector of mean payoffs of assets in a matrix $\hat{\boldsymbol{A}}$. Similarly, $COV(\hat{\boldsymbol{A}}) = (\text{cov}(\boldsymbol{A}^k, \boldsymbol{A}^j))_{k,j=1,\ldots,K}$ denotes (as before) the variance-covariance matrix associated with a matrix $\boldsymbol{A}$. Note that the variance of a portfolio of assets can be written as

$$\sigma^2(\hat{\boldsymbol{A}}\hat{\boldsymbol{\theta}}) = \hat{\boldsymbol{\theta}}'\hat{\boldsymbol{A}}'\boldsymbol{\Lambda}(\textbf{prob})\hat{\boldsymbol{A}}\hat{\boldsymbol{\theta}} - \mu(\hat{\boldsymbol{A}}\hat{\boldsymbol{\theta}})\mu(\hat{\boldsymbol{A}}\hat{\boldsymbol{\theta}})' = \hat{\boldsymbol{\theta}}' \,\text{cov}(\hat{\boldsymbol{A}})\hat{\boldsymbol{\theta}}.$$

Equipped with this notation, we analyze the decision problem of a mean-variance agent, in a setting where there is no first period consumption and endowments are spanned:

$$\max_{\hat{\boldsymbol{\theta}}^i \in \mathbb{R}^{K+1}} V^i(\mu(\boldsymbol{c}^i), \sigma^2(\boldsymbol{c}^i)) \quad \text{such that} \quad \sum_{k=0}^{K} q^k \hat{\theta}^{i,k} = \sum_{k=0}^{K} q^k \theta_A^{i,k} = w^i,$$

where $c_s^i := \sum_{k=0}^{K} A_s^k \hat{\theta}^{i,k}$, $s = 1, \ldots, S$.

Recall that we defined the risk-free rate by $q^0 := 1/R_f$. From the budget equation we can then express the units of the risk-free asset held by $\hat{\theta}^0 = R_f(w^i - \hat{\boldsymbol{q}}'\hat{\boldsymbol{\theta}})$. Hence, we can eliminate the budget restriction and re-write the maximization problem as

$$\max_{\hat{\boldsymbol{\theta}}^i \in \mathbb{R}^K} V^i\left(R_f w^i + (\mu(\hat{\boldsymbol{A}}) - R_f\hat{\boldsymbol{q}})'\hat{\boldsymbol{\theta}}^i, \sigma^2(\hat{\boldsymbol{A}}\hat{\boldsymbol{\theta}}^i)\right).$$

The first order condition is[21]: $\mu(\hat{\boldsymbol{A}}) - R_f \hat{\boldsymbol{q}} = \rho^i \operatorname{cov}(\hat{\boldsymbol{A}})\hat{\boldsymbol{\theta}}^i$, where $\rho^i := \frac{\partial_\sigma V^i}{\partial_\mu V^i}(\mu, \sigma^2)$ is the agent's degree of risk aversion.[22] Solving for the portfolio we obtain

$$\hat{\theta}^i = \frac{1}{\rho^i} COV(\hat{\boldsymbol{A}})^{-1}(\mu(\hat{\boldsymbol{A}}) - R_f \hat{\boldsymbol{q}}).$$

From the first order condition we see that any two different agents, i and i', will form portfolios whose ratio of risky assets, $\hat{\theta}^{i,k}/\hat{\theta}^{i,k'} = \hat{\theta}^{i',k}/\hat{\theta}^{i',k'}$, are identical. This is because the first order condition is a linear system of equations differing across agents only by a scalar, ρ^i. This is again the *two-fund separation property*, since every agent's portfolio is composed out of two funds, the risk-free asset and a composition of risky assets that is the same for all agents, i.e., $\hat{\boldsymbol{\theta}}^i = (\hat{\theta}^{i,0}, \hat{\theta}^{i,1}\hat{\boldsymbol{\theta}})$, $i = 1, \ldots, I$.

Dividing the first order condition by ρ^i and summing up over all agents, we obtain

$$\Big(\sum_i \frac{1}{\rho^i}\Big)\Big(\mu(\hat{\boldsymbol{A}}) - R_f \hat{\boldsymbol{q}}\Big) = \operatorname{cov}(\hat{\boldsymbol{A}}) \sum_i \hat{\boldsymbol{\theta}}^i.$$

From the equality of demand and supply of assets we know that $\sum_i \hat{\boldsymbol{\theta}}^i = \sum_i \boldsymbol{\theta}^i_A =: \hat{\boldsymbol{\theta}}^M$, where the sum of all assets available is denoted by asset M, the market portfolio. Accordingly, denote the market portfolio's payoff by $\hat{\boldsymbol{A}}^M = \hat{\boldsymbol{A}}\hat{\boldsymbol{\theta}}^M$ and let the price of the market portfolio be $\hat{q}^M = \hat{\boldsymbol{q}}'\hat{\boldsymbol{\theta}}^M$. Then we get:

$$\Big(\mu(\hat{\boldsymbol{A}}) - R_f \hat{\boldsymbol{q}}\Big) = \Big(\sum_i \frac{1}{\rho^i}\Big)^{-1} \operatorname{cov}(\hat{\boldsymbol{A}})\hat{\boldsymbol{\theta}}^M.$$

Multiplying both sides with the market portfolio yields an expression from which we can derive the harmonic mean of the agents' risk aversions:

$$\Big(\sum_i \frac{1}{\rho^i}\Big)^{-1} = \frac{\Big(\mu(\hat{\boldsymbol{A}}^M) - R_f \hat{q}^M\Big)}{\sigma^2(\hat{\boldsymbol{A}}^M)}.$$

[21]We assume that the mean-variance utility function $V^i(\mu, \sigma)$ is quasi-concave so that the first order condition is necessary and sufficient to describe the solution of the maximization problem. This is, for example, the case for the standard mean-variance function $V^i(\mu, \sigma) := \mu - \frac{\rho^i}{2}\sigma^2$, since it is even concave.

[22]Note that $\frac{\partial_\sigma V^i}{\partial_\mu V^i}$ is the slope of the indifference curve in a diagram with the mean as a function of the standard deviation.

Substituting this back into the former equation, we finally get the asset pricing rule:

$$R_f\hat{q} = \mu(\hat{A}) - \frac{\left(\mu(\hat{A}^M) - R_f\hat{q}^M\right)}{\sigma^2(\hat{A}^M)}\operatorname{cov}(\hat{A},\hat{A}^M).$$

Hence, the price of any asset k is equal to its discounted expected payoff, adjusted by the covariance of its payoffs to the market portfolio. Writing this more explicitly we have derived:

$$q^k = \frac{\mu(A^k)}{R_f} - \frac{\operatorname{cov}(A^k,A^M)}{\operatorname{var}(A^M)}\left(\frac{\mu(A^M)}{R_f} - q^M\right).$$

We see that the present price of an asset is given by its expected payoff discounted to the present minus a risk premium that increases the higher the covariance to the market portfolio. This is a nice asset pricing rule in economic terms and it is quite easy to derive the analog in finance terms. To this end multiply the resulting expression by R_f and divide it by q^k and q^M. We then obtain the by now well-known expression relating the asset excess returns to the excess return of the market portfolio:

$$\mu(R^k) - R_f = \beta^k(\mu(R^M) - R_f) \quad \text{where} \quad \beta^k = \frac{\operatorname{cov}(R^k,R^M)}{\sigma^2(R^M)},$$

which we have already seen in Sect. 3.2.

Being equipped with the economic and the finance version of the SML we can revisit the claim based on the finance SML that increasing the systematic risk of an asset is a good thing for the asset according to the SML since it increases its returns. This suggests that a hedge fund could do better than a mutual fund by simple taking more risk. The logic of the CAPM is quite the opposite: increasing the risk, the investors do *require* a higher return on the asset. The economic SML reveals that this is obviously not a good thing for the shares since the investors' demand for a higher return will be satisfied by a *decreased price*. Hence, the value of the hedge fund decreases!

What does the SML tell us about the likelihood ratio process? Recall from the general risk-return decomposition that

$$\mu(R^k) - R_f = -\operatorname{cov}(\ell,R^k), \qquad k = 1,\cdots,K.$$

Similarly the SML yields

$$\mu(R^k) - R_f = \operatorname{cov}(R^M,R^k)\frac{\mu(R^M) - R_f}{\sigma^2(R^M)}.$$

Thus we get

$$-\operatorname{cov}(\ell, R^k) = \operatorname{cov}(R^M, R^k)\frac{\mu(R^M) - R_f}{\sigma^2(R^M)}.$$

Hence, the likelihood ratio process is a linear functional of the market return $\ell = a - bR^M$ for some parameters a, b, where $b = (\mu(R^M) - R_f)/\sigma^2(R^M)$ and a is obtained from $\mu(\ell) = a - b\mu(R^M) = 1$. Thus $a = 1 + b\mu(R^M)$.[23]

4.4.2 Deriving the CAPM from the Likelihood Ratio Process

So far we have derived the SML in our general model using the specific assumptions (i)–(iv) by explicitly computing the agent's asset demand. In the following we derive it based on the likelihood ratio process. It turns out that this derivation is more easily generalizable to situations with background risk or non-standard preferences.

To begin, let us show that in the CAPM the likelihood ratio process has to be a linear combination of the risk-free asset and the market portfolio[24]: $\ell = a\mathbf{1} + bR^M$, for two scalars a and b. Here $\mathbf{1}$ denotes the risk-free payoff and R^M the market portfolio. Recall that

$$\begin{aligned} R^M &= \sum_{k=1}^{K} R^k \lambda^{M,k} = \sum_{k=1}^{K} \frac{A^k}{q^k} \frac{\sum_{i=1}^{I} q^k \theta_A^{i,k}}{\sum_{k=1}^{K} \sum_{i=1}^{I} q^k \theta_A^{i,k}} \\ &= \sum_{k=1}^{K} \frac{A^k \sum_{i=1}^{I} \theta_A^{i,k}}{\sum_{k=1}^{K} q^k \sum_{i=1}^{I} \theta_A^{i,k}} \\ &=: \frac{A^M}{q^M}. \end{aligned}$$

Hence, $\ell \in \operatorname{span}\{1, R^M\} \Leftrightarrow \ell \in \operatorname{span}\{1, A^M\}$.

Note that if we had shown $\ell = a\mathbf{1} + bR^M$ then the SML-formula does indeed follow: Inserting $a\mathbf{1} + bR^M$ for ℓ in $\mathbb{E}_p(R^k) = R_f - \operatorname{cov}_p(R^k, \ell)$ gives $\mathbb{E}_p(R^k) = R_f - b\operatorname{cov}_p(R^k, R^M)$. Applying this formula for $k = M$, one can determine b and substitute it back into the expression obtained before so that the SML follows. We have done this step already two times before in Chap. 3, so there is no point to repeat it here.

[23] Note that the linearity of the likelihood ratio process also holds in the CAPM with heterogeneous beliefs (see Sect. 3.3) on expected returns if we define the likelihood ratio process with respect to the average belief of the investors.

[24] In the exercise book we derive the CAPM in yet another way. Assume quadratic utility functions and then show that the likelihood ratio process being the marginal rates of substitution becomes proportional to a linear combination of the risk-free asset and the market portfolio.

But why should $\ell = a\mathbf{1} + bR^M$, i.e., $\ell \in \text{span}\{1, R^M\}$ or equivalently $\ell \in \text{span}\{1, A^M\}$ hold in the CAPM? Recall the optimization problem of a mean-variance consumer[25]:

$$\max_{\hat{\boldsymbol{\theta}}^i \in \mathbb{R}^{K+1}} V^i\left(c_0^i, \mu(\boldsymbol{c}_1^i), \sigma^2(\boldsymbol{c}_1^i)\right) \quad \text{such that} \quad c_0^i + \sum_{k=0}^{K} q^k \hat{\theta}^{i,k} = w_0^i + \sum_{k=0}^{K} q^k \theta_A^{i,k},$$

where $\boldsymbol{c}_1^i = \sum_{k=0}^K \boldsymbol{A}^k \hat{\theta}^{i,k}$. In terms of state prices the budget restriction can be written as[26]:

$$c_0^i + \sum_{s=1}^{S} \pi_s c_s^i = w_0^i + \sum_{s=1}^{S} \pi_s w_s^i \quad \text{and} \quad (\boldsymbol{c}_1^i - \boldsymbol{w}_1^i) \in \text{span}\{\boldsymbol{A}\},$$

where the latter is equivalent to $\boldsymbol{c}^i \in \text{span}\{\boldsymbol{A}\}$ since we assumed that endowments are spanned. Using the likelihood ratio process, the budget restriction becomes:

$$c_0^i + \sum_{s=1}^{S} p_s \frac{\ell_s}{R_f} c_s^i = w_0^i + \sum_{s=1}^{S} p_s \frac{\ell_s}{R_f} w_s^i \quad \text{and} \quad \boldsymbol{c}_1^i \in \text{span}\{\boldsymbol{A}\},$$

$$\Leftrightarrow \quad c_0^i + \frac{1}{R_f}\mathbb{E}_p(\ell c^i) = w_0^i + \frac{1}{R_f}\mathbb{E}_p(\ell w^i) \quad \text{and} \quad \boldsymbol{c}_1^i \in \text{span}\{\boldsymbol{A}\}.$$

We will show that $\boldsymbol{c}_1^i \in \text{span}\{\mathbf{1}, \ell\}$ so that aggregating over all agents we get $\ell \in \text{span}\{\mathbf{1}, A^M\}$. To this end, suppose $\boldsymbol{c}_1^i = a^i\mathbf{1} + b^i\ell + \xi^i$, where $\xi^i \notin \text{span}\{\mathbf{1}, \ell\}$. The latter means $\mathbb{E}_p(\mathbf{1}\xi^i) = \mathbb{E}_p(\ell\xi^i) = 0$. Since $\boldsymbol{c}_1^i$ is an optimal portfolio it satisfies the budget constraint and $\boldsymbol{c}_1^i \in \text{span}\{\boldsymbol{A}\}$. Since $\mathbb{E}_p(\ell\xi^i) = 0$, also $a^i\mathbf{1} + b^i\ell$ satisfies the budget constraint and can always be chosen in the span of $\boldsymbol{A}$ since any component orthogonal to the span in the sense of $\mathbb{E}_p(\ell A) = 0$ does not change the value of the assets. This is because due to the no-arbitrage condition any component that is orthogonal to span $\{\boldsymbol{A}\}$ does not contribute to $\boldsymbol{q}$, i.e., $a^i\mathbf{1} + b^i\ell \in \text{span}\{\boldsymbol{A}\}$. So is it worthwhile to include ξ^i in the consumption stream? Note that ξ^i does not increase the mean consumption, because $\mathbb{E}_p(\mathbf{1}\xi^i) = 0$. However, ξ^i increases the variance of the consumption, since

$$\text{var}_p(c^i) = \text{var}_p(a^i\mathbf{1} + b^i\ell + \xi^i) = (b^i)^2\,\text{var}_p(\ell) + \text{var}_p(\xi^i) + 2b^i\,\text{cov}_p(\ell, \xi^i)$$

[25]Note that the lower index 1 in the consumption variable denotes the period 1, i.e., $\boldsymbol{c}_1^i$ is the vector $(c_1^i, \ldots, c_S^i)$, which should not be confused with the consumption in state s: c_s^i, $s = 1$.

[26]Insert $\boldsymbol{q}' = \boldsymbol{\pi}'\boldsymbol{A}$ from the no-arbitrage condition and substitute to obtain this result.

and

$$\mathrm{cov}_p(\ell, \xi^i) = \mathbb{E}_p(\ell \xi^i) - \mathbb{E}_p(\ell)\mathbb{E}_p(1\xi^i) = 0.$$

Hence, it is best to choose $\xi^i = 0$ and we are done with the proof. Thus, the CAPM is still a special case of our model.

4.4.3 Arbitrage Pricing Theory (APT)

In the CAPM, the Beta measures the sensitivity of the security's returns to the market return. The model relies on restrictive assumptions about agents' preferences and their endowments. The Arbitrage Pricing Theory (APT) can be seen as a generalization of the CAPM in which the likelihood ratio process is a linear combination of many factors. Let $R^1, \ldots, R^F$ be the returns that the market rewards for holding the J factors $j = 1, \ldots, J$, i.e., let $\boldsymbol{\ell} \in \mathrm{span}\{\mathbf{1}, \boldsymbol{R}^1, \ldots, \boldsymbol{R}^F\}$. Following the same steps as before we get[27]

$$\mathbb{E}_p(R^k) - R_f = \sum_{j=1}^{J} b^j \left(\mathbb{E}_p(R^j) - R_f\right).$$

This gives more flexibility for an econometric regression. Seen this way, in a model with homogeneous expectations, for example, any alpha that is popping up in such a regression only indicates that the factors used in the regression did not completely explain the likelihood ratio process. Hence, there must be other factors that should have been added in the regression. This is nice from an econometric point of view, but can we give an economic foundation to it? In the following section we will do this.

4.4.4 Deriving the APT in the CAPM with Background Risk

The main idea in the following is to show that the APT can be thought of as a CAPM with background risk.

We need to prove that the likelihood ratio process is a linear combination of the risk-free asset and F mutually independent return factors i.e., $\boldsymbol{\ell} \in \mathrm{span}\left\{\mathbf{1}, \boldsymbol{R}^1, \ldots, \boldsymbol{R}^F\right\}$ with $\mathrm{cov}_p(R^f, R^{f'}) = 0$ for $f \neq f'$. Note that one of the factors may be the market itself, i.e., $f = M$ so that the APT is a true generalization of the CAPM. As before, assume that agents maximize a mean-variance utility function, but in contrast to before, we do not make the spanning assumption so that

[27] Please don't be confused: R^f denotes the return to factor f while R_f denotes the return to the risk-free asset!

consumption is also derived from exogenous wealth that is not related to the asset payoffs:

$$\max_{\hat{\theta}^i \in \mathbb{R}^{K+1}} V^i\big(c_0^i, \mu(\boldsymbol{c}_1^i), \sigma^2(\boldsymbol{c}_1^i)\big) \quad \text{such that} \quad c_0^i + \sum_{k=0}^{K} q^k \hat{\theta}^{i,k} = w_0^i + \sum_{k=0}^{K} q^k \theta_A^{i,k},$$

where $\boldsymbol{c}_1^i = \boldsymbol{w}_{\perp 1}^i + \sum_{k=0}^{K} \boldsymbol{A}^k \hat{\theta}^{i,k}$. In terms of state prices the budget restriction can be written as:

$$c_0^i + \sum_{s=1}^{S} \pi_s^* c_s^i = w_0^i + \sum_{s=1}^{S} \pi_s^* w_s^i \quad \text{and} \quad (\boldsymbol{c}_1^i - \boldsymbol{w}_{\perp 1}^i) \in \text{span}\,\{\boldsymbol{A}\}\,.$$

Using the likelihood ratio process, the budget restriction becomes:

$$c_0^i + \sum_{s=1}^{S} p_s \ell_s c_s^i = w_0^i + \sum_{s=1}^{S} p_s \ell_s w_s^i \quad \text{and} \quad (\boldsymbol{c}_1^i - \boldsymbol{w}_{\perp 1}^i) \in \text{span}\,\{\boldsymbol{A}\}\,,$$

where the first restriction can also be written as $c_0^i + \mathbb{E}_p(\ell c^i) = w_0^i + \mathbb{E}_p(\ell w^i)$. Next, we will show that $(\boldsymbol{c}_1^i - \boldsymbol{w}_{\perp 1}^i) \in \text{span}\,\{\boldsymbol{1}, \boldsymbol{\ell}\}$. To this end, suppose $(\boldsymbol{c}_1^i - \boldsymbol{w}_{\perp 1}^i) = a^i\boldsymbol{1} + b^i\boldsymbol{\ell} + \boldsymbol{\xi}^i$, where $\boldsymbol{\xi}^i \notin \text{span}\,\{\boldsymbol{1}, \boldsymbol{\ell}\}$, i.e., $\mathbb{E}_p(1\xi^i) = \mathbb{E}_p(\ell\xi^i) = 0$. Since $\boldsymbol{c}_1^i$ is an optimal portfolio it satisfies the budget and the spanning constraint. Now what would happen if we canceled $\boldsymbol{\xi}^i$ from the agent's demand? Since $\mathbb{E}_p(\ell\xi^i) = 0$, also $a^i\boldsymbol{1} + b^i\boldsymbol{\ell}$ satisfies the budget constraint and obviously $(a^i\boldsymbol{1} + b^i\boldsymbol{\ell}) \in \text{span}\,\{\boldsymbol{A}\}$ since both, the risk-free asset and the likelihood ratio process, are spanned.[28] So is it worthwhile to include $\boldsymbol{\xi}^i$ in the consumption stream? Note that $\boldsymbol{\xi}^i$ does not increase the mean consumption, because $\mathbb{E}_p(1\xi^i) = 0$. However, $\boldsymbol{\xi}^i$ increases the variance of the consumption, since

$$\text{var}_p(c^i) = \text{var}_p(a^i 1 + b^i \ell + \xi^i) = (b^i)^2\, \text{var}_p(\ell) + \text{var}_p(\xi^i) + 2b^i\, \text{cov}_p(\ell, \xi^i)$$

and

$$\text{cov}_p(\ell, \xi^i) = \mathbb{E}_p(\ell\xi^i) - \mathbb{E}_p(\ell)\mathbb{E}_p(1\xi^i) = 0.$$

Hence, it is best to choose $\boldsymbol{\xi}^i = \boldsymbol{0}$ and we are done with the main part of the proof. It remains to argue that the factors can explain the likelihood ratio process: aggregating $(\boldsymbol{c}_1^i - \boldsymbol{w}_{\perp 1}^i) = a^i\boldsymbol{1} + b^i\boldsymbol{\ell}$ over all agents gives $\boldsymbol{\ell} \in \text{span}\{\boldsymbol{1}, \boldsymbol{R}^M, \tilde{\boldsymbol{R}}^1, \ldots, \tilde{\boldsymbol{R}}^F\}$, where $\tilde{\boldsymbol{R}}^1, \ldots, \tilde{\boldsymbol{R}}^F$ are F factors that span the non-market risk embodied in the aggregate

[28] The likelihood ratio process can always be chosen in the span of $\boldsymbol{A}$ since any component orthogonal to the span in the sense of $\mathbb{E}_p(\ell A) = 0$ does not change the value of the assets. This is due to the no-arbitrage condition. Moreover, the risk-free asset is the first asset in $\boldsymbol{A}$.

wealth:

$$\sum_{i=1}^{I} w^i_{\perp 1} = \sum_{f=1}^{F} \beta^f \tilde{A}^f .$$

4.4.5 Behavioral CAPM

Finally, we want to show how Prospect Theory can be included into the CAPM to build a Behavioral CAPM, a B-CAPM, by adding behavioral aspects to the consumption based CAPM. To do so we use the C-CAPM for market aggregates and assume that the investor has the quadratic Prospect Theory utility

$$v(c_s - RP) := \begin{cases} (c_s - RP) - \frac{\alpha^+}{2}(c_s - RP)^2 & \text{, if } c_s \geq RP, \\ \lambda((c_s - RP) - \frac{\alpha^-}{2}(c_s - RP)^2 & \text{, if } c_s < RP, \end{cases}$$

and no probability weighting.

A piecewise quadratic utility is convenient because it contains the CAPM as a special case when $\alpha^+ = \alpha^-$ and $\lambda = 1$.[29] To derive the B-CAPM it is best to start from the general risk-return decomposition $\mathbb{E}(R^k) = R_f - \text{cov}(R^k, \ell)$. The likelihood ratio process for the piecewise quadratic utility is:

$$\delta^i u'(c_0)\ell(c_s) = \begin{cases} 1 - \alpha^+ c_s & \text{, if } c_s \geq RP, \\ \lambda(1 - \alpha^- c_s) & \text{, if } c_s < RP. \end{cases}$$

Now suppose that $c_s = R^M$ holds[30] and that the reference point is the risk-free rate R_f. We abbreviate $\hat{\alpha}^\pm := \alpha^\pm/(\delta^i u'(c_0))$ and denote

$$P(R^M - R_f) := \sum_{R^M_s > R_f} p_s,$$

$$\text{cov}^+(R^k, R^M) := \sum_{R^M_s > R_f} \frac{p_s}{P(R^M - F_f)}(R^k_s - \mathbb{E}(R^k))(R^M_s - \mathbb{E}(R^M)),$$

$$\text{cov}^-(R^k, R^M) := \sum_{R^M_s < R_f} \frac{p_s}{P(R^M - F_f)}(R^k_s - \mathbb{E}(R^k))(R^M_s - \mathbb{E}(R^M)).$$

---

[29] Compare Sect. 2.5 where we have seen that mean-variance preferences can be seen as a special case of EUT with quadratic utility function.

[30] See Sect. 4.6 for a justification.

Then on denoting conditional expectations by a plus sign for market returns above the risk-free rate and by a minus sign for market returns below the risk-free rate, the general risk-return decomposition is

$$P(R^M > R_f)\big(\mathbb{E}^+(R^k) - R_f + \hat{\alpha}^+ \operatorname{cov}^+(R^k, R^M)\big) \\ + (1 - P(R^M > R_f))\lambda\big(\mathbb{E}^-(R^k) - R_f + \hat{\alpha}^- \operatorname{cov}^-(R^k, R^M)\big) = 0.$$

Again, we see that if $\alpha^+ = \alpha^-$ and $\beta = 1$ then on substituting the alpha by applying the formula obtained for $k = M$, we get the CAPM. Furthermore, the B-CAPM suggests two aspects. First, that the risk factors of the CAPM may be different for up and down markets and that it may be wise to increase the returns in the loss states by the loss aversion.

4.5 Pareto Efficiency

The word efficiency has two meanings in finance. First, it is associated with *informational efficiency* of financial markets which has been postulated by Eugene Fama in his famous Efficient Market Hypothesis, EMH (see also [Ban81]). According to the EMH one cannot make excess returns based on price information, "Technical Analysis" or "Chartism", since in any point in time prices already reflect all public information. In the CAPM with heterogeneous beliefs we have seen that a learning process along which agents learn to invest actively or passively ultimately leads to a situation in which the prices are determined by the information of the best informed agent. In the short run this may not (or not yet) be the case. We will discuss informational efficiency in more details in Chap. 7.

The meaning of efficiency that we want to analyze now is different. It asks whether the allocation of assets that results in a financial market equilibrium could be improved such that nobody's utility is diminished while somebody benefits. This notion of efficiency is called *allocational efficiency*. Since it was first proposed by Vilfredo Pareto it is also called *Pareto-efficiency*. Pareto efficiency is a main subject in welfare economics. But why is this concept interesting in finance? Well, if asset allocations were Pareto-efficient then this would help to dramatically simplify our modeling of financial market equilibria. Pareto-efficiency requires that at the allocation all agents have the same marginal rates of substitution, as Fig. 4.8 already showed.[31] However, we have seen that the marginal rates of substitutions are the discount factors with which agents value future asset returns. Hence, if allocations are Pareto-efficient then all agents agree on the valuation of all possible returns, regardless whether they are already traded in the market or not. Moreover, as

[31]Strictly speaking, this is true only if efficient allocations do not lie on the boundary of the Edgeworth Box. Assumptions like that marginal utility is unbounded as consumption converges to the boundary of the Edgeworth Box are necessary here.

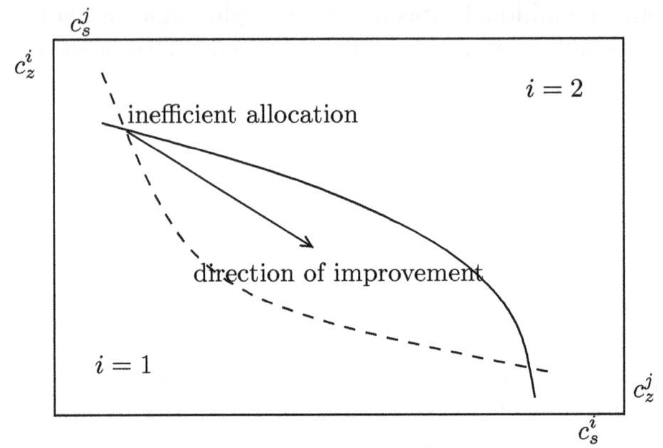

Fig. 4.11 The Edgeworth Box displays an inefficient allocation

we will see in the next section, when allocations are efficient, aggregation of the heterogeneous agent economy into a representative agent with a utility function that is of the same type as the individual agents' utilities is possible. Hence, instead of solving a system of decision problems, a single decision problem will be sufficient to determine asset prices (Fig. 4.11).

Before we can give the formal proof of the allocational efficiency of equilibria it is convenient to use the no-arbitrage condition to rewrite the decision problem in terms of state prices instead of asset prices. This will make the problem very similar to the standard general equilibrium model of microeconomics. We start with the decision problem of an investor:

$$\max_{\theta \in \mathbb{R}^{K+1}} U(c_0, \ldots, c_s) \quad \text{such that} \quad c_0 + \sum_{k=0}^{K} q^k \theta^k = w_0$$

$$\text{and} \quad c_s = \sum_{k=0}^{K} A_s^k \theta^k + w_s \geq 0, \quad s = 1, \ldots, S,$$

Substituting the asset prices from the no-arbitrage condition

$$\pi_0 q^k = \sum_{s=1}^{S} \pi_s A_s^k, \quad k = 0, \ldots, K,$$

the budget restrictions can be rewritten as:

$$\pi_0 c_0 + \sum_{s=1}^{S} \pi_s c_s = \pi_0 w_0 + \sum_{s=1}^{S} \pi_s w_s,$$

and

$$c_s - w_s = \sum_{k=0}^{K} A_s^k \theta^k, \quad s = 1, \ldots, S, \quad \text{for some } \theta.$$

The second restriction is known as the spanning constraint. It can also be written as: $(\boldsymbol{c}_1 - \boldsymbol{w}_1) \in \text{span}\,\{\boldsymbol{A}\}$.

In the notion of Pareto-efficiency one compares the equilibrium allocation with other feasible allocations. An allocation is feasible if it is compatible with the consumption sets[32] of the agents and it does not use more resources than there are available in the economy. When would we expect that equilibrium allocations are Pareto-efficient? A natural condition would be that in a certain sense agents can bet on all states of the world. Or, to put it the other way around, if some bets are *not* possible then it may happen that the marginal rates are not equalized. Hence, completeness of markets is a sufficient condition for allocational efficiency. However, as we show in the exercise book, markets may be Pareto-efficient even in the case of incomplete markets, provided utility functions are sufficiently similar to each other. The main result of this section is based on complete markets. It is stated in the following theorem that in economics is called the First Welfare Theorem:

Theorem 4.10 (First Welfare Theorem) *In a complete financial market the allocation of consumption streams, $(\boldsymbol{c}^i)_{i=1}^I$, is Pareto-efficient, i.e., there does not exist an alternative attainable allocation of consumption $(\hat{\boldsymbol{c}}^i)_{i=1}^I$ such that no consumer is worse off and some consumer is better off, i.e., $U^i(\hat{\boldsymbol{c}}^i) \geq U^i(\boldsymbol{c}^i)$ for all i and $U^i(\hat{\boldsymbol{c}}^i) > U^i(\boldsymbol{c}^i)$ for some i.*

Proof Suppose $(\hat{\boldsymbol{c}}^i)_{i=1}^I$ is an attainable allocation that is Pareto-better than the financial market allocation, i.e., $U^i(\hat{\boldsymbol{c}}^i) \geq U^i(\boldsymbol{c}^i)$ for all i and $U^i(\hat{\boldsymbol{c}}^i) > U^i(\boldsymbol{c}^i)$ for some i. Why did the agents not choose $(\hat{\boldsymbol{c}}^i)$? Because it is more expensive, i.e., $\sum_s \pi_s \hat{c}_s^i > \sum_s \pi_s c_s^i$. Adding across consumers gives:

$$\sum_{i=1}^{I} \sum_{s=0}^{S} \pi_s \hat{c}_s^i > \sum_{i=1}^{I} \sum_{s=0}^{S} \pi_s c_s^i.$$

But since $\sum_i \hat{\boldsymbol{c}}^i = \sum_i \boldsymbol{w}^i = \sum_i \boldsymbol{c}^i$, this cannot be true! □

In the exercise book we show that when markets are incomplete some version of the First Welfare Theorem is still possible. When restricting attainable allocations to those allocations that are compatible with the agents' consumption sets, that do

[32] So far we did never specify the consumption sets. This typically is the set of non-negative vectors in $\mathbb{R}^{S+1}$, since negative consumption does not have an economic interpretation. The utility functions need only be defined on the consumption sets.

not need more than the given total resources and that are attainable by trade on the given asset structure A, we can again conclude that equilibrium allocations cannot be improved in the sense of Pareto by any other allocation. This property, however, depends on the assumption of two periods, so we should not get too enthusiastic about it, since ultimately we are interested in a multi-period model.

We also remark that financial market equilibria can be Pareto-efficient even if markets are not complete. An example for this is the CAPM with homogeneous beliefs. By the two-fund separation property the utility gradients lie in a two dimensional subspace and trading mean for variance is sufficient to make them parallel. This example is however not robust since perturbing initial endowments or utility functions leads to a violation of the spanning assumption. Such perturbations of incomplete markets lead to Pareto-inefficiency (see [MQ02]).

4.6 Aggregation

Determining asset prices from the idea that heterogeneous agents trade with each other may be an intellectually plausible point of view, but for practical questions like "what drives asset prices" this may be too complicated since nobody can possibly hope to get information on every agent's utility function. If the principle of utility maximization is useful for questions of aggregate results like market prices then it would be most convenient if one had to look at one decision problem only. But then one needs to ask whom or what does this single decision problem represent. To be more precise, in this section we answer the following questions of increasing difficulty:

1. Under which conditions can prices which are market aggregates be generated by aggregate endowments (consumption) and some aggregate utility function?
2. Moreover, in this case, is it possible to find an aggregate utility function that has the same properties as the individual utility functions?
3. Finally, is it possible to use the aggregate decision problem to determine asset prices "out of sample", i.e., after some change, e.g., of the dividend payoffs?

4.6.1 Anything Goes and the Limitations of Aggregation

Figure 4.12 gives the main intuition on the aggregation problem. At the equilibrium allocation asset prices are determined by the trade of two agents, however, they could also be thought of as being derived from a single utility function that is maximized over the budget set based on aggregate endowments (the upper right corner of the Edgeworth Box).

Actually, the answer to our first question is even simpler since it does not need any information on the individual's utility functions. Any asset price vector that is arbitrage-free can also be generated by a single utility function. Hence, in a sense, anything goes!

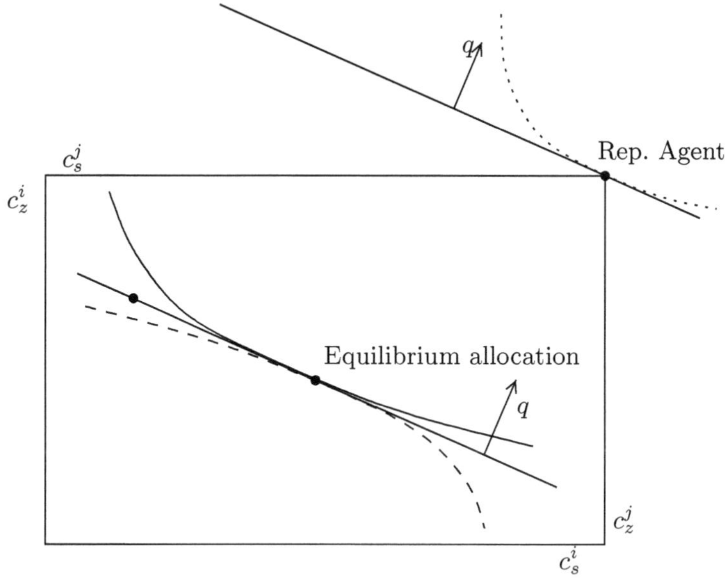

Fig. 4.12 Aggregating individual decision problems into one representative agent

Theorem 4.11 (Anything Goes Theorem) *Let $\boldsymbol{q}$ be an arbitrage-free asset price vector for the market structure $\boldsymbol{A}$. Then there exists an economy with a representative consumer maximizing an expected utility function such that $\boldsymbol{q}$ is the equilibrium price vector of this economy.*

Proof Since $\boldsymbol{q}$ is arbitrage-free there exists some risk neutral probability $\boldsymbol{\pi} \gg \boldsymbol{0}$ such that $\boldsymbol{q}' = \boldsymbol{\pi}'\boldsymbol{A}$. Choose then

$$U^R(c_0, \ldots, c_s) := c_0 + \sum_{s=1}^{S} \pi_s c_s.$$

At the prices $\boldsymbol{q}$ the representative agent will consume aggregate endowments, which can be seen immediately from the first order condition.[33] □

The argument made in this proof is the reason why the state price measure is also called the "risk neutral measure". It could be thought of as being derived in a risk neutral world, i.e., in an economy in which a single risk-neutral representative agent determines asset prices. Note, however, that every real agent in the economy might

[33] Obviously, one could also find a representative consumer with strictly concave utilities since one only needs to satisfy that his marginal rates of substitution at aggregate endowments coincide with the state prices.

be risk neutral, so that somehow the representative agent does not really *represent* the agents. This is the point of our question 2: "Is it possible to find an aggregate utility function that has the same properties as the individual utility functions?". For the answer of question 2, allocational efficiency will be quite useful, as has first been noticed by Constantinides [Con82].

Note first that Pareto-efficiency is equivalent to maximizing some *welfare function*. In other words, any Pareto efficient allocation can be obtained from the maximization of some welfare function in which the weights are chosen appropriately and the maximization of a welfare function results in a Pareto-efficient allocation. A *welfare function* assigns a social utility to each allocation. It is in a certain sense the analog of a utility function on consumption bundles in classical economies. We define the welfare function as an aggregate of individual utilities. Let $\gamma^i > 0$ be the weight of agent i in the social welfare function $\sum_{i=1}^{I} \gamma^i U^i(\boldsymbol{c}^i)$. The next argument shows that choosing the welfare weights $\gamma^i > 0$ equal to the reciprocal of the agents' marginal utility of consumption in period 0 attained in the financial market equilibrium,

$$\gamma^i = \frac{1}{\partial_0 U^i(\overset{*}{\boldsymbol{c}}{}^i)},$$

one can generate the equilibrium consumption allocation from the social welfare function: recall that under differentiability and boundary assumptions Pareto-efficiency implies

$$\frac{\nabla_1 U^1(\overset{*}{\boldsymbol{c}}{}^1)}{\partial_0 U^1(\overset{*}{\boldsymbol{c}}{}^1)} = \ldots = \frac{\nabla_1 U^I(\overset{*}{\boldsymbol{c}}{}^I)}{\partial_0 U^I(\overset{*}{\boldsymbol{c}}{}^I)} =: \pi.$$

Define

$$U^R(\boldsymbol{W}) := \sup_{c^1,\ldots,c^I} \left\{ \sum_{i=1}^{I} \gamma^i U^i(\boldsymbol{c}^i) \,\middle|\, \sum_{i=1}^{I} \boldsymbol{c}^i = \boldsymbol{W} \right\}$$

where $\gamma^i = 1/(\partial_0 U^i(\overset{*}{\boldsymbol{c}}{}^i))$. The first order condition for this maximization problem is $\gamma^1 \nabla U^1(\overset{*}{\boldsymbol{c}}{}^1) = \ldots = \gamma^I \nabla U^I(\overset{*}{\boldsymbol{c}}{}^I) =: \boldsymbol{\lambda}$ and[34] $\nabla U^R(\boldsymbol{W}) = \boldsymbol{\lambda}$, hence:

$$\nabla_1 U^R(\boldsymbol{W}) = \frac{\nabla_1 U^i(\overset{*}{\boldsymbol{c}}{}^i)}{\partial_0 U^i(\overset{*}{\boldsymbol{c}}{}^i)} \quad \text{and} \quad \partial_0 U^R(\boldsymbol{W}) = 1.$$

[34] This last claim is the so called envelope theorem.

Consider

$$\max_{\theta} U^R(\boldsymbol{c}^R) \quad \text{such that } \boldsymbol{c}^R - \boldsymbol{W} \leq \begin{pmatrix} -\boldsymbol{q}^* \\ \boldsymbol{A} \end{pmatrix} \boldsymbol{\theta}.$$

The first order condition is

$$\boldsymbol{q}^{*\prime} = \frac{\nabla_1 U^R(\overset{*}{\boldsymbol{c}}{}^R)'}{\partial_0 U^R(\overset{*}{\boldsymbol{c}}{}^R)}\boldsymbol{A} = \frac{\nabla_1 U^i(\overset{*}{\boldsymbol{c}}{}^i)'}{\partial_0 U^i(\overset{*}{\boldsymbol{c}}{}^i)}\boldsymbol{A} = \boldsymbol{\pi}'\boldsymbol{A}.$$

Note that $\boldsymbol{c}^R = \boldsymbol{W}$, the aggregate wealth of the economy.

Hence, we have found a "technique" to replace the individual utility functions by some aggregate utility function. In particular, we see that concavity of the individual utility functions is inherited by the aggregate utility function. Hence, as we argued above the likelihood ratio process should be decreasing. That is to say, postulating some utility function of the representative agent we can now test whether asset prices are in line with optimization by referring to aggregate consumption data.[35] But when does the aggregate utility function really represent the individuals? We now give a first result in this direction (others can be found in the exercise book): if all individual utility functions are of the expected utility type with common time preference and common beliefs, then the representative agent is also an expected utility maximizer with the same time preference and the same beliefs. Hence, our result shows that any heterogeneous set of risk aversions can be aggregated into one aggregate risk aversion. More precisely:

Proposition 4.12 *Assume that for all $i = 1, \ldots, I$ the utility functions u^i agree and that the time discounting δ is also independent of i. Moreover assume that the beliefs p_s, $s = 1, \ldots, S$, are homogeneous, i.e., let U^i be given by*

$$U^i(\boldsymbol{c}^i) = u^i(c_0^i) + \beta \sum_{s=1}^{S} p_s u^i(c_s^i) \quad \textit{for } i = 1, \ldots, I.$$

Then $U^R(\boldsymbol{c}^R) = u^R(\boldsymbol{c}^R) + \beta \sum_{s=1}^{S} p_s u^R(c_s^R)$, for some function $u^R: \mathbb{R} \to \mathbb{R}$.

Proof We use the definition of U^R:

$$U^R(\boldsymbol{W}) = \sup_{\boldsymbol{c}^1,\ldots,\boldsymbol{c}^I} \left\{ \sum_{i=1}^{I} \gamma^i U^i(\boldsymbol{c}^i) \,\middle|\, \sum_{i=1}^{I} \boldsymbol{c}^i = \boldsymbol{W} \right\}$$

[35] See Sect. 4.6.3 for empirical studies along this line.

where $\gamma^i = 1/(\partial_0 U^i(\overset{*}{c}{}^i))$ gives

$$U^R(\mathbf{W}) = \sup_{c^1,\dots,c^I} \left\{ \sum_{i=1}^{I} \gamma^i \left(u^i(c_0^i) + \beta \sum_{s=1}^{S} p_s u^i(c_s^i) \right) \middle| \sum_{i=1}^{I} \mathbf{c}^i = \mathbf{W} \right\}$$

$$= \sup_{c^1,\dots,c^I} \left\{ \underbrace{\textstyle\sum_{i=1}^{I} \gamma^i u^i(c_0^i)}_{u^R(W_0)} + \beta \sum_{s=1}^{S} p_s \underbrace{\textstyle\sum_{i=1}^{I} \gamma^i u^i(c_s^i)}_{u^R(W_s)} \middle| \sum_{i=1}^{I} \mathbf{c}^i = \mathbf{W} \right\}$$

$$= u^R(W_0^R) + \beta \sum_{s=1}^{S} p_s u^R(W_s^R). \qquad \square$$

Similar results are possible, for example for Prospect Theory preferences. Note that in the case of Prospect Theory the representative agent may not need to be risk loving over losses since this non-concavity of the utility gets smoothed out by the maximization as Fig. 4.13 suggests.[36]

This looks like wonderful news: taking the representative agent perspective one can even forget about non-concavities in the individual utility functions. This observation was first made in an article titled "Prospect Theory: Much Ado About Nothing!" [LL03]. So can we really forget about Prospect Theory just by aggregating the preferences of single agents? Well we should not get too enthusiastic since the representative agent technique has a natural limitation: it is generally not useful to tell us anything about asset prices that we do not know yet. More precisely, it is not useful for comparative statics, or "out of sample predictions". Indeed, as shown in the exercise book, basing one's investment decisions on the representative agents technique may result in severe losses, since asset prices would be predicted to go in the wrong direction.

Fig. 4.13 Smoothing out individual non-concavities on the aggregate

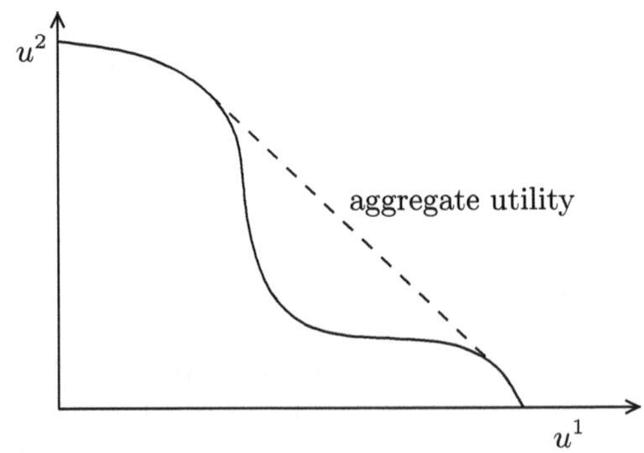

[36]For an in-depth treatment of this smoothing aggregation in general see [DDT80] and, for the case of Cumulative Prospect Theory, De Giorgi, Hens and Rieger [DGHR10].

This leads us to the final question of this section: "Is it possible to use the aggregate decision problem to determine asset prices 'out of sample', i.e., after some change, e.g., of the dividend payoffs?" If this is possible, some authors[37] say one gets "demand aggregation". This means that not only at the equilibrium point the representative agent demand function coincides with the sum of the individual demands, but it coincides for any prices. Demand aggregation is possible, however under quite restrictive assumptions. In Hens and Pilgrim [HP03] we find the following cases in which a positive answer to our third question is possible:

1. Identical utility functions and identical endowments
2. Quasi-linearity: $U^i(c_0^i, \ldots, c_S^i) = c_0^i + u^i(c_1^i, \ldots, c_s^i)$
3. Expected utility with common beliefs *and*
   (a) no-aggregate risk $\sum_{k=1}^{K} A_s^k = \sum_{k=1}^{K} A_z^k$ for all s, z *or*
   (b) complete markets *and*

   - CRRA and collinear endowments *or*
   - identical CRRA *or*
   - Quadratic utility functions

   Some of these results have been extended to incomplete markets, see [HP03].

We conclude this section by giving some example how in the representative agent utility the heterogeneous preferences get aggregated:

- Expected utility with common beliefs and no-aggregate risk:

$$U^i(c_0^i, \ldots, c_S^i) := u^i(c_0^i) + \beta^i \sum_{s=1}^{S} p_s u^i(c_s^i), \quad i = 1, \ldots, I,$$

  aggregates to

$$U^R(W_0, \ldots, W_S) = u^R(W_0) + \beta^R \sum_{s=1}^{S} p_s u^R(W_s)$$

  for any concave u^R.
- Expected utility with common beliefs and common time preference and quasi-linear quadratic preferences:

$$U^i(c_0^i, \ldots, c_S^i) := c_0^i + \beta \sum_{s=1}^{S} p_s \left(c_s^i - \tfrac{1}{2}\gamma^i (c_s^i)^2\right), \quad i = 1, \ldots, I,$$

[37] For example, Rubinstein [Rub74] or Constantinindes [Con82].

aggregates to

$$U^R(c_0^R, \ldots, c_S^R) = c_0^R + \beta \sum_{s=1}^{S} p_s \left(c_s^R - \tfrac{1}{2}\gamma^R (c_s^R)^2\right)$$

where

$$\gamma^R = \left(\sum_{i=1}^{I} \frac{1}{\gamma^i} \right)^{-1}.$$

- Expected logarithmic utility with common time preference and collinear endowments

$$U^i(c_0^i, \ldots, c_S^i) := \ln(c_0^i) + \beta \sum_{s=1}^{S} p_s^i \ln(c_s^i), \quad i = 1, \ldots, I,$$

and $\boldsymbol{w}^i = \delta^i \mathbf{W}$, $i = 1, \ldots, I$, aggregates to

$$U^R(c_0^R, \ldots, c_S^R) = \ln(c_0^R) + \beta \sum_{s=1}^{S} p_s^R \ln(c_s^R)$$

where $p_s^R = \sum_{i=1}^{I} \delta^i p_s^i$.

- Mean-Variance utilities with heterogeneous expectations on the means and common beliefs on the covariances,

$$V^i(\mu^i, \sigma) := \mu - \frac{\alpha^i}{2}\sigma^2, \quad i = 1, \ldots, I,$$

aggregates to

$$V^R(\mu^R, \sigma) = \mu^R - \frac{\alpha^R}{2}\sigma^2,$$

where $\mu^R = \sum_{i=1}^{I} a^i \mu^i$, with $a^i = \frac{r^i/\alpha^i}{\sum_i r^i/\alpha^i}$.

In each of these examples, the representative agent is of the same type as the individual agents and, moreover, he generates a mapping from the individual agents' characteristics into asset prices that also can be used for asset price predictions, i.e., after some change of the asset payoffs, for example.

4.6.2 A Model for Aggregation of Heterogeneous Beliefs, Risk- and Time Preferences

So far we have only discussed cases where the representative agent turned out to be similar to the single agents. To this end, we had to restrict ourselves to a number of special cases. In this section we will now briefly introduce a model where at the same time beliefs, risk preferences and time discounting can be heterogeneous. The resulting representative agent will then in general differ from all other agents. Moreover, "the" representative agent turns out to be non-unique: we can define infinitely many agents that represent the market. However, as in Proposition 4.12, for the construction of the representative agent one needs to know the equilibrium prices.

Proofs and further details on the result presented in this section can be found in [She08].

We consider a market with agents $i = 1, 2, \ldots$ that follow expected utility theory with power utility functions $u_i(x) = x^{1-\gamma_i}/(1-\gamma_i)$, i.e. they have CRRA preferences, but their risk aversion can be heterogeneous. The agents discount future events with classical time discounting, where their discount factors δ_i can also be heterogeneous.

Every agent has a belief p_i on the probability of the asset returns. These beliefs again may vary across agents.

In this case we can formulate the following theorem [She08, Theorem 14.1]:

Theorem 4.13 (Representative investor) *Let π be equilibrium state prices for a complete market with the investors specified above, then:*

(i) π is also the equilibrium state prices of a market with only one representative investor whose utility can be written in the form

$$\sum_t \delta_R(t) \sum_{s_t} p_R(s_t) a(s_t) c(s_t)^{1-\gamma_R(s_t)},$$

where γ_R and δ_R satisfy the equations:

$$p_R(s_t) = \frac{v(x_t)\xi(x_t)^{\gamma_R(x_t)}}{\delta^t_{R,t}} \tag{4.2}$$

$$\frac{1}{\gamma_R(s_t)} = \sum_i \theta_i(s_t)\frac{1}{\gamma_j}, \tag{4.3}$$

$$\sum_{s_t} \delta_R(t) = \pi_{s_t}\zeta(s_t)^{\gamma_R(s_t)}, \tag{4.4}$$

where $\theta_i(s_t)$ denotes the proportion of investor i in the total consumption in state s_t and $\zeta(s_t)$ denotes the value of the market portfolio.

(ii) *The representative investor is not unique: this can be seen from the fact that the two Eqs. (4.3) and (4.4) do not fully determine the three variables* γ_R, δ_R *and* p_R.

We see in particular that the representative investor can have very different properties than the other investors:

- Its time discounting does not have to be classical: δ_R depends on the state.
- Its risk aversion can also vary depending on the state: the larger the consumption share of an investor in a given state, the more he influences the risk aversion of the representative investor.

This underlines that the idea that a representative investor is "essentially" like the average investor on the market is wrong: it can have features that none of the agents on the market has!

4.6.3 Empirical Properties of the Representative Agent

In this section we *assume* that market prices are generated by an individual decision problem. The question we are interested in is which utility function is compatible with the empirical findings in asset prices. Whether this assumption is plausible is left to the reader. Some justifications on this were given in the previous section.

Some authors say that assuming the utility market hypothesis for market aggregates is maybe more plausible than assuming it for individual investors. As early as 1956 John Hicks [Hic86, page 55] wrote:

> ...the preference hypothesis only acquires a prima facie plausibility when it is applied to a statistical average... to assume that an actual person, the Mr. Brown or Mr. Jones who lives round the corner, does in fact act in such a way does not deserve a moment's consideration.

Hence, Hicks has already anticipated the psychologists' critique that describing individual decisions by utility maximization is wrong. The mystery that remains is how the individual irrationalities are washed out at the level of the market. In the next chapter, we give an evolutionary argument for this, based on market selection. From the evolutionary point of view market aggregates can be thought of as being derived from a rational utility function even though no individual ever has attempted to behave rationally.

We first have to distinguish between the implications that long-term and that short-term data of asset returns have for the utility function of the representative agent. Then we will suggest a synthesis of both. We will argue that in the long run, the utility function must have constant relative risk aversion, while in the short run it must have features of Prospect Theory.

That the utility of the representative investor must have CRRA in the long run was made pretty clear by Campbell and Viceira [CV02, page 24]:

> The long run behavior of the economy suggests that relative risk aversion cannot depend strongly on wealth. Per capita consumption and wealth increased greatly over the past two centuries. Since financial risks are multiplicative, this means that the absolute scale of financial risks has also increased while the relative scale of financial risks is unchanged. Interest rates and risk premia do not show any evidence of long-term trends in response to this long-term growth; this implies that investors are willing to pay almost the same relative costs to avoid given relative risks as they did when they were much poorer, which is possible only if relative risk aversion is almost independent of wealth.

Now supposing the utility function has CRRA, the question that remains is the magnitude of the risk aversion parameter. Let's write the utility function as $u(w) := w^{1-\alpha}/(1-\alpha)$. An upper bound for α can be found from the first order condition of utility maximization since, as we show next, the Sharpe ratio of any asset is bounded above by the volatility of the agent's consumption growth. This derivation goes back to Hansen and Jagannathan [HJ91], hence the lower bound is called the Hansen and Jagannathan bound.

Let $\zeta := \ell/R_f$ be the likelihood ratio process divided by the risk-free rate. Then the no-arbitrage condition reads $\mathbb{E}(\zeta R^k) = 1$. By the definition of the correlation we can write:

$$1 = \mathbb{E}(\zeta R^k) = \mathbb{E}(\zeta)\mathbb{E}(R^k) + \operatorname{corr}(\zeta, R^k)\sigma(\zeta)\sigma(R^k),$$

hence

$$\mathbb{E}(R^k) = R_f - \operatorname{corr}(\zeta, R^k)\frac{\sigma(\zeta)}{\mathbb{E}(\zeta)}\sigma(R^k).$$

Since the correlation is bounded between -1 and $+1$, we get the inequality:

$$\frac{\left|\mathbb{E}(R^k) - R_f\right|}{\sigma(R^k)} \leq \frac{\sigma(\zeta)}{\mathbb{E}(\zeta)}.$$

In the consumption based asset pricing model with expected utility, we have

$$q^k = u'(c_0)^{-1}\frac{1}{1+\delta}\mathbb{E}\big(u'(c_1)A^k\big),$$

hence

$$\zeta = \frac{1}{1+\delta}\frac{u'(c_1)}{u'(c_0)}.$$

Table 4.2
Hansen-Jagannathan bounds for alternative risk aversion based on annual data of the S&P 500 from 1973 to 2005

Risk aversion	Consumption SDF
1	0.0165
2.5	0.0415
5	0.0844
8	0.1376
10	0.1743
18	0.3311
20	0.3730
30	0.5987

And in the case of CRRA we get:

$$\zeta = \frac{1}{1+\delta}\left(\frac{c_1}{c_0}\right)^{-\alpha},$$

so that

$$\frac{\left|\mathbb{E}(R^k) - R_f\right|}{\sigma(R^k)} \leq \sigma\left(\frac{1}{1+\delta}\left(\frac{c_1}{c_0}\right)^{-\alpha}\right).$$

Hence, taking data on the risk-free rate, the Sharpe ratio and the volatility of consumption growth we can estimate the relative risk aversion α. In a recent paper Hens and Wöhrmann [HW07b] estimate the Sharpe ratio of the S&P 500 sampled from annual data since 1973 and compare it with the Hansen-Jagannathan bounds resulting for alternative risk aversions. The result is reported in Table 4.2.

The actual Sharpe ratio in that data is about 0.328. Hence, a relative risk aversion of about 18 would explain the risk adjusted equity premium. On data with a higher frequency the risk aversion is higher. Some authors estimate numbers in the range of 30–40. It is typically claimed that numbers above 10 are too high to be reasonable values for risk aversion. Hence, the typical finding of numbers in the area of at least 18 is puzzling to many researchers. This puzzle is called the *equity premium puzzle*. But why is a number above 18 considered to be too high? After all nobody has ever met the fictitious agent called representative investor and asked him about his degree of relative risk aversion. The idea is that the representative investor represents the individual investors and estimates of individuals' risk aversion are feasible using questionnaire techniques as explained in Sect. 2.2.4. However, we have already remarked there that such elicitations crucially depend on the assumptions on a person's wealth level.

Let us take a look at a typical question that is used to determine the relative risk aversion:

> Consider a fair lottery where you have a 50% chance of doubling your income, and a 50% chance of losing a certain percentage, say x% of your income. What is the highest loss x that you would be willing to incur to agree to taking part in this lottery?

The typical answer to this question is an x of about 23 %. Interpreting this answer based on a CRRA utility function we get an α of 3.22: we set

$$0.5\frac{(2W)^{1-\alpha}}{1-\alpha}+0.5\frac{\left((1-x)W\right)^{1-\alpha}}{1-\alpha}=\frac{W^{1-\alpha}}{1-\alpha}.$$

This gives $2^{1-\alpha}+(1-x)^{1-\alpha}=2$ or $1-\alpha=-2.22$.

Hence, the typical answer to this question is far away from the value obtained from stock market data. There is a huge literature trying to bring down the alpha obtained from stock market data. Some authors say that in the optimization of the representative agent borrowing constraints are missing. Others say that maybe the utility function is not CRRA but includes aspects like habit formation (e.g. [Abe90]). Yet others claim that consumption should be restricted to the consumption of stock holders. Finally, some researchers claim that one should calculate the equity premium based on expected stock returns that are typically smaller than their realizations. A recent book on these attempts was edited by Mehra [Meh06], the inventor of the equity premium puzzle.

There is another possible explanation: based on the observation that the background wealth plays an important role in measuring any utility function, we notice that in the above derivation of α we have implicitly assumed that the money at stake is the whole wealth of a person. However, in the question "only" the whole *salary* is at stake. Now assume that the person's background wealth is non-zero, let us say 50 % of his/her salary, then the degree of risk aversion is computed as follows:

$$0.5\frac{\left((0.5+2)w\right)^{1-\alpha}}{1-\alpha}+0.5\frac{\left((0.5+(1-x))w\right)^{1-\alpha}}{1-\alpha}=\frac{\left((0.5+1)w\right)^{1-\alpha}}{1-\alpha}.$$

If we set $x:=23\%$, we obtain $\alpha=21$, and the alpha increases even more with higher background wealth (see [HW07b]). Please keep in mind that the background wealth should reflect the total wealth of our society since the representative investor whose consumption we used to explain market data needs to own everything we can think of (land, houses, factories, cars, etc.).

In the exercise book we do similar computations based on mean-variance and on Prospect Theory. The main message is unchanged: evaluating the degree of risk aversion from market data and from experimental data under the same assumption on background wealth the equity premium puzzle may not be that puzzling.

Now we turn to the short-run properties of the representative's utility function as it can be estimated from daily data on stock market indices and their derivatives. Again we use the first order condition for optimal investment decisions as a starting point:

$$\ell_s=R_f\frac{u'(c_s)}{(1+\delta)u'(c_0)},\quad s=1,\dots,S.$$

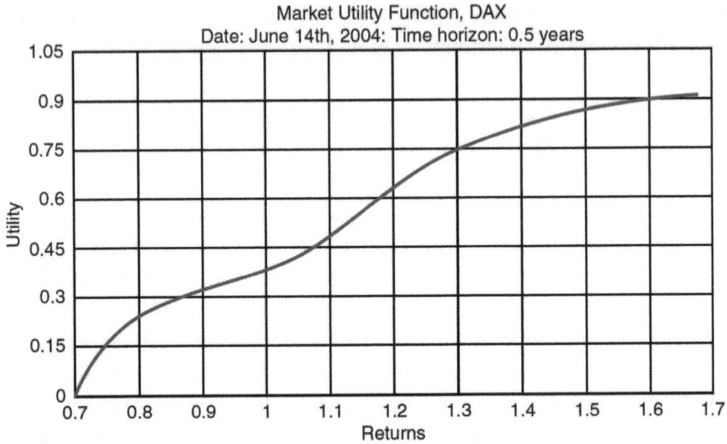

Fig. 4.14 Utility function of the representative investor estimated from daily return on the DAX for a typical trading day

Now we read this in the following way: we estimate the likelihood ratio process from observed stock market returns, which fix the p_s's and from option price data which determines the π_s's. Finally, we take the risk-free rate, the discount factor and the consumption growth as before. Following this approach Jackwerth [Jac00] showed that the ℓ is typically not a monotonic decreasing function, as is typically assumed, but instead "hump-shaped". Detlefsen, Härdle and Moro [DHM09] follow this approach and estimate the utility function on DAX-data (compare Fig. 4.14 and also [RH10]).

Note the convexity in the left part of the figure, which is reminiscent of Prospect Theory. If one were to include probability weighting, then the concavity for extreme losses which can be seen in Fig. 4.14 could be explained by the overweighting of extreme losses – compare the four fold pattern of risk from Sect. 2.4.1.

There is another possible interpretation of the "empirical" utility function: we have seen that in the long run the representative agent is a CRRA maximizer while in the short run he might be better described by elements of Prospect Theory. We can therefore propose the following synthesis of the long-run and the short-run view: define a utility function

$$u(c) := u(c_0) + \delta\big(u(c_1)(1-h) + hv(c_1 - c_0)\big)$$
$$+ \delta^2\big(u(c_2)(1-h) + hv(c_2 - c_1)\big) + \ldots,$$

where we use a new parameter, the habit formation parameter $h \in (0,1)$. For $h = 0$ we get the standard expected discounted utility function that, e.g., may have CRRA. For a positive h below 1 we can blend in the Prospect Theory value function v with reference point equal to last period's consumption. We may also include discounting by which the marginal rates of substitution between today and any future period

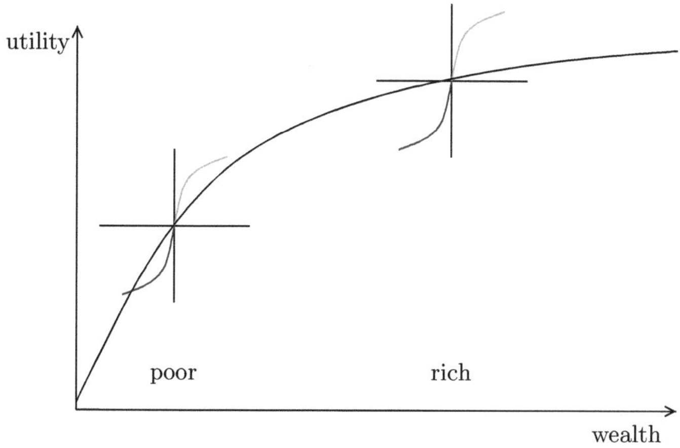

Fig. 4.15 A long-run CRRA utility with a short-run prospect-theory-overlay

are changed while those between two future periods remain as in the standard exponential discounting case (compare Sect. 2.7 and [BHS01]).

Graphically our suggested synthesis can be displayed as in Fig. 4.15. Adding both terms we arrive at a utility function which is close to the empirical one from Fig. 4.14.

It is interesting that there is also a completely different approach to the empirical pricing kernel puzzle: whereas we have so far assumed rational (or more precisely: extrapolated) expectations, but behavioral preferences, one can alternatively assume biased expectations and estimate p. This approach has been developed by Hersh Shefrin [She08] and is an extension of the representative agent results discussed in Sect. 4.6.2. In this way, it is possible to explain the observed pattern of the pricing kernel even within a standard CRRA utility framework.

It is difficult to discriminate empirically between both potential explanations for the "hump-shape" of the pricing kernel. The fact, however, that most trades on financial markets are based on heterogeneity of beliefs rather than on differences in risk preferences supports the explanation by Shefrin.

Finally, an explanation of the empirical pricing kernel puzzle based on incomplete markets is presented in the exercise book

4.7 Dynamics and Stability of Equilibria

So far we have restricted our attention to financial market equilibria. We did not give any argument how such an equilibrium is reached, starting from a non-equilibrium state. One may argue that the economy is always subject to exogenous "shocks" that will disturb the current equilibrium. Hence, if there were no forces that drive the economy back to the equilibrium then it is very unreasonable to assume that real

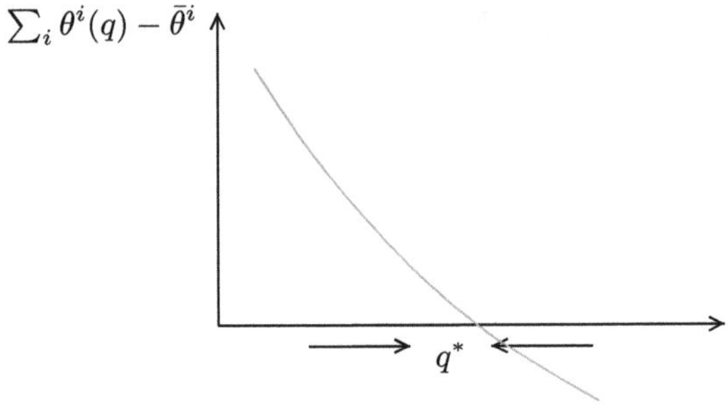

Fig. 4.16 Stability of an equilibrium in a simple one dimensional setting. The figure shows the excess demand of a single asset as a function of its price

life phenomena can be described by an equilibrium. Everything we learned so far would not have any justification! Exogenous shocks could, for example, be changes in the economic fundamentals like the payoffs of the assets or the exogenous wealth of the agents. Also they could result from changes in agents' beliefs about the states of the world (induced by natural disasters or sudden unforeseen political events). In this book we will consider three types of dynamics that can be distinguished by their speed of adjustment. We start with the fastest type, the short-run dynamics.

In the short run we look at the intraday adjustment of market prices due to excess demand or excess supply of assets. We postulate that prices move in the direction of excess demand, i.e., when demand exceeds supply, prices will increase and when supply exceeds demand, prices will decrease. This was the original idea of Adam Smith on which he based the conjecture that competitive equilibria will always be reached. While this argument is compelling for the stability of one market (see Fig. 4.16 for an illustration), it is not obvious at all if markets are linked to each other because the demand of any asset does also depend on the price of any other asset. Note that these cross-price effects naturally arise when agents have portfolio considerations like diversification. In this section we will show two results on the stability of financial market equilibria in a CAPM economy.

With simple mean-variance preferences of the form

$$U^i(c_1^i, \dots, c_S^i) := \mu^i(c_1^i, \dots, c_S^i) - \frac{\gamma^i}{2}\sigma^{2,i}(c_1^i, \dots, c_S^i),$$

we prove global stability of the unique financial market equilibrium. This case is obtained, for example, if utility functions are of the CARA-type and returns are log-normally distributed. If, however, mean-variance preferences are obtained from normally distributed returns and Prospect Theory preferences then the mean-variance utility looks more complicated than that, see [DGP08]. As an effect, CAPM equilibria may be unstable, for example because due to exogenous shocks they may jump from one possible equilibrium to another equilibrium. Intraday crashes

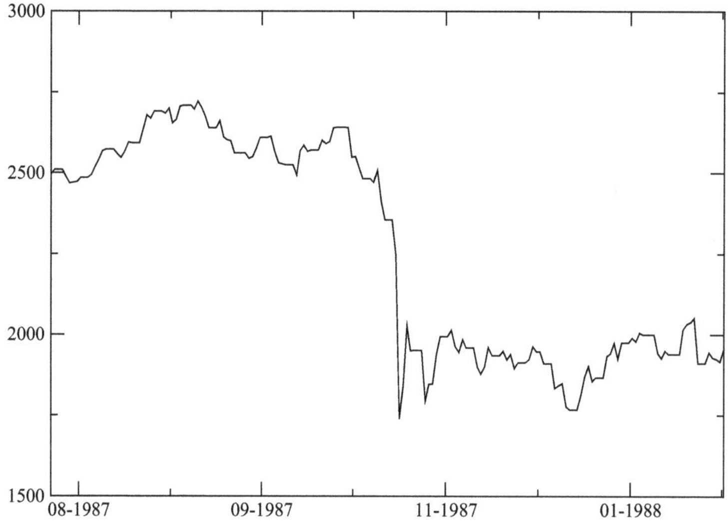

Fig. 4.17 Black Monday of October 19, 1987 in which the DJIA lost 20 % of its value in a single day!

that occur for no obvious reason like the Black Monday of October 19, 1987 (see Fig. 4.17) have been explained by this switching from one equilibrium to another.[38]

Before we conclude by proving the stability for simple mean-variance utilities, we will first give the geometric intuition for these crashes when the utilities are more complicated. Figure 4.18 shows the phenomenon of multiple equilibria in the Edgeworth Box. For a given set of endowments and preferences two different market clearing prices are obtained. One may think of one equilibrium as the "optimistic" one and one as the "pessimistic" one. In the optimistic equilibrium, the asset price is high because everybody believes the asset is attractive and, hence, prices are driven up. In the pessimistic equilibrium the reverse holds true.

Translating Fig. 4.18 into an excess demand diagram, we get a situation like that displayed in Fig. 4.19. Note that if there are multiple equilibria we actually need to have at least three of them, two of which are stable and one is unstable.

So how can it happen that on small changes of the exogenous characteristics we get drastic changes of the endogenous entities? Well, this happens if the economy is initially in one equilibrium that disappears or becomes unstable due to the exogenous changes, as the sequence in Fig. 4.20 shows.

A more compact way of showing the same phenomenon is to comprise the three parts of Fig. 4.20 into one. This can be done by looking at a mapping from the exogenous characteristics, e.g., the asset payoffs, to the endogenous asset prices (Fig. 4.21).

[38] The classical reference is Leland and Genotte [GL90].

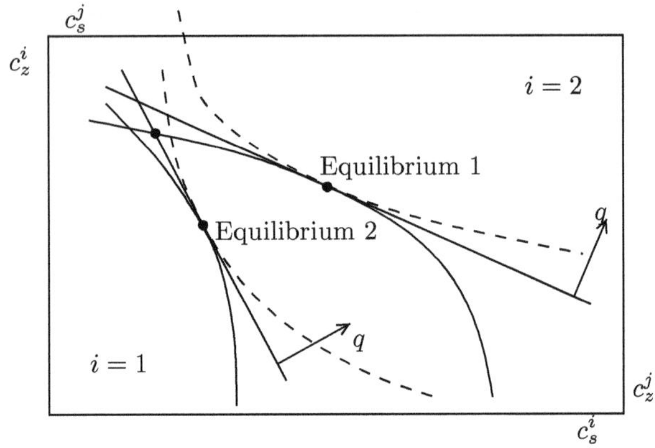

Fig. 4.18 Multiple equilibria in an Edgeworth Box

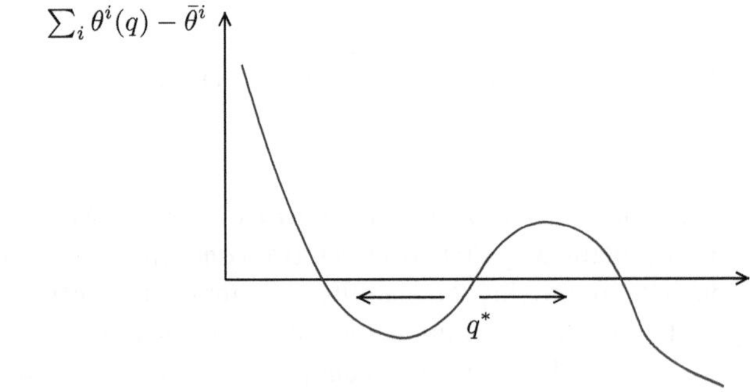

Fig. 4.19 Multiple equilibria in the excess demand diagram

We will finish this section by giving the formal argument that for simple mean-variance utilities financial market equilibria are stable. Before we do so let us mention that in the next chapter we will analyze two more adjustment processes: The medium-term adjustment in which price expectations adjust on the basis of the realized returns (see Chap. 5) and the long-run wealth adjustment in which the unsuccessful agents leave the market. The latter is also called the market selection dynamics, and is studied in evolutionary finance (see Sect. 5.7.1).

Now we prove the short-term stability of equilibria if prices are continuously adjusted in the direction of excess demand. This adjustment process is called the "Law of Demand and Supply":

Proposition 4.14 *Assume that agents have mean-variance preferences of the form*

$$U^i(c_1^i, \ldots, c_S^i) = \mu^i(c_1^i, \ldots, c_S^i) - \frac{\gamma^i}{2}\sigma^{2,i}(c_1^i, \ldots, c_S^i). \tag{4.5}$$

Then there is a unique globally stable market equilibrium.

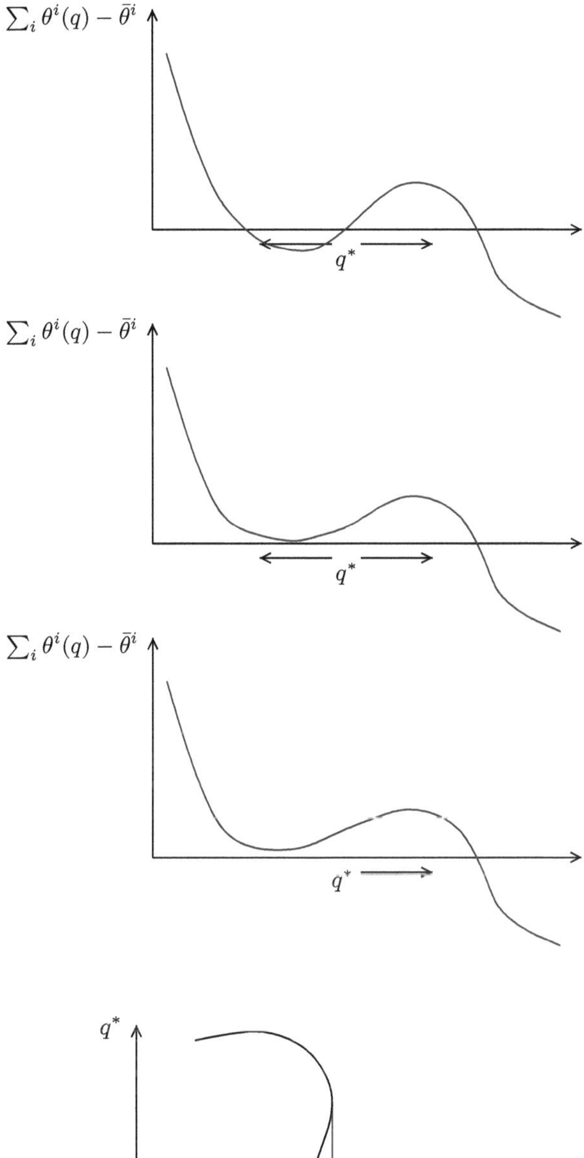

Fig. 4.20 Multiple equilibria in the excess demand diagram

Fig. 4.21 A market crash resulting from small changes in the asset payoffs

Proof Note $c_s^i = \boldsymbol{R}_s\lambda^i w_0^i$ and recall the budget constraint $\sum_{k=0}^K \lambda^{i,k} = 1$. Substitute c_s^i using (4.5), then for any portfolio $\boldsymbol{\lambda}$ we get:

$$U^i(\lambda^{i,1},\ldots,\lambda^{i,K}) = w_0^i\left(\mu^i(\boldsymbol{R}'\boldsymbol{\lambda}^i) - \frac{\gamma^i w_0^i}{2}\sigma^{2,i}(\boldsymbol{R}'\boldsymbol{\lambda}^i)\right).$$

The solution is, as before,

$$\lambda_0 = 1 - \sum_{k=1}^K \lambda^k, \quad \boldsymbol{\lambda}^i = (COV^i(\boldsymbol{R}))^{-1}\frac{\mu^i(\boldsymbol{R}) - R_f\mathbf{1}}{\gamma^i w_0^i}.$$

Written in economic terms (recall $\boldsymbol{R}^k = \frac{A^k}{q^k}$) this reads:

$$\boldsymbol{\lambda}^i = \frac{1}{\gamma^i w_0^i}\boldsymbol{A}(\boldsymbol{q})\big(COV^i(\boldsymbol{A})\big)^{-1}\big(\mu^i(\boldsymbol{A}) - R_f\boldsymbol{q}\big),$$

and finally, since $\theta^{i,k} = \frac{\lambda^{i,k}w^i}{q^k}$, we get the asset demand function

$$\theta^{i,k} = \frac{\lambda^{i,k}(1-\lambda_c^i)w_0^i}{q^k} = \frac{1}{\gamma^i}\big(COV^i(\boldsymbol{A})\big)^{-1}\big(\mu^i(\boldsymbol{A}) - R_f\boldsymbol{q}\big),$$

as in Sect. 4.4.

The continuous time dynamics (Law of Demand and Supply) guarantee that prices adjust in a way that reduces the difference between demand and supply. Mathematically, this can be expressed by

$$\dot{q}_t = a\left(\sum_{i=1}^I \theta^{i,k}(q_t) - \sum_{i=1}^I \theta_A^{i,k}\right),$$

where $a > 0$ is the speed of adjustment. Note that the market demand is monotone, i.e.

$$(\hat{q}_t - \tilde{q}_t)\left(\sum_{i=1}^I \theta^{i,k}(\hat{q}_t) - \sum_{i=1}^I \theta^{i,k}(\tilde{q}_t)\right) < 0, \quad \text{for all } \hat{q}_t, \tilde{q}_t, \tag{4.6}$$

because $\partial q_t\theta^i(q_t) = -\frac{R_f}{\gamma^i}(COV^i(\boldsymbol{A}))^{-1}$, which is a negative definite matrix if there are no redundant assets. To prove stability, we define a so called "Lyapunov function" $L(t) := \|q_t - q^*\|^2$ and show that $L(t)$ is decreasing in t. We compute

$$\begin{aligned}\dot{L}(t) &= \partial t L(t) = 2(q_t - q^*)\dot{q}_t \\ &= 2(q_t - q^*)a\left(\sum_{i=1}^I \theta^{i,k}(\hat{q}_t) - \sum_{i=1}^I \theta^{i,k}(\tilde{q}_t)\right),\end{aligned}$$

hence we can apply the monotonicity (4.6) and arrive at

$$\dot{L}(t) = 2(q_t - q^*)a\left(\sum_{i=1}^{I} \theta^{i,k}(q_t) - \theta^{i,k}(q^*)\right) < 0.$$

This proves that the market is in fact globally stable. □

4.8 Summary

In this chapter we have generalized some of the ideas from Chap. 3 by replacing the assumption of mean-variance preferences with the no-arbitrage condition. The key ideas of our model were to consider only two time steps ("investing" and "selling"), but several states of the worlds. In every state, assets can have different payoffs. A priori only the probabilities in which the states occur are known. We assumed that there is no trading strategy that gives a sure outcome above the risk-free rate (no arbitrage condition), where we distinguished slightly different ways to make this definition precise. Surprisingly, we could derive a lot of information about asset prices from the no arbitrage condition. We can, however, only price assets in a *relative* way with this method.

We then formulated a market model with several investors. Special cases of our general formulation were the CAPM, APT and the behavioral CAPM. We proved that in complete markets a market equilibrium will be Pareto efficient, i.e. no person can do better without somebody else being worse off.

Even if investors have heterogeneous preferences, the market as a whole can be described as if all investors shared the same preferences. We say that there is a "representative agent". This approach is, however, limited, since out-of-sample predictions are not possible in this way.

Finally, we discussed dynamics and stability of equilibria: while the "short-run dynamics" is bringing the market (if it satisfies certain assumptions) back to an equilibrium, it might be possible that an equilibrium is suddenly disappearing when market conditions slowly change, thus leading to a sudden transition to a different equilibrium. This can be an explanation of crashes on financial markets.

Test

1. What cannot be said about complete markets?

   - □ In a complete market, all period consumptions can be achieved by asset trades.
   - □ When the states of the world are defined by asset returns themselves, then the variation of returns is less then the number assets.

- ☐ If the income is exogenous, there do not have to be sufficient assets to hedge all the risks in the exogenous income.
- ☐ The rank of the return matrix needs to be equal to the number of states.

2. What do agents trade?

- ☐ Consumption.
- ☐ Opinions.
- ☐ Financial assets.
- ☐ Risk Factors.

3. What can be said about arbitrage?

- ☐ In a arbitrage-free market, the price of the derivative asset price must be equal to its duplicating portfolio.
- ☐ The definition of arbitrage certainly depends on the qualitative properties of the investor's utility function.
- ☐ Arbitrage strategy, more specifically, depends on the monotonicity of the investor's utility function.
- ☐ The absence of arbitrage is equivalent to the existence of state prices.

4. What are risk-neutral probabilities?

- ☐ The probabilities representing the homogenous beliefs of the investor.
- ☐ The physical probabilities observed in the market.
- ☐ The probabilities representing the heterogenous beliefs of the investor.
- ☐ The probabilities used in the pricing equation, called "risk adjusted probabilities".

5. What can be said about financial market equilibria?

- ☐ There cannot be any arbitrage opportunities in the financial market equilibria.
- ☐ A financial market equilibrium is a system of asset prices and an allocation of assets such that every agent optimizes his decision problem and markets clear.
- ☐ With simple mean-variance preferences, the financial market equilibrium exists uniquely and globally stable.
- ☐ Deriving the asset prices from an equilibrium does not necessarily lead to arbitrage-free prices.

6. What is the likelihood ratio process?

   ☐ The ratio of the state price measure and the physical probabilities.
   ☐ Risk neutral probabilities.
   ☐ Normalized state price process.

7. What is the general formula of the Arbitrage Pricing Theory?

   ☐ $\mathbb{E}_p(R^k) - R_f = \sum_{f=1}^{F} b^f(\mathbb{E}_p(R^f))$
   ☐ $\mathbb{E}_p(R^k) - R_f = \sum_{f=1}^{F} b^f(\mathbb{E}_p(R^f) - R_f)$
   ☐ $\mathbb{E}_p(R^k) = \sum_{f=1}^{F} b^f(\mathbb{E}_p(R^f) - R_f)$

8. What are the main implications of the Behavioral CAPM with quadratic value function?

   ☐ B-CAPM can be considered as a more general version of CAPM since CAPM is a special case of the model.
   ☐ Differently from the standard CAPM, the B-CAPM with its risk-return decomposition captures the possible asymmetries among risk factors between up and down markets.
   ☐ The CAPM model can be still considered as good in many practical applications.

9. Which meanings of efficiency in finance refers to Efficient Market Hypotheses while the other refers to Pareto Efficiency?

   ☐ Informational efficiency, referring that at any time, the prices reflect already all public information.
   ☐ Allocational efficiency, referring that in a financial equilibrium, allocations of assets are such that nobody's utility can decrease or increase by changing the allocations.

10. What are the empirical properties of the representative agent?

   ☐ The utility function of an representative investor is not concave everywhere.
   ☐ For the short-run return dynamics, the representative agent is better modeled with Prospect Theory.
   ☐ The representative agent is simply the CRRA maximizer.
   ☐ The representative agent is CRRA maximizer for the long run dynamics.

Multiple-Periods Model 5

"It will fluctuate." John P. Morgan's reply, when asked what the stock market will do.

In the previous two chapters, we have restricted ourselves to the case of two time periods, one for investing and one for receiving payoffs. For many applications it is, however, necessary to allow for models with more than two time periods. In particular one can then study re-trading on the arrival of new information. Nevertheless we will see that many of the insights we have won for the two-period model will be useful also for multi-period models.

5.1 The General Equilibrium Model

To find the equilibrium in a system that runs over more than two-periods, it is necessary to define first the uncertainty associated with time and information. We follow the approach of Lucas [LJ78] and define a model over discrete time, i.e., $t = 0, 1, 2, \ldots, T$, by a tree-like extension of our two-period model (see also [Con82]). The information structure of this "Lucas tree model" is given by a finite set of *states* $\omega_t \in \Omega_t$ in each t. A path of state realizations over time is denoted by the vector $\omega^t = (\omega_0, \omega_1, \ldots, \omega_t)$. The uncertainty with respect to information decreases with the time since the paths are not recombining. The time-uncertainty can be described graphically by an *event tree*[1] consisting of an initial date ($t = 0$)

[1]The mathematical term for an event tree is a filtration, $\mathcal{F}_0, \mathcal{F}_1, \ldots, \mathcal{F}_T$, i.e. a sequence of partitions of a set $\{1, \ldots, s\}$ such that $\mathcal{F}_0 = \{\{1, \ldots, s\}, \emptyset\}$, $\mathcal{F}_T = \{\{1\}, \{2\}, \ldots, \{s\}\}$ and for all $e_t \in \mathcal{F}_t$ there exists $e_{t-1} \in \mathcal{F}_{t-1}$ such that $e_t \subseteq e_{t-1}$.

T. Hens, M.O. Rieger, *Financial Economics*, Springer Texts in Business and Economics, DOI 10.1007/978-3-662-49688-6_5

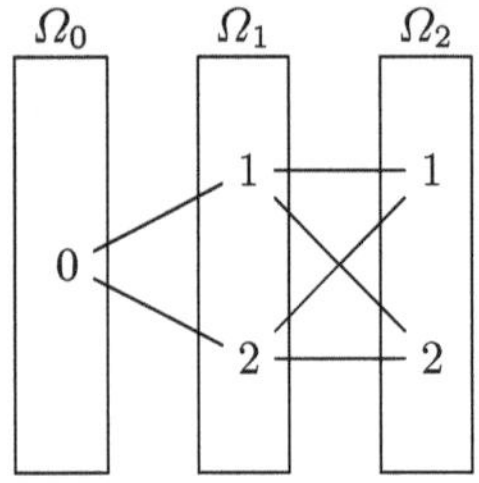

Fig. 5.1 Generating paths by drawing states $\omega_t \in \Omega_t$

(0,11)	(0,11)	(0,1,1)
(0,12)	(0,12)	(0,1,2)
(0,21)	(0,21)	(0,2,1)
(0,22)	(0,22)	(0,2,2)

Fig. 5.2 Associated event tree

and a set of paths ω^t at time t. In any intermediate time period t the event tree consists of a partition of the set of paths so that two paths that cannot yet be distinguished at t belong to the same subset. See Fig. 5.1 for an example.

The probability measure determining the occurrence of the states is denoted by P. We will define P over the set of paths. We call P the *physical measure*, since it is exogenously given by nature, and use it to model the exogenous dividends process. If the realizations are independent over time, P can be calculated as a product of the probabilities associated with the realizations building the vector ω^t. For example, the probability of getting two times "head" by throwing a fair coin is equal to the probability of getting "head" once (equal to 0.5) multiplied with the probability of getting "head" in the second run (equal to 0.5) (Fig. 5.2).

In the Lucas [LJ78] model the payoffs are determined by the dividend payments and capital gains in every period. Let $i = 1, \ldots, I$ denote the investors and $k = 1, \ldots, K$ some long-lived assets in unit supply that enable wealth transfers over different periods of time. In addition, there is a consumption good. This good is perishable, i.e., it cannot be used to transfer wealth over time. All assets pay off in terms of the consumption good. This clear distinction between means to transfer wealth over time and means to consume is one of the important modeling assumptions of Lucas [LJ78].

Ultimately, we are of course interested in the evolution of payoffs of the assets, but not every node in the tree has to result in a payoff. It could well be that other events than payoffs trigger trades and, therefore, have to be represented in the tree. As an example, think of a company paying out dividends only once a year, but having quarterly earning reports: obviously, the earning reports will lead to changes in the probability distribution of the dividend payoffs, thus we have to include them into our tree model although there is no payment connected with them. Moreover, even events that are unrelated to the company's dividend payoff have sometimes to be considered. An example would be substantial changes in the investor population or preferences: they will result in trades that can change the stock prices of the company.

Following Lucas [LJ78] once more, we assume *perfect foresight*, which means that *conditionally* on the events, all investors agree on the prices. Although this seems to be a strong assumption, our model is still flexible enough to accommodate different opinions: we just have to split states into sub-states whenever some investors disagree about the prices in the original state. Then, we assume the same price expectations in each state and allow agents to hold different probabilities of the occurrence of the states.

In a *competitive equilibrium* with perfect foresight, every investor decides about his portfolio strategy according to his consumption preferences[2] over time. Investors may disagree on the probability distribution of the states, but by construction, they agree on the prices conditionally on the states. This leads to the following definition of an equilibrium extending Definition 4.8 to the multi-period case.[3]

Definition 5.1 A competitive equilibrium with perfect foresight is a list of portfolio strategies θ_t^i and a sequence of prices q_t^k, $t = 0, 1, \ldots, T$, such that for all $i = 1, \ldots, I$

$$(\theta_0^i, \ldots, \theta_T^i) \in \underset{\substack{\theta_t \in \mathbb{R}^{K+1} \\ t=0,1,\ldots,T}}{\arg\max}\ U^i\left(\theta^{\text{cons}}\right) \quad \text{s.t.} \quad \theta_t^{\text{i,cons}} + \sum_{k=1}^{K} q_t^k \theta^k \overset{!}{=} W_t^i,\ \theta_t^{\text{i,cons}} \geq 0,$$

$$\text{with} \quad W_t^i := \sum_{k=1}^{K} \left(D_t^k + q_t^k\right) \theta_{t-1}^{i,k} + w_t^i,$$

$$\text{for all } t = 0, 1, \ldots, T,$$

where D_t^k are the total dividend payments of asset k, w_t^i is the endowment of investor i in period t and markets clear [4]:

$$\sum_{i=1}^{I} \theta_t^{i,k} \overset{!}{=} 1 \qquad \text{for all } k \text{ and all } t.$$

[2]Note that investors' preferences are defined over consumption and not over the depot value. The utility function representing the investors' time preferences and risk attitude determines the consumption, which is smoothed over the realized states.

[3]Note that in contrast to Chap. 4, $k = 0$ does not denote the risk-free asset but consumption. This is because long-lived assets that are re-traded are rarely risk-free since their prices might fluctuate.

[4]In principle, all quantities will be dependent on the entire history/path ω^t – or at least on the realized state ω_t – and we should write, e.g., $q_t^k(\omega^t)$, as there might be a different path ω'^t for which $q_t^k(\omega'^t)$ is a different value. Thus, in writing q_t^k above we not only name a function where instead its value is meant, but we are not precise on which value we actually mean, either. Doing so is therefore – if not stated differently – understood just as an abbreviation for ease of reading. Please observe that we cannot write $q^k(t)$, because there is not "one" function q or q^k which gets evaluated at two different points in time: q_t and q_{t+1} are two different functions, as they are defined on two different domains ($\Omega^t := \Omega_0 \times \ldots \times \Omega_t$ and $\Omega^{t+1} = \Omega^t \times \Omega_{t+1}$ respectively).

Note that $\theta_t^{i,k}(\omega^t) \in \mathbb{R}$ is the amount of asset k, respectively for $k = 0$ the amount of the consumption good, that agent i has in period t given the path ω^t and θ_t^i is the portfolio strategy along the set of paths. θ_t^i is accordingly this amount of all assets for all paths ω^t and θ^i is the list of portfolio strategies over time. Thus, the objective function U^i is a function on the space of all consumption paths. Note also that we normalized the price of the consumption good to one and that we used Walras law to exclude the market clearing condition for the consumption good.

To be more concrete, investigate the decision problem of an expected utility maximizer in the situation/*node* ω^t restricted to the decision variables in that and the adjacent *nodes*:

$$\max_{\theta_t^{i,k},\theta_t^{i,\text{cons}},\theta_{t+1}^{i,\text{cons}}} u^i(\theta_t^{i,\text{cons}}) + \delta_t^i \sum_{\omega_{t+1}\in\Omega_{t+1}} \text{prob}(\omega_{t+1}) u^i\left(\theta_{t+1}^{i,\text{cons}}(\omega^t\omega_{t+1})\right),$$

where u^i is the utility function of investor i (compare Sect. 4.1.3). The budget constraints are:

$$\begin{aligned}\theta_t^{i,\text{cons}} &+ \sum_{k=1}^K q_t^k \theta_t^{i,k} \\ &= \sum_{k=1}^K (D_t^k + q_t^k)\theta_{t-1}^{i,k} + w_t^i, \qquad \theta_t^{i,\text{cons}} \geq 0\end{aligned}$$

and for all $w_{t+1} \in \Omega_{t+1}$

$$\begin{aligned}\theta_{t+1}^{i,\text{cons}}(\omega^t\omega_{t+1}) &+ \sum_{k=1}^K q_{t+1}^k(\omega^t\omega_{t+1})\,\theta_{t+1}^{i,k}(\omega^t\omega_{t+1}) \\ = \sum_{k=1}^K &(D_{t+1}^k(\omega^t\omega_{t+1}) + q_{t+1}^k(\omega^t\omega_{t+1}))\,\theta_t^{i,k} + w_{t+1}^i(\omega^t\omega_{t+1}), \quad \theta_t^{i,\text{cons}} \geq 0.\end{aligned}$$

We start the economy with some initial endowment of assets θ_{-1}^i such that $\sum_i \theta_{-1}^i = 1$. Assets start paying dividends in $t = 0$, i.e., the budget constraint at the beginning is

$$\theta_0^{i,\text{cons}} + \sum_{k=1}^K q_0^k \theta_0^{i,k} = \sum_{k=1}^K (D_0^k + q_0^k)\theta_{-1}^{i,k} + w_0^i.$$

We can think of $t = 0$ as the starting point of our analysis, i.e., θ_{-1}^i can be interpreted as the allocation of assets that we inherit from a previous period ("the past"). Hence, in a sense the economy can be thought of as restarted at $t = 0$.

Instead of using the *amount* of asset k held in the portfolio of investor i in time t, the investors' demand can be expressed in terms of the asset allocation or percentage of the budget value. We define $\lambda_t^{i,k} = (q_t^k \theta_t^{i,k})/W_t^i$ (in analogy to the two-period model definition from Sect. 4.1.3). Thereby we get $\theta_t^{i,k} = \lambda_t^{i,k} W_t^i / q_t^k$. Equalizing demand with supply, i.e.,

$$\sum_{i=1}^{I} \frac{\lambda_t^{i,k} W_t^i}{q_t^k} \stackrel{!}{=} 1 \quad \text{for all } k \text{ and all } t,$$

gives the following result:

Proposition 5.2 *The price of asset k is the average wealth of the traders' asset allocation for asset k, i.e.,*

$$q_t^k = \sum_{i=1}^{I} \lambda_t^{i,k} W_t^i .$$

This pricing rule is equivalent to the simple equilibrium condition that demand is equal to supply. We have made no other assumptions to derive this result!

We are interested in optimal asset allocation strategies, thus it is more useful to formulate the investments in terms of percentage of wealth rather than in terms of the absolute number of assets. Therefore, we rewrite the model in terms of $\lambda_t^{i,k}$ (i.e., strategy as a percentage of wealth) instead of $\theta_t^{i,k}$ (i.e., strategy in terms of asset units) which leads to the following reformulation of Definition 5.1:

Definition 5.3 A *competitive equilibrium* with perfect foresight is a list of portfolio strategies λ_t^i, and a sequence of prices q_t^k for all $t = 0, 1, \ldots, T$, such that for all $i = 1, \ldots, I$

$$\lambda^i \in \mathop{\arg\max}_{\substack{\lambda_t^i \in \Delta^{K+1} \\ t=0,1,\ldots,T}} U^i(\lambda^{\text{cons}} W^i) \quad \text{s.t.} \quad W_t^i = \left(\sum_{k=1}^{K} \frac{D_t^k + q_t^k}{q_{t-1}} \lambda_{t-1}^{i,k} \right) W_{t-1}^i + w_t^i,$$

$$\text{for all } t = 0, 1, \ldots, T$$

and markets clear:

$$\sum_i \lambda_t^{i,k} W_t^i = q_t^k , \quad \text{for all } k \text{ and all } t.$$

Again, for simplicity we have omitted the dependence of the decisions on the paths. In other words, in a competitive equilibrium all investors choose an *asset allocation* $\lambda_{t=0\ldots T}^i$ that maximizes their utility over time under the restriction of a budget constraint with a stochastic compound interest rate.[5] As in Definition 5.1, the compatibility of these decision problems today and in all later periods and events is assured by the assumption of perfect foresight. This equilibrium is therefore also called equilibrium in plans and price expectations.

5.2 Complete and Incomplete Markets

The model can considerably be simplified if markets are complete, i.e., if there are sufficiently many assets to hedge all risks. With complete markets – as we will show below – the sequence of budget constraints can be reduced to a single budget constraint.

Definition 5.4 A financial market (D, q) is said to be *complete* if any consumption stream $\{\theta^{\text{cons}}\}$ can be *attained* with at least one initial wealth w_0, i.e., it is possible to find some trading strategy θ such that for all periods $t = 1, 2, \ldots, T$,

$$\theta_t^{\text{cons}} + \sum_k q_t^k \theta_t^k = \sum_k \left(D_t^k + q_t^k\right) \theta_{t-1}^k, \quad \text{and} \quad \theta_0^{\text{cons}} + \sum_k q_0^k \theta_0^k = w_0.$$

A financial market is said to be *incomplete* if there are some consumption streams that cannot be achieved whatever the initial wealth is.[6]

The necessary and sufficient condition for a financial market to be complete is:

$$\operatorname{rank} A_t(\omega^{t-1}\omega_t) = \left|\Omega_t(\omega^{t-1})\right| \quad \text{for all } \omega^t, t = 1, 2, \ldots, T,$$

where

$$A_t(\omega^{t-1}\omega_t) := \left[D_t^k(\omega^{t-1}\omega_t) + q_t^k(\omega^{t-1}\omega_t)\right]_{\omega_t \in \Omega_t}^{k=1,\ldots,K}$$

and $\left|\Omega_t(\omega^{t-1})\right|$ is the number of states that can be reached from ω^{t-1}. Hence, if $K < |\Omega_t(\omega^t)|$ for some ω^t, then markets are incomplete.

For example, in a symmetric tree model, where the set of possible states is equal to S, a necessary condition for market completeness is that the number of assets is not smaller than the number of possible states, i.e., $K \geq S$.

[5]The budget constraint is defined over the wealth in period t and $t-1$, where the sum $\sum_k \left(\frac{D_t^k + q_t^k}{q_{t-1}^k}\right) \lambda_{t-1}^{i,k}$ is the compound interest rate.

[6]This may arise if there are more states than insurance possibilities.

To illustrate the concept of market completeness, let us consider the following example:

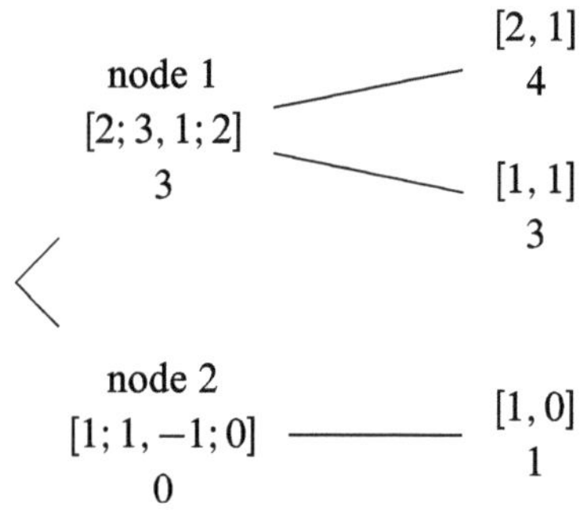

Asset payoffs (dividend payments and prices) are given in brackets over every possible state, e.g. in node 1 asset 1 pays a dividend of 2 and is priced at 3. Note that in terminal nodes asset prices are zero. Thus, we excluded the prices from the description of asset payoffs. The number below indicates the target consumption θ^{cons} over the time. There are two assets, i.e., $k = 1, 2$. An investor has to decide how many units of asset 1 and 2 to buy at $t = 0$ and $t = 1$ in order to achieve the target consumption path.

Starting at the end of the period, an investor has to solve the following problem for node 1:

$$2\theta_1^1(1) + \theta_1^2(1) = 4,$$
$$\theta_1^1(1) + \theta_1^2(1) = 3.$$

In other words, the sum of the payoffs of asset 1 and 2 multiplied by the number of assets held in the two possible states in node 1 must be equal to the target consumption in these states. The solution to this system of equations is $\theta_1^1(1) = 1$, $\theta_1^2(1) = 2$, i.e., in $t = 1$ given that node 1 is realized an investor has to hold 1 unit of asset 1 and 2 units of asset 2. This portfolio costs $3 \cdot 1 + 2 \cdot 2 = 7$.

Applying the same calculation procedure for node 2, we get

$$1\theta_1^1(2) + 0\theta_1^2(2) = 1.$$

Thus, $\theta_1^1(2) = 1$, $\theta_1^2(2) = 0$, i.e., in $t = 1$ given that node 2 is realized an investor has to hold 1 unit of asset 1 and no asset 2. The portfolio costs are $1 \cdot 1 + (-1) \cdot 0 = 1$.

Applying the above procedure for $t = 0$, we get:

$$(2+3)\theta_0^1 + \quad (1+2)\theta_0^2 = 3+7$$
$$(1+1)\theta_0^1 + (-1+0)\theta_0^2 = 0+1$$

which gives $\theta_0^1 = 13/11, \theta_0^2 = 15/11$.

The same argument can be applied for any other target consumption, since in this example markets are complete. An alternative way to define market completeness is by saying that every new asset is redundant to the already existing assets. Redundancy is defined analogously to the two-period case: in that case it is possible to find prices for the assets such that the introduction of the asset does not change the set of attainable consumption streams. Respectively, any target consumption stream can be also achieved with other assets.

Hence, an asset is redundant if it has payoffs $\{D_t^{K+1}(\omega)\}_{\omega \in \Omega_t}$ which are a (positive) linear combination of the existing assets $k = 1, 2, \ldots, K$, i.e., for some α^k:

$$D_t^{K+1}(\omega) = \sum_{k=1}^{K} \alpha^k(t) D_t^k(\omega) \quad \text{for all } \omega \in \Omega_t\,.$$

Choosing prices according to the linear rule $q^{K+1} = \sum_k \alpha^k q^k$ in every event ω^t we have:

$$\operatorname{rank}\left[A_t(\omega^{t-1}\omega_t) \,\middle|\, D_t^{K+1}(\omega^{t-1}\omega_t) + q_t^{K+1}(\omega^{t-1}\omega_t)\right] = \operatorname{rank} A_t(\omega^{t-1}\omega_t)\,,$$

i.e., the rank of the payoff matrix that includes the payoffs of the redundant assets does not change, since the additional column in the payoff matrix is a linear combination of other columns.

If assets are redundant, they do not create additional insurance possibilities, the efficient frontier in the mean-variance framework and the rank of the payoff matrix in the state-preference model do not change. If there are non-redundant assets, i.e., assets that change the rank of the payoff matrix and the efficient frontier, their inclusion creates additional insurance possibilities. Thus the market cannot have been complete.

5.3 Term Structure of Interest

Now we first want to apply the multi-period model to fixed-income markets. To this end let r_{t_0,t_n} denote the annual interest rate applied for borrowing and lending money between t_0 and t_n. The collection of interest rates $r_{t_0,t_1}, \ldots, r_{t_0,t_T}$ is called the spot rate curve or the term structure of interest rates, which is usually increasing and concave, i.e., borrowing/lending over longer time periods gives higher interest at a decreasing rate. Whereas in the two-period model fixed interest investments were trivial to describe, even adding one more period can reveal interesting effects like the so-called *forward rate bias*. Let us first define the *forward rate*: the forward rate f_{t_0,t_1,t_2} is the (annual) interest rate between t_1 and t_2 that is agreed today for the borrowing and lending between t_1 and t_2, i.e., if I promise today to give to a friend an amount of money in a year and he pays it back in two years together with no interest, then $f_{t_0,t_1,t_2} = 0\,\%$. Since the spot interest rates for the maturities t_1 and t_2 are known at t_0, f_{t_0,t_1,t_2} can be determined by the No-arbitrage Principle. To invest a certain amount of money between t_0 and t_2 in a bond with maturity t_2 is the same as to invest the same amount of money into a bond with maturity $t_1 < t_2$ and into the forward f_{t_0,t_1,t_2}. Therefore, the interest rate over the whole period must be the same in both cases, or in mathematical terms

$$(1 + r_{t_0,t_1})^{t_1}\,(1 + f_{t_0,t_1,t_2})^{t_2-t_1} \stackrel{!}{=} (1 + r_{t_0,t_2})^{t_2}\,,$$

where r_{t_0,t_1} is the annual interest rate of a bond in t_0 which has a maturity of t_1. For the forward rate we get:

$$1 + f_{t_0,t_1,t_2} = \frac{(1 + r_{t_0,t_2})^{\frac{t_2}{t_2-t_1}}}{(1 + r_{t_0,t_1})^{\frac{t_1}{t_2-t_1}}}\,.$$

The forward rate is often seen as the interest rate we expect today for tomorrow. In other words, the realized interest rate between t_1 and t_2, $r_{t_1t_2}$ should be on average the same as the forward rate. The empirical data shows, however, a different picture: the difference between the forward rate and realized interest rate, the so-called forward rate bias, is quite persistent. The forward rate bias is either positive or negative for long periods of time. Even more it seems that if the interest rates are rising, we have a negative forward rate bias, and we have a positive forward rate bias if the interest rates are falling (Figs. 5.3, 5.4 and 5.5). Can these empirical facts (increasing concave term structure leading to a forward rate bias) be explained by an economic model?

5.3.1 Term Structure Without Risk

We consider a three-period economy with $t = 0, 1, 2$ and a representative investor with the utility function

$$U(c) = \ln(c_0) + \frac{1}{1+\delta}\mathbb{E}\left(\ln(c_1)\right) + \frac{1}{(1+\delta)^2}\mathbb{E}\left(\ln(c_2)\right)\,.$$

The agent can trade bonds in $t = 0$ with a time to maturity of 1 or 2 and at the forward rate $f_{0,1,2}$, which we further on denote by f_{12}. His exogenous income (including initial endowment) is w_0, w_1 and w_2. Furthermore, the prices of the

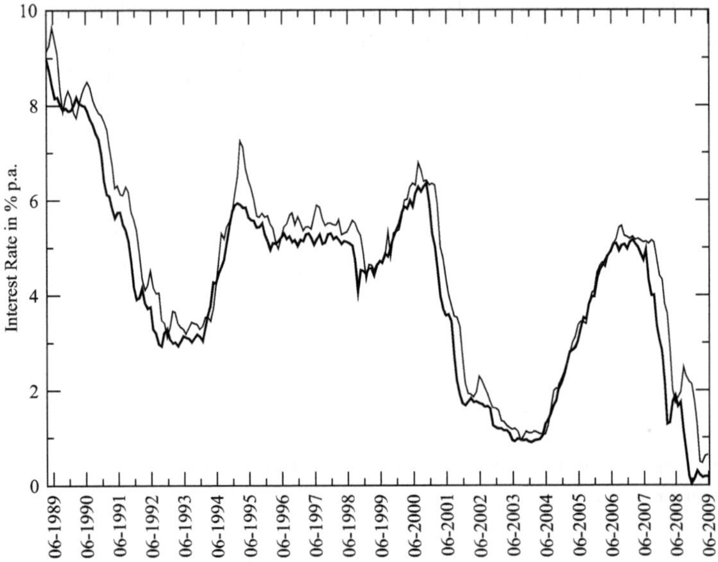

Fig. 5.3 Three-month forward interest rate (*thin line*) vs. three-month spot interest rate (*thick line*) (USA)

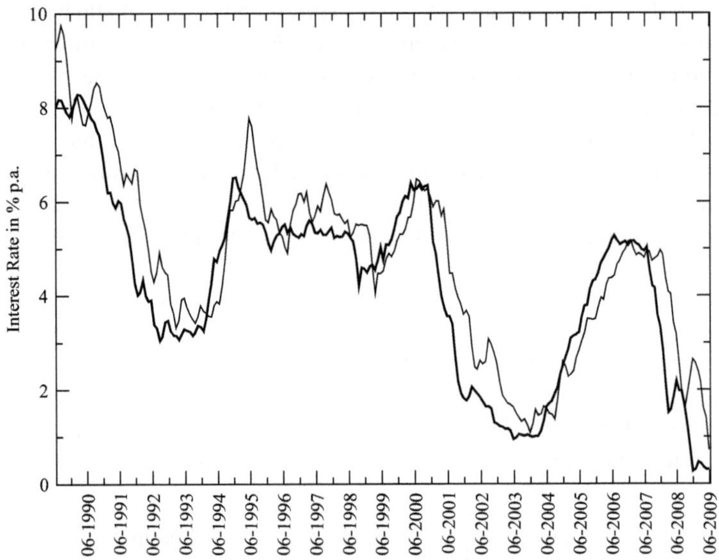

Fig. 5.4 Six-month forward interest rate (*thin line*) vs. six-month spot interest rate (*thick line*) (USA)

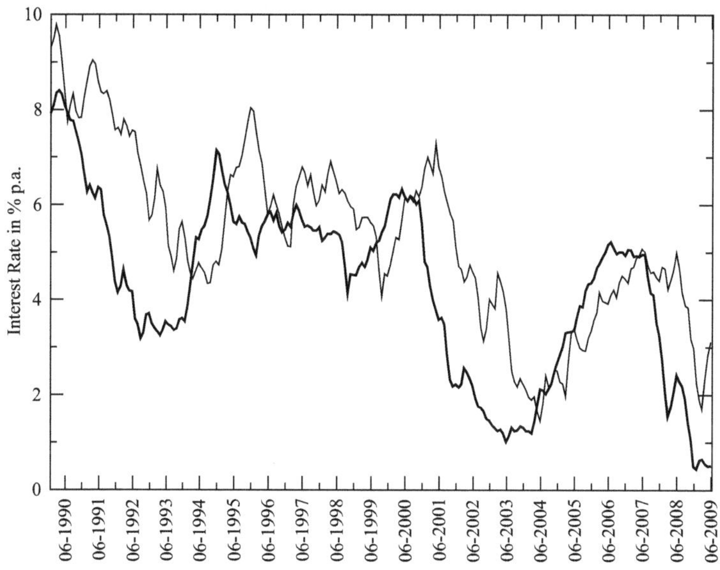

Fig. 5.5 One-year forward interest rate (*thin line*) vs. one-year spot interest rate (*thick line*) (USA)

consumption goods are p_0, p_1 and p_2. We want to determine the term structure and the forward rate at $t = 0$ and the realized spot rate at $t = 1$. Here we denote the amount borrowed/lent in a spot market and in the forward market by s.

The utility maximization problem of the representative investor is:

$$\max_{\substack{c_0,c_1,c_2\\ s_{01},s_{02},s_{12}}} \sum_{t=0}^{2} \frac{1}{(1+\delta)^t} \ln(c_t)\,, \qquad c_t \geq 0,$$

$$\begin{aligned} \text{s.t. } p_0c_0 + s_{01} + s_{02} &= p_0w_0\,, \\ p_1c_1 + s_{12} &= p_1w_1 + (1+r_{01})s_{01}\,, \\ p_2c_2 &= p_2w_2 + (1+f_{12})s_{12} + (1+r_{02})^2 s_{02}\,, \end{aligned}$$

where s_{01}, s_{02} and s_{12} are the investments into the bonds and the forward. The market clearing conditions are $c_0 = w_0$, $c_1 = w_1$ and $c_2 = w_2$. Solving this, we obtain[7]

$$1 + r_{01} = (1+\delta)(1+g_{01})\,, \qquad 1 + r_{02} = (1+\delta)\sqrt{1+g_{02}}\,,$$

$$1 + f_{12} = (1+\delta)(1+g_{12})\,,$$

[7]Re-arrange the budget constraints for c, insert the result in the utility function and differentiate with respect to the bond holdings s. Finally, evaluate the marginal utilities at $c = w$.

where $1+g_{t\,t+1} = p_{t+1}w_{t+1}/(p_t w_t)$ is the nominal growth rate between t and $t+1$. We see that if the growth rate between period 1 and 2 is larger than between 0 and 1, we have an increasing term structure. If the economy is shrinking, the term structure is falling.[8]

Arriving at $t = 1$ the maximization problem of the representative agent is:

$$\max_{c_1,c_2,s_1} \ln(c_1) + \frac{1}{1+\delta}\ln(c_2),$$
$$\text{s.t. } p_1c_1 + s_1 = p_1w_1,$$
$$p_2c_2 = p_2w_2 + (1+r_{12})s_1,$$

where s_1 is the amount of money which the representative investor is saving in 1. By solving the budget constraints for the consumption the optimization problem becomes:

$$\max_{s_1} \ln\left(w_1 - \frac{s_1}{p_1}\right) + \frac{1}{1+\delta}\ln\left(w_2 + (1+r_{1,2})\frac{s_1}{p_2}\right).$$

The first order condition is:

$$\frac{1}{c_1}\frac{-1}{p_1} + \frac{1}{1+\delta}\frac{1}{c_2}\frac{1+r_{1,2}}{p_2} = 0\,.$$

We plug in the market clearing conditions (i.e. $c_1 = w_1$ and $c_2 = w_2$) and then solve for $1+r_{12}$ to get:

$$1+r_{1,2} = (1+\delta)\frac{w_2p_2}{w_1p_1} = (1+\delta)(1+g_{1,2})\,,$$

i.e. $1+r_{12} = (1+\delta)(1+g_{12}) = 1+f_{12}$. In other words, the forward rate is exactly the realized interest rate and there is no forward rate bias. We see that our model is too simple to capture the effects causing a forward rate bias. One idea to make it more realistic would be to consider quasi-hyperbolic instead of exponential time discounting (compare Sect. 2.7). The utility of the representative investor then becomes

$$U^H(c) = \ln(c_0) + \frac{1}{1+\beta}\left(\frac{1}{1+\delta}\mathbb{E}\left(\ln(c_1)\right) + \frac{1}{(1+\delta)^2}\mathbb{E}\left(\ln(c_2)\right)\right),$$

where $\beta > 0$ describes the degree of quasi-hyperbolic discounting.

[8]Note that in reality, the growth rate is unknown while the spot rate curve can be observed at all times. Thus, a falling term structure is typically an indicator for a recession (see [CS09]).

The utility maximization problem of the representative investor in a world without risk is then:

$$\max_{\substack{c_0,c_1,c_2\\ s_{01},s_{02},s_{12}}} \ln(c_0) + \frac{1}{1+\beta}\left(\frac{1}{1+\delta}\ln(c_1) + \frac{1}{(1+\delta)^2}\ln(c_2)\right),$$

$$\begin{aligned}
\text{s.t. } p_0c_0 + s_{01} + s_{02} &= p_0w_0\,,\\
p_1c_1 + s_{12} &= p_1w_1 + (1+r_{01})s_{01}\,,\\
p_2c_2 &= p_2w_2 + (1+f_{12})s_{12} + (1+r_{02})^2 s_{02}\,.
\end{aligned}$$

The market clearing conditions are $c_0 = w_0$, $c_1 = w_1$ and $c_2 = w_2$. This problem can be solved in the same way as before. For the interest rates we obtain

$$\begin{aligned}
&1 + r_{01} = (1+\beta)(1+\delta)(1+g_{01})\,, \quad 1 + r_{02} = \sqrt{1+\beta}\,(1+\delta)\sqrt{1+g_{02}}\,,\\
&1 + f_{12} = (1+\delta)(1+g_{12})\,.
\end{aligned}$$

In $t = 1$, the utility function of the representative investor is $\ln(c_1) + \frac{1}{1+\beta}\frac{1}{1+\delta}\ln(c_2)$. Analogously to before the realized interest rate becomes

$$1 + r_{12} = (1+\beta)(1+\delta)(1+g_{12})\,.$$

With $\beta > 0$, we obtain a negative forward rate bias. Thus, hyperbolic discounting implies a negative forward rate bias, but in reality positive and negative forward rate biases can be observed. Moreover, hyperbolic discounting would lead to a decreasing term structure, contrary to what we usually observe on the market. Thus, we need to extend our model into a different direction.

5.3.2 Term Structure with Risk

So far we have only considered a world without risk. Let us now include risk into our model so that in $t = 1$ the economy can develop better or worse, i.e. we have two states, an up and a down state. In $t = 1, 2$ the exogenous income and the prices of the consumption good depend on the state occurred.

The utility maximization problem of the representative investor with beliefs for the occurrence of the up sate is therefore

$$\max_{\substack{c_0,c_{1u},c_{1d},c_{2u},c_{2d}\\ s_{01},s_{02},s_{12}}} \ln(c_0) + \text{prob}\left(\frac{1}{1+\delta}\ln(c_{1u}) + \frac{1}{(1+\delta)^2}\ln(c_{2u})\right)$$
$$+ (1-\text{prob})\left(\frac{1}{1+\delta}\ln(c_{1d}) + \frac{1}{(1+\delta)^2}\ln(c_{2u})\right),$$

$$\begin{aligned}
\text{s.t. } p_0c_0 + s_{01} + s_{02} &= p_0w_0\,,\\
p_{1u}c_{1u} + s_{12} &= p_{1u}w_{1u} + (1+r_{01})s_{01}\,,\\
p_{1d}c_{1d} + s_{12} &= p_{1d}w_{1d} + (1+r_{01})s_{01}\,,\\
p_{2u}c_{2u} &= p_{2u}w_{2u} + (1+f_{12})s_{12} + (1+r_{02})^2 s_{02}\,,\\
p_{2d}c_{2d} &= p_{2d}w_{2d} + (1+f_{12})s_{12} + (1+r_{02})^2 s_{02}\,,
\end{aligned}$$

where s_{01}, s_{02} and s_{12} are the investments into the bonds and the forward.[9] The market clearing conditions are $c_0 = w_0$, $c_{1u} = w_1$, $c_{1d} = w_{1d}$, $c_{2u} = w_{2u}$ and $c_{2d} = w_{2d}$. Denoting by $\mathbb{E}(x) = \text{prob}\, x_u + (1-\text{prob})\, x_d$ and solving this problem gives

$$1 + r_{01} = (1+\delta)\frac{1}{\mathbb{E}\left(\frac{1}{1+g_{01}}\right)}\,, \qquad 1 + r_{02} = (1+\delta)\frac{1}{\sqrt{\mathbb{E}\left(\frac{1}{1+g_{02}}\right)}}\,,$$

$$1 + f_{12} = (1+\delta)\frac{\mathbb{E}\left(\frac{1}{1+g_{01}}\right)}{\mathbb{E}\left(\frac{1}{1+g_{02}}\right)}\,,$$

where g_{tt+1} is the nominal growth rate between t and $t+1$. r_{12s}, the interest rate realized in $t = 1$, depends on the state s of the economy: in the up state we have

$$1 + r_{12u} = (1+\delta)(1+g_{12u})\,,$$

and in the down state we have

$$1 + r_{12d} = (1+\delta)(1+g_{12d})\,.$$

[9] Note that s_{12} does not depend on the state u or d. This is because the forward contract is written in $t = 0$ without conditioning on the state realized in $t = 1$. The amount borrowed/lent will be effective for the budget constraint in $t = 1$.

The expected return is therefore

$$\mathbb{E}(1 + r_{12}) = (1 + \delta)\mathbb{E}(1 + g_{12}) .$$

It is quite difficult to analyze this problem in general. Therefore, we look at an example with specific parameters: the nominal growth rate is $g_{t,t+1,s} = (w_{t+1,s}\,p_{t+1,s})/(w_{t,s}\,p_{t,s}) - 1$ and we set $\delta = 0.1$, prob $= 0.5$, $g_{0,1,u}(u) = g_{1,2,u} = \frac{1}{9}$, $g_{0,1,d} = -\frac{1}{21}$ and $g_{0,2,d} = 0$.

With these parameters, we get for the interest rates in $t = 0$: $1 + r_{01} = 1.128$, $1 + r_{02} = 1.156$ and $1 + f_{12} = 1.185$. The interest rates realized in $t = 1$ depend on the state: $1 + r_{12u} = 1.222$, $1 + r_{12d} = 1.155$ and $\mathbb{E}\,(1 + r_{12}) = 1.189$.

The shape of the term structure can be explained similarly as in the economy without risk. The numerical example shows that in the upper state the realized interest rate rises and we have a negative forward rate bias. In the down state just the opposite happens. This is in line with the empirical observations.

5.4 Arbitrage in the Multi-period Model

We have defined in Chap. 4 an *arbitrage opportunity* as a riskless profit and we have seen that in equilibrium, there cannot be any arbitrage opportunity, since otherwise adding an arbitrage opportunity would improve the portfolio of an investor and thus contradict the optimality assumption in the definition of a market equilibrium. The same holds for the multi-period case, and we can extend the definition of arbitrage as follows:

Definition 5.5 An *arbitrage* is a self-financing trading strategy, i.e., there is some strategy θ_t with $\theta_{-1} = 0$ such that for all $t = 0, 1, 2, \ldots, T$,

$$\theta_t^{\text{cons}} + \sum_{k=1}^{K} q_t^k \theta_t^k = \sum_{k=1}^{K} (D_t^k + q_t^k)\theta_{t-1}^k$$

for which the resulting consumption is positive: $\theta_t^{\text{cons}} > 0$, i.e., $\theta_t^{\text{cons}}(\omega^t) \geq 0$ for all ω^t and all $t = 0, 1, 2, \ldots, T$, and $\theta^{\text{cons}} \neq 0$.[10]

5.4.1 Fundamental Theorem of Asset Pricing

The non-existence of arbitrage opportunities is fundamental for asset pricing. Since these prices must be consistent with the optimization calculus of agents with different utility functions and different risk aversions, we use the so-called

[10]We give only the strict monotonic variant of arbitrage (compare Chap. 4 for other definitions).

risk-neutral measures $\pi_{t=1,2,...,T}$ to build expected values. The existence of this measure is guaranteed by the absence of arbitrage according to the Fundamental Theorem of Asset Pricing (FTAP). In the multi-period case this theorem becomes:

Theorem 5.6 (FTAP) *There is no arbitrage opportunity if and only if there is a state price process* $\pi_{t=1,2,...,T} \gg 0$ *such that for all* ω^t

$$q_{t-1}^k(\omega^{t-1}) = \frac{1}{\pi_{t-1}(\omega^{t-1})} \sum_{\omega_t \in \Omega_t} \pi_t(\omega^t)(D_t^k(\omega^t) + q_t^k(\omega^t)) \tag{5.1}$$

where $\omega^t = \omega^{t-1}\omega_t$.

In other words, we can price assets by discounting their payoffs with respect to the state prices, which depend on agent's preferences only indirectly through the dependence on asset prices. Again, state prices are not equivalent to the so-called physical (or objective) probability measure P. The state prices are a theoretical construct that helps to find the fair[11] price of payoffs. In particular, $\pi_{t=1,2,...,T}$ is the price of an elementary security paying 1 only in state ω_t of the path ω^t. The proof of Theorem 5.6 is a straightforward extension of the two-period case (compare Theorem 4.2):

Proof The proof showing that the existence of a positive state price process rules out arbitrage opportunities is straightforward. Consider any self-financing strategy, i.e., suppose that for all $t = 0, 1, 2$, and for all ω^t the budget constraints are satisfied:

$$\theta_t^{\text{cons}} + \sum_{k=1}^{K} q_t^k \theta_t^k = \sum_{k=1}^{K} (D_t^k + q_t^k)\theta_{t-1}^k .$$

Multiplying both sides by $\pi_t(\omega^t)$ and summing up over ω^t allows us to use (5.1):

$$\sum_{\omega_t} \pi_t(\omega^t) \left(\theta_t^{\text{cons}} + \sum_{k=1}^{K} q_t^k \theta_t^k \right) = \sum_{\omega_t} \pi_t(\omega^t) \left(\sum_{k=1}^{K} (D_t^k + q_t^k)\theta_{t-1}^k \right)$$
$$= \sum_{k=1}^{K} q_{t-1}^k(\omega^{t-1})\theta_{t-1}^k(\omega^{t-1}).$$

[11] In the insurance context, "fair" means that the insurance premium must be equal to the expected damages.

Adding this along all paths $\omega^t = (\omega_0, \omega_1, \ldots, \omega_t)$ gives:

$$\sum_t \sum_{\omega_t} \pi_t(\omega^t)\theta_t^{\text{cons}}(\omega^t) = \sum_{k=1}^{K} q_0^k \theta_{-1}^k .$$

Now, if $\theta_t^{\text{cons}}(\omega^t) > 0$ and $\pi_t(\omega^t) \gg 0$ this would require $\theta > 0$, saying that a positive payoff incurs positive costs, ruling out arbitrage opportunities.

For the other direction of the proof we refer to [MQ02] (the key idea is to use the separation theorem as in the two-period model). □

Note that in this proof we reduced the sequence of budget constraints to a single budget constraint in terms of present values of all future expenditures and incomes. This single budget constraint always needs to hold. If markets are incomplete, then in addition one needs to make sure that the process of gaps between an agent's expenditure and his income can be achieved with the set of available assets.

5.4.2 Consequences of No-Arbitrage

Similar to the two-period case, we can find two important consequences of the absence of arbitrage: first, the Law of One Price says that if from period t onwards two assets have identical dividend processes, then in period $t-1$ they must have the same price.

To see this, suppose that $q_{t-1}^1 < q_{t-1}^2$ and $D_\tau^1 = D_\tau^2$ for $\tau \geq t$. Buying θ_{t-1}^1 units of the cheaper asset and selling the same amount of the expensive asset gives $(q_{t-1}^2 - q_{t-1}^1)\theta_{t-1}^1 > 0$ in $t-1$, and in all other periods where the portfolio is hold, i.e. $\theta_t^k = \theta_{t-1}^k$ the portfolio is hedged, i.e.,

$$\sum_{k=1}^{K} q_t^k \theta_t^k = \sum_{k=1}^{K} (D_t^k + q_t^k)\theta_{t-1}^k ,$$

since $\theta_t^k = 0$ for $k \neq 1, 2$ and $\theta_t^k = \theta_{t-1}^k$ for $k = 1, 2$ and $\theta_t^1 = -\theta_t^2$.

The second consequence of the No-arbitrage Principle is the concept of linear pricing: it says that if in period $t-1$ one buys a portfolio $\hat{\theta}_{t-1}$ and later on holds it fixed, then in $t-1$ the price of the portfolio must be a linear combination of the prices of its components:

$$q_{t-1}(\hat{\theta}) = \hat{\theta}_{t-1} q_{t-1} = \sum_{k=1}^{K} \hat{\theta}_{t-1}^k q_{t-1}^k .$$

To see this, suppose that $q_{t-1}(\hat{\theta}) > \hat{\theta}_{t-1} q_{t-1}$. Then, buying $\hat{\theta}_{t-1}^k$ units of asset k and selling the portfolio $\hat{\theta}_{t-1}$ gives $q_{t-1}(\hat{\theta}) - \hat{\theta}_{t-1} q_{t-1} > 0$ in $t-1$. Otherwise the position is hedged:

$$q_t(\hat{\theta})\hat{\theta}_t + \sum_{k=1}^{K} q_t^k \hat{\theta}_t^k = (D_t(\hat{\theta}) + q_t(\hat{\theta}))\theta_{t-1}(\hat{\theta}) + \sum_{k=1}^{K} (D_t^k + q_t^k)\hat{\theta}_{t-1}^k ,$$

because $D_t(\hat{\theta}) = \sum_{k=1}^{K} D_t^k \hat{\theta}_t^k$ and $\hat{\theta}_t^k(\omega^t) = \hat{\theta}_{t-1}^k(\omega^{t-1})$.

5.4.3 Applications to Option Pricing

The Fundamental Theorem of Asset Pricing is essential for the valuation of redundant assets such as derivatives. We have seen already in Chap. 4 that there are two possible ways to determine the value of a derivative. The first approach is based on determining the value of a *hedge portfolio*. This is a portfolio of assets that delivers the same payoff as the derivative. The second approach uses the *risk-neutral probabilities* in order to determine the current value of the derivative's payoff. To value an option in a *multi-period* setting, we use the binomial lattice model (see Fig. 5.6). Note that the risk-neutral probability is a stationary measure,

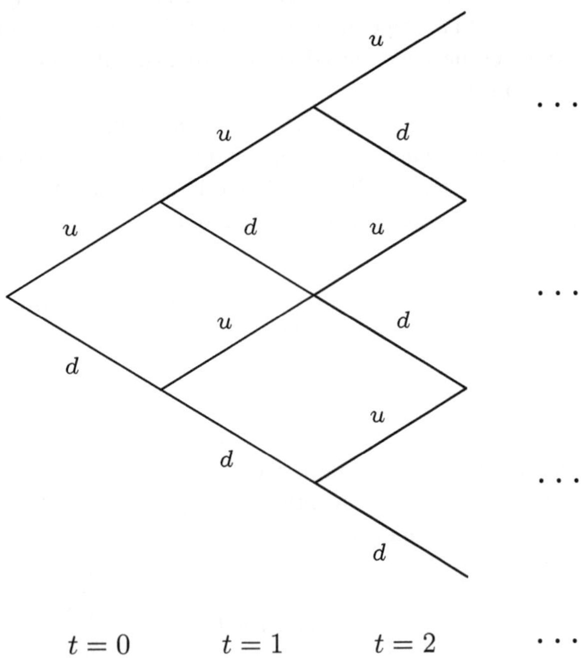

Fig. 5.6 Binomial lattice model

i.e., it remains the same at every node. To see this,[12] suppose that at some node of the binomial lattice the stock price is S. Then, its expected value after one period is $\mathbb{E}_{\pi^*}(S) = \pi^* Su + (1-\pi^*)Sd$, where u, d are the gross returns in the up and the down state. In a riskless world this value must be equal to SR, in other words $\mathbb{E}_{\pi^*}(S) = \pi^* Su + (1-\pi^*)Sd = SR$. Since S cancels, we get that the risk-neutral probability is constant over time and depends only on the size and the frequency of "up" and "down" movements. Consider an example of a call option over the periods $t = 0, 1, 2$. The value of this option in $t = 1$ depends on the realized state ("up" or "down"), i.e.:

$$C_u := \frac{1}{R}\left(\pi^* C_{uu} + (1-\pi^*)C_{ud}\right),$$

$$C_d := \frac{1}{R}\left(\pi^* C_{ud} + (1-\pi^*)C_{dd}\right).$$

The value of the call at $t = 0$ is

$$C = \frac{1}{R^2}\left((\pi^*)^2 C_{uu} + 2\pi^*(1-\pi^*)C_{ud} + (1-\pi^*)^2 C_{dd}\right),$$

i.e.,

$$C = \frac{1}{R^2}\Big((\pi^*)^2 \max\{0, u^2S - K\} + 2\pi^*(1-\pi^*)\max\{0, udS - K\} \\ + (1-\pi^*)^2 \max\{0, d^2S - K\}\Big).$$

We can continue this argument for more and more periods to obtain the hypergeometric distribution, which in the limit gives the normal distribution (see [Var78]).

5.4.4 Stock Prices as Discounted Expected Payoffs

Suppose we have two assets where the first asset is short-lived and risk-free (e.g., a one-period loan-saving contract) and the second is risky. Then, applying (5.1) gives:

$$q^1_{t-1}(\omega^{t-1}) = \frac{1}{\pi_{t-1}(\omega^{t-1})} \sum_{\omega_t \in \Omega_t} \pi_t(\omega^{t-1}\omega_t)(\underbrace{D^1_t(\omega^{t-1}\omega_t)}_{1} + \underbrace{q^1_t(\omega^{t-1}\omega_t)}_{0}). \tag{5.2}$$

Since the security is riskless, its payoff $D^1_t(\omega^t)$ is equal to 1 in all states. Under the assumption that the asset lives only one period, there is no price for it after this period, i.e., $q^1_t(\omega^t)$ is equal to 0.

[12] We use the same notation as for the similar example given in Chap. 4.

This is equivalent to:

$$q_{t-1}^1(\omega^{t-1}) = \frac{1}{\pi_{t-1}(\omega^{t-1})} \sum_{\omega_t \in \Omega_t} \pi_t(\omega^{t-1}\omega_t) \equiv \frac{1}{1 + r_{ft-1}(\omega^{t-1})}.$$

Using this and the condition (5.2), we get

$$q_{t-1}^k(\omega^{t-1}) = \frac{1}{1 + r_{ft-1}(\omega^{t-1})} \sum_{\omega_t \in \Omega_t} \pi_t^*(\omega^{t-1}\omega_t)(D_t^k(\omega^{t-1}\omega_t) + q_t^k(\omega^{t-1}\omega_t)), \tag{5.3}$$

where

$$\pi_t^*(\omega^t) = \frac{\pi_t(\omega^t)}{\sum_{\omega_t \in \Omega_t} \pi_t(\omega^{t-1}\omega_t)} > 0$$

is indeed a (risk-neutral) probability measure based on the information of period $t-1$. Hence, asset prices can be presented as discounted expected payoffs, conditional on the information available at the time of valuation. This is a sequence of events (or a path) realized from the beginning until $t-1$.

Forward iteration of (5.3) along paths yields the discounted dividends model:

$$q_{t-1}^k(\omega^{t-1}) = \mathbb{E}_{\pi_{t-1}^*(\omega^{t-1})} \sum_{\tau=t}^{\infty} \frac{D_\tau^k(\omega^\tau)}{\Pi_{\tau'=t-1}^{T-1}(1 + r_{f\tau'}(\omega^{\tau'}))}$$

Thus, if a market is rational, price movements will depend only on movements of the risk-free interest rate and the expected dividend payments. Assuming that the dividend process follows a random walk, we can conclude that perfectly anticipated prices must be random, i.e.,

$$\mathbb{E}_{\pi_t^*}(q_{t+1}^k - (1 + r_t)q_t^k) = -\mathbb{E}_{\pi_t^*}(D_{t+1}^k).$$

This random character of stock prices has been summarized by Cootner [Coo64, page 232]:

> The only price changes that would occur [in a financial market] are those that result from new information. Since there is no reason to expect information to be non-random in appearance, the period-to-period price changes of stock should be random movements, statistically independent of one other.

Expressing the no-arbitrage condition in terms of excess returns (returns exceeding the riskless return) we get $\mathbb{E}_{\pi_t^*}(R_{t+1}^k - R_{ft}) = 0$. In other words, the net present value of a strategy with respect to the risk-neutral probability must be equal to 0. Positive net present values are possible only if one uses a probability measure different from π^*, however in this case the probability measure used does not include all possible risks.

5.4.5 Equivalent Formulations of the No-Arbitrage Principle

According to the Fundamental Theorem of Asset Pricing, if a price process is arbitrage free, there exists no strategy θ_t, $t = 0, 1, 2, \ldots, T$, that generates risk-free excess returns.

This is equivalent to the existence of a market expectation or a risk-neutral probability such that

$$q_t^k = \frac{1}{1+r_t}\mathbb{E}_{\pi_t^*}(D_{t+1}^k + q_{t+1}^k).$$

Applying forward iteration to the expression above in order to get a result which is not dependent on future realizations, we get the Dividend Discount Model (DDM)[13]:

$$q_t^k = \mathbb{E}_{\pi_t^*}\left(\sum_{\tau=t+1}^{\infty}\left(\frac{1}{1+r}\right)^{\tau-t} D_\tau^k\right), \qquad t = 0, 1, 2, \ldots, T.$$

There are no expected gains $E_{\pi_t^*}(G_{t+1} - G_t) = 0$, i.e., the gains process is a *martingale*, where $G_t = \sum_{\tau=1}^{t} g_\tau \theta_{t-1}$ for some portfolio strategy θ and

$$g_{t+1}^k = \left(\frac{1}{1+r}\right)^t \left[\frac{1}{1+r}(D_{t+1}^k + q_{t+1}^k) - q_t^k\right]$$

is the discounted gain from holding asset k from t till $t+1$. Hence, the cumulative expected gains are zero:

$$E_{\pi_t^*}\left(\sum_{\tau=t+1}^{\infty} g_\tau \theta_{\tau-1}\right) = 0$$

That is to say: "Nobody can beat the market", i.e., you cannot beat a martingale.

To summarize, the absence of arbitrage is equivalent to the conclusion that gains are martingales, prices are random and do not generate excess returns with respect to the risk-neutral probability π^*.[14]

[13] To simplify expressions we have assumed a constant interest rate.

[14] See [HK78] for more details.

5.4.6 Ponzi Schemes and Bubbles

Up to now we have left the choice of time horizon in our multi-period model open: there could have been finitely many periods ($T < \infty$) or infinitely many ("$T = \infty$"). There are, however, some interesting theoretical issues arising in the infinite time horizon settings that we discuss in this section.

The first is the so-called *Ponzi scheme*. A Ponzi scheme is probably the only theoretical concept in economics that is named after an outright criminal, Charles Ponzi. In 1920 he attracted enormous investments and paid huge interest for it – but only by using newly arriving investments, thus getting deeper and deeper into debt. At the same time he financed a luxurious life from this and even bought a private bank. Finally, the whole scheme broke down, and he ended in prison for many years.

The idea of using the money collected from subsequent investors to repay the initial investors and their interest was not invented by him, but since his was probably the most famous case, the scheme was named after him. In economics, Ponzi schemes are defined in a more abstract way: borrowing money from the future to repay debts now and, at the same time, finance consumption. To model this more precisely, we consider a set of short-lived bonds D^t with payoff

$$D^t(\omega) = \begin{cases} 1 & \text{if } \omega \in \Omega_t\,, \\ 0 & \text{otherwise.} \end{cases}$$

With their help we can construct the following consumption path:

$$c_t = w_t + 1 \text{ for all } t = 0, 1, 2, \ldots, T,$$

where w_t is the exogenous income. Taking a closer look, this looks very attractive: we get something extra in *every* period in time! It seems impossible to finance such a consumption stream without a large initial endowment, but in fact we can do this without any initial endowment: define the investment strategy $\theta_{t=0,1,\ldots,T}$ as

$$\theta_0 = (-R, 0, \ldots),\ \ldots,\ \theta_t = (0, \ldots, 0, -\sum_{i=1}^{t} R^i, 0, \ldots)\,,$$

where θ_t is different from zero in its t-th component (Fig. 5.7).

This means nothing else than that consumption in period $t = 0$ is financed by going short on the first bond with a total amount of R. To finance consumption in period $t = 1$ and to pay back the first bond one issues a second bond with a total amount of $R + R^2$, etc. Formally, we get in period $t = 0$:

$$\theta_0^1 q_0^1 + c_0 = -R\frac{1}{R} + c_0 = w_0\,.$$

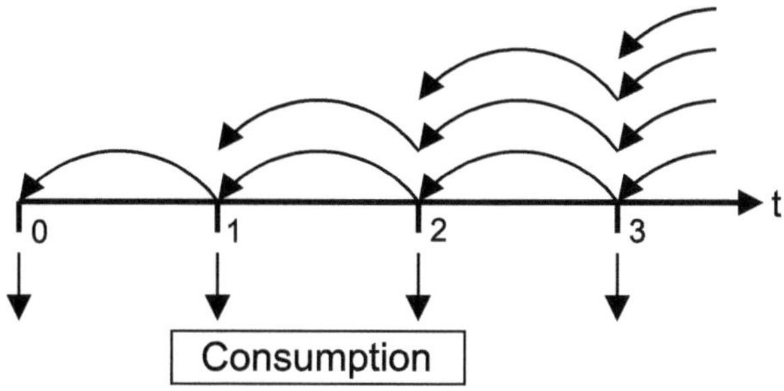

Fig. 5.7 Illustration of the Ponzi scheme

Analogously for any arbitrary period t,

$$\theta_t^{t+1} q_t^{t+1} + c_t = -\sum_{i=1}^{t+1} R^i \frac{1}{R} + c_t = \theta_{t-1}^t + w_t = -\sum_{i=1}^{t} R^i + w_t \,.$$

Thereby we get $c_t = w_t + 1$ for all t as mentioned above. This leads to immediate problems. For instance, there does not exist any maximum utility, since the choice set becomes unbounded. Whatever we have, we can still get more. An arbitrage opportunity arises out of seemingly innocent short-term bonds!

There are two ways of solving this problem:

- We can restrict our multi-period model to a finite time horizon $T < \infty$.
- We can impose a *transversality condition*:

$$\lim_{t\to\infty} R^t \left(\sum_{k=1}^{K} \theta_t^k q_t^k \right) > 0 \,,$$

where $\sum_{k=1}^{K} \theta_t^k q_t^k$ is the total value of the portfolio.

Are Ponzi schemes nowadays extinct? Not really: there are still many fraudulent schemes like this, disguised under various, often creative names. How long can a Ponzi scheme run? Since the input of money is bounded in reality, such schemes are doomed to collapse. Charles Ponzi's scheme broke down after several months – the interest rates he promised were just too high. The lower the interest rates promised, the longer such a scheme can live.

The most recent example was the now infamous hedge fund run by Bernard Madoff, former chairman of the NASDAQ stock exchange, from the 1980s to December 2008. His returns were always around 10 %. Nothing outrageously high, but still substantially above the risk-free rate and even above the average performance of stocks, yet seemingly without much variance. Even in times of stock market declines like after the end of the internet bubble (2000–2002) or during the financial crisis (2007–2008), his hedge fund still generated steady returns. From December 1990 to May 2005 there were in fact only seven months (out of 174) with negative returns!

That Madoff was quite silent about *how* he generated these returns was not remarkable: no hedge fund would share its secrets and thus destroy the sole base of its business. What was maybe more remarkable is that even though the hedge fund got checked a few times by the SEC and other regulatory commissions, there seemed to have been found nothing wrong with it. In the end, 50 billion US dollars were invested into the fund, some of it directly, some of it via a network of hedge funds and investment companies worldwide that either knew or didn't know what was really going on (we might never know!): the biggest Ponzi scheme of all times!

The way Madoff had marketed his fund was fundamentally different from Ponzi's. This allowed the fund to grow over such a long time and to such a large size. The end, however, was similar: some investors wanted to withdraw their money, but there was none left, and Madoff had to admit that everything had been a giant Ponzi scheme. Not only rich private investors that knew Madoff personally and trusted him as a respectable person had lost their money (and the money of several charity funds), but also various hedge funds and even larger banks had lost billions.

Are all Ponzi schemes designed by criminals collecting money of investors by pretending to invest it? Or are there other phenomena that can be interpreted as Ponzi schemes in the economic sense? Let us consider the following prominent real life example: credit cards. To finance consumption some people obtain a credit by getting a credit card. A second credit card is later used to repay the debt on the first credit card and also to finance new consumption, etc. In praxis such strategies are used by a substantial number of people, in particular in the USA where the average credit card debts is particularly high.

But not only Ponzi schemes can pose a problem for an infinite horizon model. *Bubbles* are a further challenge. Similar to a Ponzi scheme, a bubble relies on expectations of large profits in the future. However in this case no debts are accumulated. Consider an asset with $D^t = 1$ for all t (a "console"). We can compute its price at $t = 0$:

$$q_0 = \sum_{t=1}^{\infty} \pi(\omega_t)\,.$$

Using

$$\pi(\omega_t) = \delta^t \frac{u'(c_t)}{u'(c_0)}\,,$$

which we know from the two-period model (Sect. 4.1), we get:

$$q_0 = \sum_{t=1}^{\infty} \delta^t \frac{u'(c_t)}{u'(c_0)}\,.$$

Analogously we compute at $t = 1$:

$$
\begin{aligned}
q_1 &= \sum_{t=2}^{\infty} \pi(\omega_t) = \sum_{t=2}^{\infty} \delta^{t-1} \frac{u'(c_t)}{u'(c_1)} \\
&= \delta^{-1} \frac{u'(c_0)}{u'(c_1)} \sum_{t=2}^{\infty} \delta^t \frac{u'(c_t)}{u'(c_0)} \\
&= \delta^{-1} \frac{u'(c_0)}{u'(c_1)} \left(q_0 - \delta \frac{u'(c_1)}{u'(c_0)} \right) .
\end{aligned}
$$

For an arbitrary t, we get

$$
q_{t+1} = \delta^{-1} \frac{u'(c_t)}{u'(c_{t+1})} \left(q_t - \delta \frac{u'(c_{t+1})}{u'(c_t)} \right) .
$$

Hence,

$$
\begin{aligned}
q_t &= \delta \frac{u'(c_{t+1})}{u'(c_t)} + \delta \frac{u'(c_{t+1})}{u'(c_t)} q_{t+1} \\
&=: M_{t+1} + M_{t+1} q_{t+1} .
\end{aligned}
$$

We define $M'_t := \prod_{\tau=1}^{t} M_\tau$ and obtain by iteration

$$
\begin{aligned}
q_0 &= M_1 + M_1 q_1 = M_1 + M_1(M_2 + M_2 q_2) = \cdots = \\
&= \underbrace{\sum_{t=1}^{\infty} M'_t}_{\text{fundamental}} + \underbrace{\lim_{T\to\infty} M'_T q_T}_{\text{bubble}} .
\end{aligned}
$$

Whereas the first component is the classical "fundamental value" price of the asset, the second component only depends on expected later prices and is therefore in fact a "bubble" component. This decomposition implies in particular that the price is not uniquely determined anymore. But even worse, a bubble can lead to a positive value for an asset without any dividend or payment! Let $D_t = 0$ for all t, then

$$
q_0 = \lim_{T\to\infty} M'_T q_T
$$

can be positive (compare [Bew80]). One could argue that money, having no dividend payment nor interest, has just a value because it is a bubble, but maybe this point of view is taking it a bit too far...

For many applications, we would like to exclude bubbles. This can be achieved by either restricting the analysis to a finite time horizon or by imposing the condition

$$\lim_{T\to\infty} M_T' q_T = 0 \,.$$

It is, however, interesting to know that bubbles occur naturally in an infinite horizon model, given that they also form on real financial markets.

5.5 Pareto Efficiency

5.5.1 First Welfare Theorem

As in the two-period case (see Sect. 4.1), we can prove that market equilibria in complete markets are Pareto-optimal, i.e., there is no improvement for investors possible that don't make some other investors worse off. We first have to generalize the necessary definitions to the multi-period case:

Definition 5.7 (Attainability) An allocation of consumption streams $\{\{\theta_t^{i,\text{cons}}(\omega^t)\}_{t=0}^T\}_{i=1}^I$ is attainable if each component is in the consumption set of the agent and if it does not use more consumption than is available from the dividend process D and exogenous endowments:

$$\sum_{i=1}^{I} \theta_t^{i,\text{cons}}(\omega^t) = \sum_{k=1}^{K} D_t^k(\omega^t) + \sum_{i=1}^{I} w_t^i(\omega^t) \quad \text{for every } \omega^t,\ t = 0,1,2,\ldots,T.$$

Accordingly, Pareto efficiency needs to be adapted.

Definition 5.8 (Pareto efficiency) In a financial market the allocation of consumption streams $\{\{\theta_t^{i,\text{cons}}(\omega^t)\}_{t=0}^T\}_{i=1}^I$ is *Pareto efficient* if and only if it is attainable and there does not exist an alternative attainable allocation of consumption streams $\{\{\theta_t^{i,\text{cons}}(\omega^t)\}_{t=0}^T\}_{i=1}^I$, such that no consumer is worse off and some consumer is better off:

$$U^i\big(\{\theta_t^{i,\text{cons}}(\omega^t)\}_{t=0}^T\big) \geq U^i\big(\{\theta_t^{i,\text{cons}}(\omega^t)\}_{t=0}^T\big) \text{ for all } i, \text{ and } > \text{ for some } i.$$

Now we are in a position to state the First Welfare Theorem.

Theorem 5.9 (First Welfare Theorem) *In a complete financial market, the allocation of consumption streams in any market equilibrium is Pareto efficient.*

Note that market efficiency does not rule out that some agents consume much more than others. From the perspective of fairness, this might not be optimal.

Proof (First Welfare Theorem) Suppose $\{\{\theta_t^{i,\text{cons}}(\omega^t)\}_{t=0}^T\}_{i=1}^I$ is an attainable allocation that is Pareto-better than the financial market allocation, i.e.,

$$U^i\left(\{\theta_t^{i,\text{cons}}(\omega^t)\}_{t=0}^T\right) \geq U^i\left(\{\theta_t^{i,\text{cons}}(\omega^t)\}_{t=0}^T\right) \text{ for all } i, \text{ and } > \text{ for some } i.$$

Why did the agents not choose $\{\{\theta_t^{i,\text{cons}}(\omega^t)\}_{t=0}^T\}_{i=1}^I$?

Since markets are complete the alternative allocation must have been too expensive:

$$\sum_t \sum_{\omega_t} \pi_t(\omega^{t-1}\omega_t)\theta_t^{i,\text{cons}}(\omega^{t-1}\omega_t) > \sum_t \sum_{\omega_t} \pi_t(\omega^{t-1}\omega_t)\theta_t^{i,\text{cons}}(\omega^{t-1}\omega_t).$$

Summing up over consumers gives:

$$\sum_i \sum_t \sum_{\omega_t} \pi_t(\omega^{t-1}\omega_t)\theta_t^{i,\text{cons}}(\omega^{t-1}\omega_t) > \sum_i \sum_t \sum_{\omega_t} \pi_t(\omega^{t-1}\omega_t)\theta_t^{i,\text{cons}}(\omega^{t-1}\omega_t),$$

therefore (by market clearing)

$$\sum_t \sum_{\omega_t} \pi_t(\omega^{t-1}\omega_t)\Bigg[\underbrace{\sum_i \theta_t^{i,\text{cons}}(\omega^{t-1}\omega_t) + \sum_i w_t^i(\omega^{t-1}\omega_t)}_{\sum_k D_t^k(\omega^{t-1}\omega_t)}\Bigg]$$

$$> \sum_t \sum_{\omega_i} \pi_t(\omega^{t-1}\omega_t)\Bigg[\underbrace{\sum_i \theta_t^{i,\text{cons}}(\omega^{t-1}\omega_t) + \sum_i w_t^i(\omega^{t-1}\omega_t)}_{\sum_k D_t^k(\omega^{t-1}\omega_t)}\Bigg]$$

which is a contradiction. □

The result so far only uses that the agents' utility functions are increasing in $\theta_t^{i,\text{cons}}$. Moreover, if the functions are smooth, i.e. continuously differentiable, and if boundary solutions are ruled out, Pareto efficiency can be defined using the consumers' marginal rates of substitution and we have for all investors i, j:

$$MRS_{s,z}^i = \frac{\partial_{\theta_s^{\text{cons}}} U^i(c^i)}{\partial_{\theta_z^{\text{cons}}} U^i(c^i)} = \frac{\partial_{\theta_s^{\text{cons}}} U^j(c^j)}{\partial_{\theta_z^{\text{cons}}} U^j(c^j)} = MRS_{s,z}^j. \tag{5.4}$$

The graphical representation of this efficiency concept can also be illustrated in the Edgeworth Box (compare Fig. 4.8).

5.5.2 Aggregation

As in the two-period model we can now use Pareto efficiency to aggregate a heterogeneous investor economy into a single representative investor decision problem. As a result, asset prices can be described by marginal rates of substitution based on aggregate consumption. If the representative agent's utility is of the expected utility type, then in any point in time we get

$$q_{t-1}^k(\omega^{t-1}) = \frac{\delta_{t-1}(\omega^{t-1})}{u'(c(\omega^{t-1}))} \sum_{\omega_t \in \Omega_t} p_{t-1}(\omega^{t-1}) u'(c(\omega^t))[D_t^k(\omega^t) + q_t^k(\omega^t)],$$

where δ is the discount factor process and u is the von Neumann-Morgenstern risk utility of the representative investor.

5.6 Dynamics of Price Expectations

In the previous chapter we studied the short-run (intraday) price adjustment dynamics. Now we are in a position to develop a simple model in which we can study the medium-run (months or quarters) dynamics of price expectations. Finally, at the end of this chapter we will then study the long-term (years or decades) dynamics of wealth evolution. As mentioned before, whenever we study one type of dynamics we blend out the other two types of dynamics by either keeping fixed those variables that are the subject of the other dynamics, or by assuming an infinitely quick adjustment in the other variables.[15] For the case of price expectation dynamics this means assuming that in the short-run prices adjust perfectly to the demand and supply while we keep the wealth of the agents fixed.

In the model of price expectation dynamics we consider two types of input data (previous prices used by chartists and fundamental values used by fundamentalists) on which the price expectations are formed and we combine this with two types of expectation rules: momentum and reversal. Hence, we study the interaction of four types of expectation formations: momentum or reversal rules for chartists and fundamentalists.

5.6.1 What Is Momentum?

We first have to add a little excursion into the topic of momentum to see that the idea to base expectations solely on previous prices is not as way off as one might think: at first glance the efficient market hypothesis would say that information is always already incorporated into the current price – at least when we consider time

[15] Considering different speeds of adjustment in economic dynamics was first suggested by Samuelson [Sam64].

scales above split-seconds, since in times of algorithmic trading it takes only as long to trade on newly arriving information.[16] The prices should therefore completely be driven by news. Since the prices do not effect the news, we would not see any useful pattern in them. So how come that then there is still information in the prices?

This is indeed a puzzling observation. One of the most prominent and well-established examples for the effect is the *momentum effect*: investing in assets that had previously outperformed the market leads on average to excess returns, even when controlling for risk. This phenomenon has been observed by De Bondt and Thaler [DBT85] and more detailed by Jegadeesh and Titman [JT93] and by Lakonishok, Shleifer and Vishny [LSV92, LSV94] and Chan, Jegadesh and Lakonishok [CKL96]. The result has, as usual, first been studied for the US market, but in the meanwhile it has been found also on many other markets [Rou98, SDBW99], although there seem to be cultural differences in its intensity, see Chui and Titman [CTW10]. The connection to trading volume has been studied by Lee and Swaminathan [LS00].

The momentum effect disappears on long time scales (years), where instead a reversal is observable.

There are several competing explanations for momentum and why it is so persistent. Barberis, Shleifer and Vishny [BSV98] use an underreaction explanation for short-term momentum, and an overreaction explanation for long term reversals. They postulate that investors believe earnings growth always to be in one of two regimes, a mean-reverting regime that applies most of the time and (occasionally) a trend regime. Since a representative investor never knows exactly in which regime the market is, he can only try to infer the likelihood of the prevailing regime from the price history. The real earnings in the model are random, so an investor who observes a number of earnings surprises in one direction will conclude that the regime has switched to a trend regime and will react to this, causing a momentum effect.

Daniel, Hirshleifer and Subrahmanyam [DHS98] provide an alternative explanation based on the existence of private signals and overconfident reaction to them and afterwards a too slow adjustment to "reality".

Finally, we want to mention the work by Hong and Stein [HS99]. Their model comes close to what we want to discuss in the next section: they consider a market with two investor types, "newswatchers" (fundamentalists) and "momentum traders" (chartists). In their model news travel slowly, leading initially to an underreaction of prices and thus to some price trend. This trend is picked up by momentum traders who want to participate from the resulting price drift.

[16] In fact, trading is so fast nowadays, that even the physical distance from the stock exchange plays a role, as buying and selling orders are transmitted via cables and since the speed of light and particularly the processing times of signals at switching points are limited, a trading computer on Wall Street can react so much faster than its colleague a few blocks away that the latter does not stand a chance in competing when trying to exploit new information on the short-run.

This reinforces the price drift and an overreaction takes place. The result is a strong momentum effect on the medium-run, but finally, when it comes to a correction, a price reversal occurs.

For a nice summary of momentum-reversal theories and further insights we refer to [She00].

5.6.2 Dynamical Model of Chartists and Fundamentalists

Let us model more closely price expectation dynamics with two investor groups: chartists and fundamentalists, where chartists use previous price data while fundamentalists rely on fundamental values.

The steady state of the dynamical system is characterized by the discounted dividends rule, and the stability of it will depend on the relative proportions of investors with different types of price expectation. To keep things simple, we assume that given his price expectations every trader maximizes a mean-variance utility for one period ahead. The model we develop is similar to the famous Brock and Homes [BH97] model.

As before, we assume that the economy follows a discrete time tree model with a finite number of states in each period. Since the maximization problems span two-periods only, we switch back to the notation of Chap. 4. The following then defines the maximization problem for any trader $i = 1, \ldots, I$ in any period of time t with $s = 1, \ldots, S$ states of the world[17]:

The utility of agent i is of the mean-variance type[18]:

$$U^i(c_1^i, \ldots, c_S^i) = \mu(c_1^i, \ldots, c_S^i) - \frac{\gamma^i}{2}\sigma^2(c_1^i, \ldots, c_S^i).$$

Note that his consumption in state s is given by $c_s^i = \lambda^{i,c} R_s^i \lambda^i w^{i,f}$ and recall the budget constraint[19] $\sum_{k=0}^{K} \hat{\lambda}^{i,k} = 1$. Since we did not state the time dependence explicitly, let us point out that the consumption in state s is given by the consumption rate of that period applied to the wealth in state s, which is obtained from the previous period financial wealth[20] multiplied by the gross return obtained in state

[17] To simplify matters we first suppress all time and uncertainty dependence in this generic one step ahead optimization problem.

[18] $R_s^{i,k}$ is the return agent i expects to get from asset k if state s occurs.

[19] Recall that in Chap. 4 $k = 0$ was the risk-free asset.

[20] I.e. the wealth not consumed, but spent on financial assets.

s: $R^i_s\hat{\lambda}^i = \sum_{k=0}^{K} R^{i,k}(s)\hat{\lambda}^{i,k}$. Hence, for any portfolio $\hat{\lambda}$ we get a utility from that portfolio[21]:

$$U^i(\hat{\lambda}^{i,0},\ldots,\hat{\lambda}^{i,K}) = w^{i,f}\left(\mu(\lambda^{i,c}R^i\hat{\lambda}^i) - \frac{\gamma^i w^{i,f}}{2}\sigma^2(\lambda^{i,c}R^i\hat{\lambda}^i)\right).$$

The solution is as before[22]

$$\hat{\lambda}^i = \left(COV(R^i)\right)^{-1}\frac{\mu(R^i) - R_f}{\gamma^i\lambda^{i,c}w^{i,f}},$$

which we can write in economic terms as

$$\theta^i = \frac{1}{\gamma^i}\left(COV(A^i)\right)^{-1}\left(\mu(A^i) - R_f q\right).$$

Recall that in the multi-period model payoffs are given by dividends and resale prices, i.e.

$$A^i(\omega^t) = D^i(\omega^t) + q^i(\omega^t).$$

We assume that conditionally on ω^t agents agree on the dividends, and we assume point expectations,[23] i.e. q^i_t is independent of (ω^t). Then, the demand in period t given the expectations for the next period is

$$\hat{\lambda}^i_t = \frac{1}{\gamma^i\lambda^{i,c}_t w^{i,f}_t}\Lambda(q_t)\left(COV(D_{t+1})\right)^{-1}\left(\mu(D_{t+1}) + q^i_{t+1} - R_f q_t\right).$$

Normalizing the supply of each asset to 1, assuming short-run equilibrium and defining

$$\bar{\gamma}^{-1} = \sum_{i=1}^{I}(\gamma^i)^{-1}$$

---

[21]To be mathematically correct one should introduce a different symbol for a utility function if it depends on different variables than before. However, adding more notation will be confusing to many readers.

[22]Again, the notation is used: $\mu(R^i)$ is a vector in $\mathbb{R}^k$ but R_f is a scalar. A more correct notation would be $\mu(\tilde{R}^i) - R_f\mathbb{1}$, where $\tilde{R}^i$ denotes the matrix of risk assets and $\mathbb{1} \in \mathbb{R}^k$ is a vector with 1 in each entry.

[23]The first assumption is not restrictive, since in case two agents were to disagree on the dividends in one of the states, one might introduce more states and let the agents disagree over the occurrence of the states. The second assumption is strong. The only excuse we have is that it is sufficient to generate interesting dynamics – and that it is convenient in the Brock-Homes-Model.

gives

$$\sum_{i=1}^{I} \lambda_t^{i,c} w_t^{i,f} = q_t = \Lambda(q_t)\,(COV(D_{t+1}))^{-1}\left(\frac{\mu(D_{t+1})}{\bar{\gamma}} + \sum_{i=1}^{I} \frac{q_{t+1}^i}{\gamma^i} - \frac{R_f q_t}{\bar{\gamma}}\right).$$

Multiplying both sides by $\Lambda(q)_k^{-1}$ and $COV(D_{t+1})$ and defining $D^M = \sum_{k=1}^K D^k$, for any asset k gives

$$q_t^k = \frac{\mu(D_{t+1}^k) - \bar{\gamma}_t COV(D_{t+1}^k, D_{t+1}^M) + \sum_{i=1}^I \delta^i q_{t+1}^{i,k}}{R_f},$$

where

$$\delta^i = \frac{\bar{\gamma}}{\gamma^i}.$$

Hence, the price of any asset k in period t is given by the discounted expected dividends minus the risk of those dividends relative to the market dividends plus the average expected price for the next period.

Before we analyze the dynamics of the model we first characterize its steady state stationary solution. To this end we assume that the trading strategies $\hat{\lambda}^i$, the consumption rates $\lambda^{i,c}$, the expected dividends $\mu(D)$ and the covariance $COV(D)$ are all stationary. Then, the price equation reduces to:

$$\bar{q}^k = \frac{\mu(D^k) - \bar{\gamma} COV(D^k, D^M)}{r_f},$$

which is equal to the discounted expected dividends of the constant payoff

$$\mu(D^k) - \bar{\gamma} COV(D^k, D^M)$$

discounted at $R_f = 1 + r_f$. Recall $R_f = 1 + r_f$ and substitute

$$\bar{\gamma} = \frac{\mu(D^M) - r_f q^M}{\sigma^2(D^M)},$$

from summing the above formula over k. Then we obtain

$$\left(\bar{q}^k - \mu\left(\frac{D^k}{r_f}\right)\right) = \beta^k\left(\bar{q}^M - \mu\left(\frac{D^M}{r_f}\right)\right),$$

where

$$\beta^k = \frac{COV(D^k, D^M)}{\sigma^2(D^M)}.$$

Hence, we have derived a Security Market Line formula similar to that of the static CAPM, but in terms of first principles: dividends and the risk-free rate!

Now we model some structure on the price expectations more explicitly. There are two types of traders: chartists $i \in C$ and fundamentalists $i \in F$. Chartists only use price data as an input for forming their price expectations while fundamentalists compare current prices to the long term steady state prices. Chartists form the price expectations

$$q_{t+1}^{i,k} = q_t^k + a^{i,k}(q_t^k - q_{t-1}^k)$$

with $a^{i,k} > 0$ being a momentum chartist and $a^{i,k} < 0$ being a reversal chartist. Fundamentalists form the price expectations

$$q_{t+1}^{i,k} = q_t^k + b^{i,k}(\bar{q}_t^k - q_t^k)$$

with $b^{i,k} > 0$ being value investors and $b^{i,k} < 0$ being growth investors.

Note that the price dynamics developed above is an inhomogeneous first order difference equation. Such a dynamical system converges to its steady state $\bar{q}$ iff the absolute value of the coefficient in front of the price variable $\sum_{i=1}^{I} \delta^i q_{t+1}^{i,k}$ is smaller than one.[24] Thus we can ignore those forms that do not depend on prices. Inserting the expectation functions we get:

$$R_f q_t^k = \sum_{i \in C} \delta^i \left(q_t^k + a^{i,k}(q_t^k - q_{t-1}^k)\right) + \sum_{i \in F} \delta^i \left(q_t^k + b^{i,k}(\bar{q}_t^k - q_t^k)\right).$$

Rearranging while ignoring constant terms we get:

$$\left(\sum_{i \in C} \delta_t^i (1 + a^{i,k}) + \sum_{i \in F} \delta_t^i (1 - b^{i,k}) - R_f\right) q_t^k = \left(\sum_{i \in C} \delta_t^i a^{i,k}\right) q_{t-1}^k$$

$$\text{or} \quad q_t^k - \frac{\left(\sum_{i \in C} \delta_t^i a^{i,k}\right)}{\left(\sum_{i \in C} \delta_t^i a^{i,k} - \sum_{i \in F} \delta_t^i b^{i,k} - r_f\right)} q_{t-1}^k.$$

[24]For a proof, we show that the iteration $X_{n+1} := aX_n + b$ always converges if $|a| < 1$. To see this, we compute the fixed point of the iteration as $X = b/(1-a)$, i.e. if $X_n = b/(1-a)$ then $X_{n+1} = X_n$. Then we consider the squared difference of X_n to the fixed point and show with a small computation that it is decreasing. From this we can deduce that X_n indeed converges. We can apply this auxiliary result to the dynamic system above.

Thus, in the stability analysis of the dynamical system we need to consider four cases in which we would get stability of the steady state:

Case 1
(numerator and denominator positive)
This happens, e.g. with strong momentum and weak value. Consequently stability occurs iff

$$\sum_{i \in F} \delta^i b^{i,k} + R_f < 0,$$

which is unlikely since $R_f > 1$.
Case 2
(numerator positive and denominator negative)
This happens, e.g. with medium momentum and strong growth. Consequently stability occurs iff

$$2 \sum_{i \in C} \delta^i a^{i,k} < \sum_{i \in F} \delta^i b^{i,k} + r_f,$$

which cannot be since in this case

$$\sum_{i \in C} \delta^i_t a^{i,k} > 0 \quad \text{and} \quad \sum_{i \in F} \delta^i_t b^{i,k} < 0.$$

Case 3
(numerator negative and denominator positive)
This happens, e.g. with reversal and strong growth. Consequently stability occurs iff

$$\sum_{i \in F} \delta^i b^{i,k} + r_f < 2 \sum_{i \in C} \delta^i a^{i,k},$$

which is well possible.
Case 4
(numerator and denominator negative)
This happens, e.g. with reversal and value or weak growth. Stability occurs iff

$$-2 \sum_{i \in C} \delta^i a^{i,k} < \sum_{i \in F} \delta^i b^{i,k} + r_f,$$

which is possible if the reversal is not too strong relative to value.

Thus, reversal chartists and value fundamentalists stabilize the dynamic system while momentum and growth destabilize it. This conclusion is quite intuitive, but the model makes clear that it depends on quite some assumptions: two-period mean-variance optimization with homogeneous beliefs on dividends and point expectations for prices.

Brock and Hommes [BH98] enrich the dynamics of this model by changing the number of traders of each type according to the performance along the paths of the model. Interesting new phenomena like chaotic dynamics may occur. But the general conclusion that reversal chartists and value fundamentalists stabilize the economy is still true in the extended model.

5.7 Survival of the Fittest on Wall Street

We finish this section by analyzing the long term dynamics of our model, the evolution of wealth over time and uncertainty. A first result shows that assuming complete markets, perfect foresight and intertemporal utility maximization, the wealth of investors with rational expectations will grow fastest in a financial market equilibrium.

5.7.1 Market Selection Hypothesis with Rational Expectations

We can use the Pareto efficiency property of competitive equilibria to formulate a Market Selection Hypothesis that determines which investor survives best in the dynamics of the market in terms of (relative) wealth over time. If every investor has some expected utility function with the same time preferences, but possibly different risk attitude, and if payoffs are stochastic, then investor i will dominate investor j if his *beliefs* on the occurrence of the states are more accurate. Note that the investor's dominance is not defined over his strategy, but on his ability to make good estimates.

To get some intuition, let investors maximize their expected utilities:

$$\mathbb{E}_{u^i} = \sum_{t=0}^{T} \delta^t \sum_{\omega_t \in \Omega_t} p_t^i(\omega) u^i(\theta_t^i(\omega^t)) .$$

They may differ with respect to the risk preferences $u^i(\cdot)$ and their personal beliefs $p^i(\omega)$ for the occurrence of a particular event ω. But they have the same degree of time discounting. Consider the marginal rate of substitution of two expected utility investors between two states s and z. Pareto efficiency requires them to equalize, i.e. (5.4) needs to hold. If investors differ in their beliefs (expectations) for state s, e.g., $p_s^i > p_s^j$, there must also be some states such that $p_z^i < p_z^j$. Then $c_s^i > c_s^j$

and $c^i_z < c^j_z$ to ensure equality (5.4).[25] Thus, the better the agents' beliefs, the more those agents get in the more likely states. Hence, their wealth will grow faster. Note that the only one requirement for our fitness criteria to hold is decreasing marginal utilities as in the expected utility framework.

The result that investors get more wealth in those states to which they assign a higher probability goes back to Sandroni [San00] and (under more restrictive assumptions) to Blume and Easley [BE82]. They formulate a theorem for competitive equilibria with perfect foresight that selects for consumers with correct beliefs. It says that in dynamically complete markets with two expected utility investors i and j with different beliefs such that $P^i = P$ (investor i has correct beliefs) and $P^j \neq P$, where the dividend process is i.i.d., the wealth relation $W^i_t / W^j_t \rightarrow \infty$ (P-almost surely), as $t \rightarrow \infty$, with W^i_t (W^j_t) as the relative wealth of investor i (j). In other words, excluding "impossible" events in the long-run, investor i is wealthier than investor j if he makes better predictions. As Blume and Easley [BE92] have recently shown, the result does however not hold for incomplete markets.

5.7.2 Evolutionary Portfolio Theory

In the following we want to derive market selection results for any complete or incomplete market and without assuming that investment strategies are generated by intertemporal expected utility maximization with perfect foresight and rational expectations. To this end we use an analogy to the biological evolution discovered by Charles Darwin. The first point Charles Darwin made was that for the evolution we do not need to model the interaction of the individuals, but the interaction of the strategies played by individuals, i.e., the interaction of the species. From an evolutionary point of view the fate of a single individual animal counts nothing as compared to the relative size of the population of its species. Hence, we suggest to stratify the financial markets not in terms of individuals but in terms of strategies. For the market it is totally irrelevant who is investing according to, say, mean-variance or Prospect Theory or rules of thumb. The only thing that matters is how much money is invested according to such a criterion. In biology strategies fight for resources like food. In finance strategies fight for market capital.

In an evolutionary model there are two forces at play: the selection force reducing the set of strategies and the mutation force enriching it. You see the selection force in financial markets when you realize that every loss some strategy made by buying at high prices and selling at low prices must have generated an equally sized gain for a set of counterparties. The mutation force is clearly seen if you look back a bit in history and observe that previously popular strategies like trying to corner a market are no longer so frequent while new strategies like those followed by hedge funds have emerged. Ultimately, what the evolutionary finance model tries to explain is

[25] This result follows from the decreasing marginal utility. The more wealth an investor receives the lower is his marginal utility.

how the ecology of the market evolves over time, i.e., how the distribution of wealth across strategies changes over time. In the following we will outline the theoretical foundation for it, compare also [HSH09, DLW91, LAP99].

Note that the evolutionary model is based on observables like strategies. We claim that for the vast majority of capital the strategy according to which it is managed is in principle observable. This is because most capital is managed by delegation and in this process the principal (the investor) wants the agents (the wealth manager) to commit to some strategy in order to simplify monitoring. Indeed many banks compete for investors' money by advertising strategies they want to commit to. Moreover, the evolutionary approach takes a "flow of funds" perspective. It claims that understanding according to which principles wealth flows into strategies is the key to understand where asset prices are going. The flow of funds consideration is very popular in practice – even regarding daily price levels. This is because we have now reliable data on flows of funds, between mutual funds and hedge funds, for example. We are more skeptical that the flows approach is useful for daily data – within one day (or even within minutes) prices can change drastically on the occurrence of strong news without any high volume of trade. On the other hand one can get good predictions on the longer run, on monthly data for example, from the flows approach.

Let us now try to make these vague ideas a bit more precise.

5.7.3 Evolutionary Portfolio Model

We base our evolutionary model again on the Lucas [LJ78] asset pricing model. This has two advantages. First, the Lucas model is a model that makes sense from an economic point of view. The time uncertainty structure is simple enough to penetrate, budget equations and homogeneity properties are satisfied. Moreover, displaying our new ideas in this traditional model will help to asses the differences of the two. Recall from the traditional finance model that the Lucas model is defined in discrete time, i.e., $t = 0, 1, 2, \ldots, T$. The *information* structure $(\{\Omega_t\}_t)$ and probability measure (P), the payoffs, the long-lived assets $(k = 1, \ldots, K)$ in unit supply (enabling wealth transfers over different periods of time) as well as the perishable consumption good ("$k = 0$") are still given as in Sect. 5.1, but instead of individual investors let $i = 1, \ldots, I$ now denote investment strategies.

All assets pay off in terms of the consumption good. This clear distinction between means to transfer wealth over time and means to consume is taken from Lucas [LJ78]. Hens and Schenk-Hoppé [HSH05] show that this assumption is essential when taking the evolutionary perspective: if the consumption good were non-perishable and hence could also be used to transfer wealth over time, then every strategy that tries to save using the consumption good will be driven out of the market by the strategy that does not use the consumption good to transfer wealth over time, but otherwise uses the same investment strategy. As in the traditional model, we use the consumption good as the numeraire of the system. Note that one of the long-lived assets could be a bond paying a risk-free dividend. Bonds may however be risky in terms of their resale values, i.e., in terms of the prices $q^k_{t+1}\ (\equiv q^k_{t+1}(\omega^{t+1}))$.

The evolutionary portfolio model we propose studies the evolution of wealth over time as a random dynamical system (RDS). A dynamical system is an iterative mapping on a compact set that describes how a particle (in our model the relative wealth among strategies) moves from one position to the next. In a random dynamical system this movement is not deterministic, but subject to random shocks. An example of a random dynamical system is the famous Brownian motion describing the stochastic movement of a particle in a fluid. Note that in contrast to traditional finance we assume that in order to "predict" the next position of the particle one is not allowed to know the realizations of future data of the system, i.e., we do not allow for perfect foresight of prices! Yet it may happen that under certain nice conditions the limit position of the RDS could also be described "as if" agents were maximizing expected utility under perfect foresight. For an example see the exercise book.

As in the traditional model, we start from the fundamental equation of wealth dynamics:

$$W^i_{t+1} = \sum_{k=1}^{K} \frac{D^k_{t+1} + q^k_{t+1}}{q^k_t} \lambda^{i,k}_t W^i_t,$$

with $\sum_{k=1}^{K} \lambda^{i,k}_t = 1 - \lambda^{i,c}_t$ for all i and t. From one period to the next the wealth of any strategy i is multiplied by the gross return it has generated by its portfolio strategy $\lambda^{i,k}_t\ (\equiv \lambda^{i,k}_t(\omega^t))$ executed in the previous period.[26] Returns come from two sources, dividends and capital gains, but there is no exogenous wealth w^i_t.

In every period asset prices are determined by the equilibrium between demand and supply within that period. Since D^k_{t+1} was meant to denote the total dividends of asset k, we have normalized the supply to one, and as before equilibrium prices are given by:

$$q^k_t = \sum_{i=1}^{I} \lambda^{i,k}_t W^i_t$$

In other words, the price of asset k is the wealth-average of the strategies' portfolio share for asset k.

Note that wealth, dividends and prices may all be subject to some growth rate like the rate at which nominal GDP is growing. However, for analyzing what is the best way of splitting your wealth among the long-lived assets, we can restrict attention to *relative* wealth, *relative* dividends and *relative* prices getting rid of the absolute growth rates. To do so we assume that all strategies have the same time

[26] Note that up to now we did not make any assumption on how the portfolio strategy $\lambda^{i,k}_t(\omega^t)$ executed at ω^t is determined!

independent consumption rate λ^c. The fundamental equation of wealth evolution written in relative terms is given by:

$$r_{t+1}^i = \sum_k \frac{\lambda^c d_{t+1}^k + \hat{q}_{t+1}^k}{\hat{q}_t^k} \lambda_t^{i,k} r_t^i,$$

where

$$\hat{q}_{t+1}^k = \frac{q_t^k}{\sum_i W_t^i}, \qquad d_{t+1}^k = \frac{D_{t+1}^k}{\sum_{k'} D_{t+1}^{k'}}, \qquad r_t^i = \frac{W_t^i}{\sum_i W_t^i}.$$

Here we used that only dividends are consumed[27]

$$\lambda^c \sum_{i=1}^{I} W_t^i = \sum_k D_t^k.$$

In deriving the fundamental equation of wealth evolution written in relative terms we did however make one important assumption: all strategies have the same consumption rate λ^c. The justification of this assumption is that we are searching for the best allocation of wealth among the long-lived assets. It is clear that among two strategies with otherwise equal allocation of wealth among long-lived assets the one with a smaller consumption rate will eventually dominate. Written in relative terms the asset pricing equation keeps its nice form.The relative asset prices are simply the convex combination of the strategies in the market:

$$\hat{q}_t^k = \sum_i \lambda_t^{i,k} r_t^i.$$

In terms of evolutionary game theory this means that strategies are "playing the field", i.e., one strategy has an impact on any other strategy only via the average of the strategies. Careful reading of the wealth flow equation reveals that the flow of wealth is not already described by a dynamical system. We would like to have a mapping from the relative wealth in one period $r_t := (r_t^1, \ldots, r_t^I)$ to the relative wealth in the next period $r_{t+1} = (r_{t+1}^1, \ldots, r_{t+1}^I)$. But r_{t+1}^i also enters on the right hand side because capital gains do depend on the strategies played. Fortunately, the dependence of capital gains of strategies wealth is linear so that we can solve the

[27]Formally, this identity follows from aggregating $W_t^i = \sum_k (D_t^k + q_t^k)\theta_{t-1}^{i,k}$ over all agents noting that $\sum_k q_t^k = (1 - \lambda^c) \sum_k W_t^i$.

wealth flow equation in the RDS-form. In the resulting equation, the inverse matrix captures the capital gains. Note that the $I \times K$ matrix $\Lambda_{t+1} := (\hat{\lambda}^{i,k}_{t+1})_{i,k}$ is the matrix of portfolio strategies,

$$r_{t+1} = \lambda^c \left(\mathrm{Id} - \left[\frac{\hat{\lambda}^{i,k}_t r^i_t}{\hat{\lambda}^{\cdot,k}_t r^{\cdot}_t} \right]_{i,k} \Lambda^{\mathrm{T}}_{t+1} \right)^{-1} \left[\sum_k d^k_{t+1} \frac{\hat{\lambda}^{i,k}_t r^i_t}{\hat{\lambda}^{\cdot,k}_t r^{\cdot}_t} \right]_i ,$$

where as before $\hat{\lambda}^{i,k}_t = \lambda^{i,k}/(1-\lambda^c)$, so that $\sum_k \hat{\lambda}^{i,k}_t = 1$.

Note that this equation is a first order stochastic difference equation describing a mapping from the simplex Δ into itself.

$$r_{t+1}(\omega^{t+1}) = F_t(\omega^{t+1}, r_t)$$

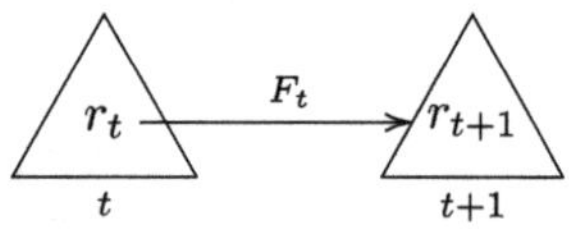

Let us summarize the assumptions made so far. We used Lucas' [LJ78] distinction between means to transfer wealth over time and means to consume, we assumed that all strategies have the same consumption rate and by writing the Lucas model as a RDS we assumed that strategies are not allowed to use information that is not available at the time when executed. Every of these assumptions seems well justified to us. Note that in contrast to many other economic models generating dynamics we did not make any simplifying assumptions like linear demand functions, usually justified by first order Taylor approximations or by mean-variance optimization. One has to be very careful when making these seemingly innocuous assumptions. When iterating a dynamical system terms of higher order may accumulate so that the real dynamics of the system looks quite different from the dynamics of the system based on the simplifying assumptions.

It might be instructive to variegate our model at this point with a little simulation. Figure 5.8 shows a typical run of a simulation with two strategies, a strategy generated from mean-variance analysis (solid line) and the naive diversification rule of fixing equal weights in the portfolio (dotted line). Even though initially the wealth of the mean-variance rule accounts for 90 % of the market wealth after a few iterations the situation has reversed and the $1/n$ rule has 90 % of the market wealth. This wealth dynamics is reflected in the asset prices: they initially reflect the mean-variance rule but rapidly converge to the $1/n$ rule.

This shows that seemingly rational portfolio rules like mean-variance can do quite poorly against seemingly irrational rules like $1/n$ – a result that was first pointed at by DeLong, Shleifer, Summers and Waldman [DLSSW90] and recently found empirical support in DeMiguel, Garlappi and Uppal [MGU09].

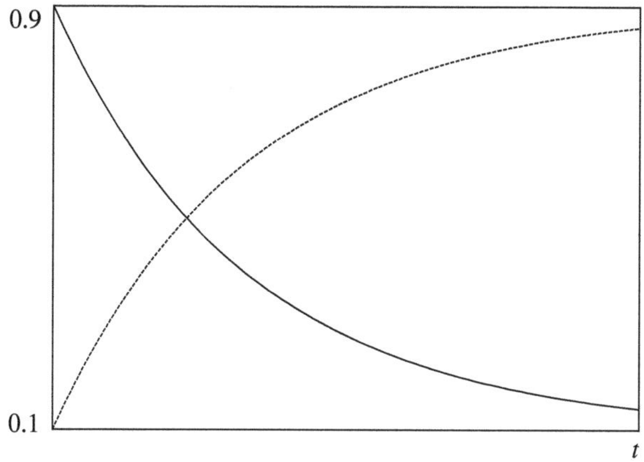

Fig. 5.8 Market shares (mean-variance rule vs. $1/n$ rule)

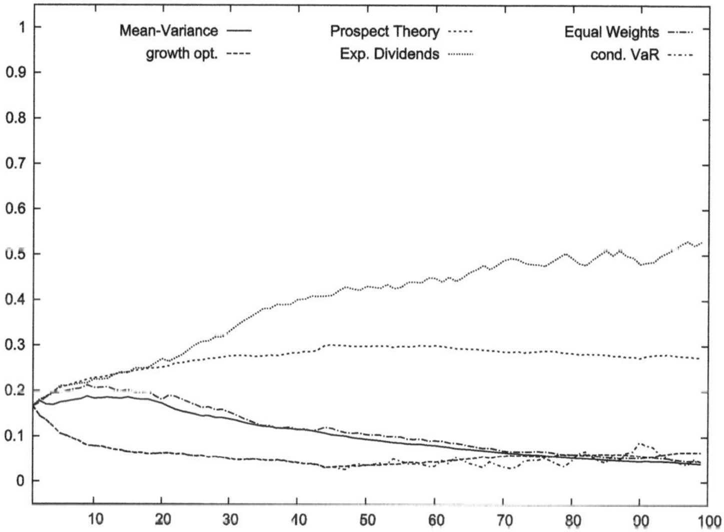

Fig. 5.9 Sample run of a simulation with six strategies

5.7.4 The Unique Survivor: λ^*

We conclude our simulation analysis by including the expected relative dividends portfolio $\hat{\lambda}^{*,k} = \mathbb{E}_P d^k$, $k = 1, \ldots, K$, in the market selection process. As a result the process always converges to the situation in which all the market wealth is concentrated at the strategy $\hat{\lambda}^*$. Figure 5.9 shows a typical run of the simulation. An analysis of the standard deviation bands of this simulation would show that they do not widen but get tighter as time goes on, indicating that the process converges. We displayed the expected dividends rule in competition with a mean-variance and conditional value-at-risk (VaR) rule, as well as an approximation to the growth optimal rule maximizing the expected logarithm of returns by means of Cover's algorithm [Cov84], the equal weights or naive diversification portfolio $1/n$, and a portfolio based on Prospect Theory.

Our conjecture from these simulations is: *starting from any initial distribution of wealth, on P-almost*[28] *all paths the market selection process converges to* $\hat{\lambda}^*$ *if the dividend process d is i.i.d.*

Indeed, it can be even shown analytically, that if dividends d follow an i.i.d. process and we only consider stationary adapted strategies, then

$$\lambda^{\star,k} = (1-\lambda^c)\mathbb{E}_p d^k_{(\omega)}.$$

is the unique evolutionary stable strategy. The key to understand this result is the notion of expected growth rate of wealth of any strategy $\hat{\lambda}$ in a market governed by strategy λ^M:

$$g(\hat{\lambda}, \hat{\lambda}^M) = \mathbb{E}_p \ln\left(1 - \lambda^c + \lambda^c \sum_{k=1}^K \frac{\mathrm{d}^k \hat{\lambda}^k}{\hat{\lambda}^{M,k}}\right)$$

Note that $g(\hat{\lambda}^M, \hat{\lambda}^M) = 0$, i.e. if all strategies are identical then none can grow at the cost of others. The evolutionary stability property is $g(\hat{\lambda}, \hat{\lambda}^\star) < 0$, hence all strategies die out in a market governed by $\lambda^\star$. Moreover, it is apparent that under-diversified strategies have no chance to survive [EHSH06].

Over recent years even more general statements have been proven – but for those, as well as for further applications, stability analysis or empirical tests as well as for alternative formulations of evolutionary models, we refer to [Sch09].

Summary

The multi-period model is an important generalization of the two-period case. While traditional finance highlights the similarities of this model to the two-period model, no-arbitrage pricing, financial market equilibria, Pareto efficiency and aggregation, the analysis of dynamics like price expectations and wealth dynamics explores the new insights that are then possible.

There are a number of textbooks that cover the standard theory of this chapter. For further reading we recommend [Coc01, Duf96, MQ02].

[28] That is to say on all paths except for those that are highly unlikely, i.e. those that have measure zero according to the probability measure P. For example, if P is i.i.d., every infinite sequence in which some state is not visited infinitely often has measure zero.

Test

1. We want to compare the multi-period model with the two-period model of the previous chapter. How can we describe a two-period model with the state s_0 in $t = 0$ and three possible states s_1, s_2, s_3 at time $t = 1$ in the new terminology?

   ☐ We define $\Omega_t = s_t$.
   ☐ We define $\Omega_0 = s_0$ and Ω_1 as s_1, s_2 or s_3, depending on which state occurs at $t = 1$.
   ☐ We define $\Omega_0 = \{s_0\}$ and $\Omega_1 = \{s_1, s_2, s_3\}$.
   ☐ We define $\Omega_0 = \{s_0\}$ and $\Omega_1 = \{s_0, s_1, s_2, s_3\}$.

2. Which of the following statements on the Ponzi scheme are correct?

   ☐ It works only in an infinite horizon model.
   ☐ It can be understood as a long-term bubble.
   ☐ It means taking larger and larger loans for repaying old loans and financing consumption.

3. Which of the following statements on wealth dynamics are correct on a complete market with two expected utility investors, where the first one estimates probabilities correctly, the second one does not?

   ☐ Depending on their risk attitudes, any of them can survive on the market in the long-run.
   ☐ The first investor will survive on the market in the long-run.
   ☐ The second investor will be driven out of the market almost surely in the long run.

4. What can we say about momentum?

   ☐ Investing in assets that have performed well in the last six months is a strategy that is successful on average.
   ☐ Momentum is a behavioral bias that explains why people buy stocks that are already overpriced.
   ☐ Explanations of the momentum effect often involve under- and overreaction of stock market participants.

5. In evolutionary finance...

   ☐ ...investors fight for survival by competing on the market.
   ☐ ...investment strategies fight for survival by competing on the market.
   ☐ ...only the asymptotical behavior for $t \to \infty$ is of interest.

6. What can we say about the strategy λ^*?

- ☐ It is the asymptotical strategy of a certain market environment.
- ☐ It means investing into the expected relative dividend portfolio.
- ☐ In very general cases, λ^* is the only evolutionary stable strategy.

Part III

Advanced Topics

Theory of the Firm*

6

The great difference between the industry of today as compared to that of yesterday is what might be referred to as the necessity of the scientific approach, the elimination of operation by hunches.

ALFRED P. SLOAN

We will now extend the financial economy $\mathcal{E}_F$ to cover problems of production and production units, i.e. firms. Among other things, this allows conclusions about the behaviour of firms in markets. So far we assumed bond payoffs to be exogenous, ignoring the decision-making process of the bonds' issuers. A precise theory of the firm will analyze this process, resulting in bonds whose payoff structure is determined by various economic parameters.

6.1 Basic Model

The model builds on the two-period model introduced in Sect. 4.1. In time $t = 0$ a set of bonds $\mathcal{K} := \{1, \ldots, K\}$ can be traded; in time $t - 1$ they payoff in dependence of the state of the world. Formally, there are $S + 1$ states of the world, where $s = 0$ corresponds to $t = 0$ and in $t = 1$ one state $s \in \{1, \ldots, S\}$ is realized.

6.1.1 Households and Firms

In this chapter we consider two types of economic agents: a set of households $\mathcal{I} := \{1, \ldots, I\}$ (or agents in the narrower sense) and a set of firms $\mathcal{J} := \{1, \ldots, J\}$. A household's main interest lies in consumption (in the sense that utility only depends on it). Let $\mathcal{X}^i \subset \mathbb{R}^{S+1}$ be a set of consumption plans, where $x^i \in \mathcal{X}^i$ describes agent i's consumption for each state s. Let for a household i the mapping $U^i : \mathcal{X}^i \to \mathbb{R}$

T. Hens, M.O. Rieger, *Financial Economics*, Springer Texts in Business and Economics, DOI 10.1007/978-3-662-49688-6_6

represent its utility. The firms' genuine task is (not to consume, but) to produce. Each firm $j \in \mathcal{J}$ is characterised by its exogenous production technology $\mathcal{Y}^j \subseteq \mathbb{R}^{S+1}$, which can be interpreted as

> the outcomes arising from alternative modes of specialization and organization [...] expressed in reduced form. [MQ02]

The net output is denoted by $y^j := (y_0^j, \ldots, y_S^j) \in \mathcal{Y}^j$, where $y_s^j > 0$ is the net output of firm j in state s. $y_s^j < 0$ is interpreted as the net input. Let $\mathcal{Y} := \bigcup_{j \in \mathcal{J}} \mathcal{Y}^j$ be the set of all production capabilities and let $Y \in \mathbb{R}^{(S+1) \times J}$ be a production matrix of the entire economy.

When working with production economies, one frequently has to make assumptions about the production technology set $\mathcal{Y}^j$. These assumptions are intended to simplify various problems and some of them are rather technical in nature; still, we give a short interpretation:

Assumption 6.1 (Production technology set) *We make the following assumptions on the production technology set $\mathcal{Y}^j$ of a firm:*

(i) $\mathcal{Y}^j \subset \mathbb{R}^{S+1}$ *is closed,*
(ii) $\mathcal{Y}^j$ *is convex,*
(iii) $\mathcal{Y}^j \supset \mathbb{R}^{S+1}_{\leq 0}$,
(iv) $\mathcal{Y}^j \cap \mathbb{R}^{S+1}_{\geq 0} = \{0\}$,
(v) for all $\omega \in \mathbb{R}^{S+1}_{\geq 0}$ *we have* $(\omega + \mathcal{Y}) \cap \mathbb{R}^{S+1}_{\geq 0}$ *compact.*

(i) has only technical character. (ii) implies that the technologies have non-increasing returns to scale. (iii) means on the one hand that $0 \in \mathbb{R}^{S+1}$ belongs to the technology of every firm; i.e., it has the option of producing nothing. On the other hand every firm can freely dispose its resources: a given output does not have to be produced with minimal possible input. (iv) is a kind of no arbitrage condition (NAC) for the production: a positive output in one state can only be produced by investing a strictly positive input in another. Since we are looking at a one-good economy, this can also be interpreted as: there can be no investment projects that have positive cash-flow in at least one state, but non-negative cash-flow in every state. Finally, (v) formalizes the limitations of production: given the resources in an economy, production is bounded; which implies that for any price system q cash-flows are bounded.

6.1.2 Financial Market

Both types of economic agents actively participate in the bond market. Let $\mathcal{Z}^i$ be the set of a household's possible portfolios, i.e., its portfolio space. An element $z^i := (z_1^i, \ldots, z_K^i) \in \mathcal{Z}^i$ describes its position on the bond market, where z_k^i denotes the position for the k-th bond. Let $\mathcal{Z} := \bigcup_{i \in \mathcal{I}} Z^i$ be the set of all portfolio spaces.

Similarly, we denote by $\mathcal{P}^j \subset \mathbb{R}^K$ a firm's portfolio space, with an element $\xi^j := (\xi_1^j, \ldots, \xi_K^j) \in \mathcal{P}^j$. We call a firm's portfolio ξ^j its financial policy to distinguish it from a household's portfolio. Let $\mathcal{P} := \bigcup_{j \in \mathcal{J}} P^j$ denote the set of all portfolio spaces.

In time $t = 0$ the set of bonds $\mathcal{K} := \{1, \ldots, K\}$ are traded at some prices $q \in \mathbb{R}^K$. Each bond's payoff can be described by a $S \times K$ matrix with entries A_s^k denoting the payoff of bond k in state s.

We will distinguish between *non-incorporated* and *incorporated companies.* In the context of non-incorporated companies, the agents exogenously hold firm shares, and cannot trade them. On the other hand, in an economy with incorporated companies these shares are tradeable on a market. This distinction leads to two variants of the set up. In the benchmark case the financial market A only consists of bonds. Firm shares are held by households who do not sell them. Let $\bar{\delta}^i = (\bar{\delta}_1^i, \ldots, \bar{\delta}_J^i) \in \mathbb{R}_{\geq 0}^J$ denote the initial distribution of firm shares of agent i, where $\bar{\delta}_j^i$ is its share of firm j.

In the case of incorporated companies firm shares are tradeable (in a stock market). The market for firm shares is assumed to be open to both, households and firms. Let $\mathcal{H}^i \subset \mathbb{R}^J$ be a household's portfolio space on this market. An element $h^i = (h_1^i, \ldots, h_J^i)$ means that the i-th agent holds h_j^i shares in the j-th firm. As in the benchmark case, we assume that any household is endowed with some portfolio $\bar{\delta}^i$, which can be sold now. Since there are no transaction costs, we let the agents sell their stocks $\bar{\delta}^i$, and then demand δ^i on the market. The firms now may also buy shares of other firms, so that they are connected by cross ownership. We assume they do not hold any such shares at the beginning. Let $\mathcal{T}^j$ be the set of possible firm shares. As before, $\mathcal{T}^j \subseteq \mathbb{R}^J$. Let $\mathcal{T} := \bigcup_{j \in \mathcal{J}} \mathcal{T}^j$ be the set of firm interdependencies. The dependencies can be described from a firm's point of view by vectors $\tau^j = (\tau_1^j, \ldots, \tau_J^j)$, meaning the j-th firm holds τ_l^j shares in the l-th firm.

In period $t = 0$ firm shares are traded at some prices $p \in \mathbb{R}^J$. If traded or not, holding firm shares leads to dividends in both periods. Let the matrix $D := (d^1, \ldots, d^J) \in \mathbb{R}^{(S+1) \times J}$ describe the dividend of all firms (in any state).[1] Dividends can also be negative which would mean a liability (to subsequent payment).

Example 6.2 (Dividend policy) Consider a firm j and only two states $s = \{0, 1\}$. For a given production decision $y^j \in \mathcal{Y}^j$ and a given financial policy $\xi^j \in \mathcal{P}^j$ it might fix the following dividends[2]:

$$d^j = y^j + \begin{pmatrix} -q' \\ A \end{pmatrix} \xi^j$$

[1]Its structure depends on the concrete production decision y^j and the financial policy ξ^j of each firm $j \in \{1, \ldots, J\}$, so we write $D(Y, \xi)$.

[2]The optimal choice of dividends, financial policy, production will be derived later.

that is $d_0^j = y_0^j - \sum_{k \in \mathcal{K}} q^k \xi_k^j$ and $d_1^j = y_1^j - \sum_{k \in \mathcal{K}} A_s^k \xi_k^j$. Differing financial policies determine different dividends, as can be seen in the following cases:

- 100 % equity finance: $\xi^j = 0$

$$d^j = y^j \quad \rightarrow \quad d_0^j = y_0^j \quad \text{complete equity finance}$$
$$\rightarrow \quad d_1^j = y_1^j \quad \text{full risk}$$

- 100 % bonded capital finance: $-q'\xi^j = y_0^j$

$$d^j = y^j + \begin{pmatrix} -q' \\ A \end{pmatrix} \xi^j \quad \rightarrow \quad d_0^j = 0$$
$$\rightarrow \quad d_1^j = y_1^j + A\xi^j$$

- 100 % risk coverage: $A\xi^j = -y_1^j$

$$d^j \quad \rightarrow \quad d_0^j = y_0^j - q'\xi^j$$
$$\rightarrow \quad d_1^j = 0.$$

6.1.3 Financial Economy with Production

To wrap up the model consider all the events in the two periods: In $t = 0$ a household i is endowed with some assets w_0^i and a portfolio of firm shares $\bar{\delta}^i$. It consumes x_0^i and buys a portfolio of bonds z^i at prices q. In the general case (of incorporated companies), it can also buy firm shares δ^i and sell its initial firm shares at some prices p. In the end of this period the household gets a dividend $\sum_{j \in \mathcal{J}} d_0^j h_j^i$ for the owned firms. At the same time $t = 0$ a firm j produces a net output of y_0^j. It can buy a portfolio of bonds ξ^j at prices q. It has to determine its dividend d_0^j. If firm shares are tradeable, a firm can also buy a portfolio τ^j, which leads to dividends $\sum_{l \in \mathcal{J}} d_0^l \tau_l^j$ at the end of the first period.

In $t = 1$ one state $s \in \{1, \ldots, S\}$ is realized. A household i is endowed with some assets w_s^i, gets dividends of its firm shares ($\sum_{j \in \mathcal{J}} d_s^j h_j^i$) and receives payoffs of its bond portfolio $\sum_{k \in \mathcal{K}} A_s^k z_k^i$. It can use these sources of income to consume x_s^i. In the same period $t = 1$, a firm j produces a net output of y_s^j. It earns the payoffs $\sum_{k \in \mathcal{K}} A_s^k \xi_k^j$ by its position in the bond market. It has to give dividends d_s^j, but also receives dividends when owning shares of other firms $\sum_{l \in \mathcal{J}} d_s^l \tau_l^j$.

Definition 6.3 (Financial Economy with Production $\mathcal{E}_F^P$) A financial economy with production $\mathcal{E}_F^P(\mathcal{I}, \mathcal{X}, U, \omega, \mathcal{Z}, \bar{\delta}, \mathcal{J}, \mathcal{Y}, \mathcal{P}, [\mathcal{H}, \mathcal{T}], A)$ is an economy with

(i) I economical agents, their consumption spaces $\mathcal{X}$, utility functions U, initial distributions of goods ω, initial distribution of firm shares $\bar{\delta}$, portfolio spaces $\mathcal{Z}$ (and corresponding firm share portfolios[3] $\mathcal{H}$);
(ii) J firms with technologies $\mathcal{Y}^j$, portfolio spaces $\mathcal{P}$ (and corresponding firm share portfolios $\mathcal{T}$);
(iii) and a financial market A for bonds (and shares).

We need to restate a few already known concepts in the setting of a financial economy with production. Let $\phi_{\mathcal{Y}}, \phi_{\mathcal{P}}$, and $\phi_{\mathbb{T}}$ be the corresponding production allocation, financial policy allocation and share allocation (respectively) of the firms,[4] and $\phi_{\mathcal{H}}$ the households' allocation of firm shares.[5]

Definition 6.4 (Achievability) An allocation of goods $\phi_{\mathcal{X}}$ and a production allocation $\phi_{\mathcal{Y}}$ are *achievable*, if

(i) $x^i \in \mathcal{X}^i$ for all i,
(ii) $y^j \in \mathcal{Y}^j$ for all j,
(iii) $\sum_{i\in\mathcal{I}}(x^i - \omega^i) \leq \sum_{j\in\mathcal{J}} y^j$.

6.1.4 Budget Restriction/Households' Decisions and Firms' Decisions

At the beginning of the first period, the agents and firms plan all traded and produced quantities for all $S + 1$ states. This means that at $t = 0$, not only the production plans for $t = 1$ have to be set, but in particular their financing has to be guaranteed in advance. This set-up leads to a list of budget restrictions. We will keep the distinction between incorporated and non-incorporated firms, because it plays an important role in the description of the budget restrictions.

6.1.4.1 Non-incorporated Companies

Since firm shares are not tradeable, households' hold just their initial firm shares $h^i = \bar{\delta}^i$. At time $t = 0$, agent i's expenses for consumption x_0^i and investment in bonds z_k^i may not exceed the value of its initial assets ω_0^i and its dividends $\sum_{j=1}^J d_0^j \bar{\delta}_j^i$

[3]If firm shares are tradeable for households (or firms, respectively).

[4]In the scope of firms, an allocation is a map $\phi : \mathcal{J} \to \mathcal{M}$, i.e., from the set of firms to a corresponding set $\mathcal{M}$. In particular, it is different from an allocation in the scope of households. However, as the distinction is always clear from context, we will not differentiate between them.

[5]$\phi_{\mathcal{H}} : \mathcal{I} \to [0, 1]$ with $\sum_{i=1}^I \phi_\delta(i) = 1$.

from its firm shares, i.e. at $t = 0$,

$$x_0^i + \sum_{k\in\mathcal{K}} q^k z_k^i \leq \omega_0^i + \sum_{j\in\mathcal{J}} d_0^j \bar{\delta}_j^i. \tag{6.1}$$

At $t = 1$, in a given state s, the agent gains the payoff of its portfolio and the dividends which can all be used for consumption. Hence the budget restriction for $t = 1$ are

$$x_s^i \leq w_s^i + \sum_{k\in\mathcal{K}} A_s^k z_k^i + \sum_{j\in\mathcal{J}} d_s^j \bar{\delta}_j^i. \tag{6.2}$$

We combine the two conditions (6.1) and (6.2), where the latter shall hold for all states $s \geq 1$:

$$\mathbb{B}^i(q, \omega^i, \bar{\delta}^i, A, D)$$

$$:= \left\{ \begin{array}{c|ll} (x^i, z^i) & x_0^i + \sum_{k\in\mathcal{K}} q^k z_k^i \leq \omega_0^i + \sum_{j\in\mathcal{J}} d_0^j \bar{\delta}_j^i & \\ \in \mathcal{X}^i \times \mathcal{Z}^i & x_s^i \leq w_s^i + \sum_{k\in\mathcal{K}} A_s^k z_k^i + \sum_{j\in\mathcal{J}} d_s^j \bar{\delta}_j^i, & \text{for all } s \geq 1 \end{array} \right\}.$$

The budget set can also be written in matrix form:

$$\mathbb{B}^i(q, \omega^i, \bar{\delta}^i, A, D) = \left\{ \begin{array}{c|l} (x^i, z^i) & \\ \in \mathcal{X}^i \times \mathcal{Z}^i & x^i \leq \omega^i + \begin{pmatrix} -q' \\ A \end{pmatrix} z^i + D\bar{\delta}^i. \end{array} \right\}$$

Therefore, an economic agent i is confronted with the following utility maximization problem:

$$\max_{\substack{x^i \in \mathcal{X}^i \\ z^i \in \mathcal{Z}^i}} U^i(x) \quad \text{such that} \quad (x^i, z^i) \in \mathbb{B}^i(q, \omega^i, \bar{\delta}^i, A, D).$$

A firm, on the other hand, should maximize its total profit. Since profits also arise at $t = 1$ and in different states, they have to be discounted or weighted. In the following, we assign a vector $\pi^j \in \mathbb{R}^{S+1}$ to every firm j, for which the NAC holds ($q_k = \sum_{s=1}^S A_s^k \pi_s^j$, for all k; see Theorem 4.2).[6] In $t = 0$, a firm's dividends are restricted by its net production and its actions on the bond market. In $t = 1$ for any state $s \geq 1$, the dividends may not exceed the net output and the payoffs from the

[6] If we restrict ourselves to the case of complete markets, common state prices π^N may be used to value the profits.

bond market. Thus, we consider the following maximization problem:

$$\max_{\substack{d^j \in \mathbb{R}^{S+1} \\ y^j \in \mathcal{Y}^j \\ \xi^j \in \mathcal{P}^j}} \pi^{j\prime} d^j \quad \text{such that} \quad d^j \leq y^j + \begin{pmatrix} -q' \\ A \end{pmatrix} \xi^i. \tag{6.3}$$

6.1.4.2 Incorporated Companies

Now let the households and firms be allowed to trade firm shares at some prices p. The households obtain a new source of income through stock trade by selling their initial allocation of stock. Conversely, they can also use their resources to buy shares. This means the budget restriction of a household i needs to be changed to account for this possibility. At $t = 0$, agent i has a demand for goods x_0^i and for bonds z_k^i and shares δ_j^i.[7] Apart from the initial distribution of goods ω_0^i, the sales revenue from the initial distribution and the dividends of its shares limit a household's decisions, i.e.,

$$x_0^i + \sum_{k \in \mathcal{K}} q^k z_k^i + \sum_{j \in \mathcal{J}} p^j \delta_j^i \leq \omega_0^i + \sum_{j \in \mathcal{J}} d_0^j \delta_j^i + \sum_{j \in \mathcal{J}} p^j \bar{\delta}_j^i. \tag{6.4}$$

The revenues and expenses at $t = 1$ in a given state s are:

$$x_s^i \leq \omega_s^i + \sum_{k \in \mathcal{K}} A_s^k z_k^i + \sum_{j \in \mathcal{J}} d_s^j \delta_j^i. \tag{6.5}$$

We combine the two conditions of the budget restriction, where the latter shall hold for all states $s \geq 1$:

$$\mathbb{B}^i(q, p, \omega^i, \bar{\delta}^i, A, D) = \left\{(x^i, z^i, \delta^i) \in \mathcal{X}^i \times \mathcal{Z}^i \times \mathcal{H}^i \mid \text{(6.4) and (6.5) hold}\right\}$$

which can be written more compactly as

$$\mathbb{B}^i(q, p, \omega^i, \bar{\delta}^i, A, D) = \left\{(x^i, z^i, \delta^i) \in \mathcal{X}^i \times \mathcal{Z}^i \times \mathcal{H}^i \mid \text{(6.6) holds}\right\}$$

where

$$x^i \leq \omega^i + \begin{pmatrix} -q' \\ A \end{pmatrix} z^i + \begin{pmatrix} (D_0 - p)' \\ D_1 \end{pmatrix} \delta^i + \begin{pmatrix} p'\bar{\delta}^i \\ 0 \end{pmatrix}. \tag{6.6}$$

$D_0 \in \mathbb{R}^J$ is the vector of dividends of all J firms at $t = 0$, and $D_1 \in \mathbb{R}^{S \times J}$ is the matrix of dividends of all firms in all states at $t = 1$.

[7] Mind the notation: δ_j^i is the new demand for shares, after selling $\bar{\delta}_j^i$, the initially distributed shares! We presume that shares are traded first, and dividends are paid only afterwards.

An economic agent i is therefore confronted with the following utility maximization problem:

$$\max_{\substack{x^i \in \mathcal{X}^i \\ z^i \in \mathcal{Z}^i \\ \delta^i \in \mathcal{H}^i}} U^i(x) \quad \text{such that} \quad (x^i, z^i, \delta^i) \in \mathbb{B}^i(q, p, \omega^i, \bar{\delta}^i, A, D)$$

Now consider the budget restriction of an incorporated firm. For a given production decision y^j and a given financial policy (ξ^j, τ^j), the maximal dividends of firm j are:

$$d^j = y^j + \begin{pmatrix} -q' \\ A \end{pmatrix} \xi^j + \begin{pmatrix} (D_0 - p)' \\ D_1 \end{pmatrix} \tau^j.$$

Note that the dividends of firm j depend on the dividends of all firms. Each firm chooses a production plan, a financial policy and firm shares in order to maximize profits. Thus, a firm solves the following maximization problem (where the weights π^j, again serve to evaluate different dividend vectors):

$$\max_{\substack{d^j \in \mathbb{R}^{S+1} \\ y^j \in \mathcal{Y}^j \\ \xi^j \in \mathcal{P}^j \\ \tau^j \in \mathcal{T}^j}} \pi^{j\prime} d^j \quad \text{such that} \quad d^j \leq y^j + \begin{pmatrix} -q' \\ A \end{pmatrix} \xi^j + \begin{pmatrix} (D_0 - p)' \\ D_1 \end{pmatrix} \tau^j.$$

The budget restriction requires that in any state s the dividends (and the investment in bonds and stocks in $t = 0$) do not exceed the net output plus the payoffs from the financial market. Having a concise notation for the "budget restriction" of a firm will often come in handy.

Let

$$\mathbb{W}^j(q, p, A, D) := \left\{ \begin{matrix} (d^j, y^j, \xi^j, \tau^j) \\ \in \mathbb{R}^{S+1} \times \mathcal{Y}^j \times \mathcal{P}^j \times \mathcal{T}^j \end{matrix} \;\middle|\; (6.7) \text{ holds} \right\}$$

where

$$d^j \leq y^j + \begin{pmatrix} -q' \\ A \end{pmatrix} \xi^j + \begin{pmatrix} (D_0 - p)' \\ D_1 \end{pmatrix} \tau^j. \tag{6.7}$$

Now we can formulate the firm's decision problem as

$$\max_{\substack{d^j \in \mathbb{R}^{S+1} \\ y^j \in \mathcal{Y}^j \\ \xi^j \in \mathcal{P}^j \\ \tau^j \in \mathcal{T}^j}} \pi^{j\prime} d^j \quad \text{such that} \quad (d^j, y^j, \xi^j, \tau^j) \in \mathbb{W}^j(q, p, A, D).$$

6.2 Modigliani-Miller Theorem

In this section we consider financial market equilibria and ask to what extent they depend on the firms' financial policies. A seminal contribution to this topic was made by Modigliani and Miller [MM58]. While in the traditional view, the shareholder value of a firm depends on its debt-equity ratio, Modigliani and Miller show that under well defined conditions the funding of a firm is irrelevant. Their proof is based on an intuitive arbitrage argument.

Consider two firms with identical production vector y^j, $j = 1, 2$, but different debt-equity ratios. If the market value were different, investors would have an arbitrage opportunity: they could buy one of the firms, change the ratio, and resell it with profits.

The irrelevance theorem of Modigliani and Miller not only addresses the valuation of companies, but also the average cost of capital, and decisions on investment projects. Their important contribution lies in the precise formulation of the conditions under which financial decisions must be irrelevant.

In the analysis of financial market equilibria, we distinguish between incorporated and non-incorporated firms.

6.2.1 The Modigliani-Miller Theorem with Non-incorporated Companies

Consider a financial economy with production $\mathcal{E}_F^P$ according to Definition 6.3. An $\mathcal{E}_F^P$ is considered to be in a state of equilibrium, if four conditions are met: (i) Consumers maximize utility within their budget constraints, (ii) firms maximize profits given their constraints, without allowing for arbitrage, (iii) the allocations must be achievable, and (iv) financial markets clear. Given the model of non-incorporated companies introduced in the last subsection, this leads to the following definition:

Definition 6.5 (Financial Market Equilibrium with Endogenous Production) A FME for a financial economy $\mathcal{E}_F^P$ with endogenous production are allocations $(\overset{*}{\phi}_{\mathcal{X}}, \overset{*}{\phi}_{\mathcal{Z}}, \overset{*}{\phi}_{\mathcal{H}}, \overset{*}{\phi}_{\mathcal{Y}}, \overset{*}{\phi}_{\mathcal{P}})$ and a pricing system $(\overset{*}{q}, \{\pi^j\}_{j\in\mathcal{J}})$, such that

(i) $(\overset{*}{x}{}^i, \overset{*}{z}{}^i, \overset{*}{\delta}{}^i) \in \mathbb{B}^i(\overset{*}{q}, \omega^i, \bar{\delta}^i, A, \overset{*}{D})$ and $\overset{*}{x}{}^i \in \arg\max U^i(x^i)$ for all $i \in \mathcal{I}$,

(ii) $(\overset{*}{y}{}^j, \overset{*}{\xi}{}^j, \overset{*}{d}{}^j)$ satisfies $\overset{*}{d}{}^j \leq \overset{*}{y}{}^j + \begin{pmatrix} -q' \\ A \end{pmatrix} \overset{*}{\xi}{}^j$ and $\overset{*}{d}{}^j \in \arg\max \pi^{j\prime} d^j$ for all $j \in \mathcal{J}$,

(iii) $\sum_{i\in\mathcal{I}} \overset{*}{x}{}^i \leq \sum_{i\in\mathcal{I}} \omega^i + \sum_{j\in\mathcal{J}} \overset{*}{y}{}^j$,

(iv) $\sum_{i\in\mathcal{I}} \overset{*}{z}{}^i + \sum_{j\in\mathcal{J}} \overset{*}{\xi}{}^j = 0$,

where π^j satisfies the NAC for all j, i.e., the firms see no opportunity for arbitrage.

If we do not require condition (ii) to hold, then we speak of a financial market equilibrium with exogenous production decisions. For existence of financial market equilibria (with exogenous or endogenous production decisions) we refer the reader to [MQ02]. In the following we shall examine how a given equilibrium changes under alternative financial policies of the firms. The answer is given by the Modigliani-Miller theorem (MMT).

Theorem 6.6 (MMT with non-incorporated companies) *Let* $(\overset{*}{\phi}_{\mathcal{X}}, \overset{*}{\phi}_{\mathcal{Z}}, \overset{*}{\phi}_{\mathcal{H}}, \overset{*}{\phi}_{\mathcal{Y}}, \overset{*}{\phi}_{\mathcal{P}})$ *and* $(\overset{*}{q}, \{\pi^j\}_{j\in\mathcal{J}})$ *be a FME with endogenous production decision and let* $\hat{\phi}_{\mathcal{P}}$ *be any financial policy allocation of the firms, then* $(\overset{*}{\phi}_{\mathcal{X}}, \hat{\phi}_{\mathcal{Z}}, \overset{*}{\phi}_{\mathcal{H}}, \overset{*}{\phi}_{\mathcal{Y}}, \hat{\phi}_{\mathcal{P}})$ *and* $(\overset{*}{q}, \{\pi^j\}_{j\in\mathcal{J}})$ *is also an FME with exogenous/endogenous production decision, where*

$$\hat{z}^i = \overset{*}{z}{}^i + \sum_{j\in\mathcal{J}} \left(\overset{*}{\xi}{}^j - \hat{\xi}^j\right) \bar{\delta}^i_j, \text{ for all } i \in \mathcal{I}.$$

This shows that the funding of optimal production plans is irrelevant, as the good allocation $\overset{*}{\phi}_{\mathcal{X}}$, the production $\overset{*}{\phi}_{\mathcal{Y}}$, and the price vector $\overset{*}{q}$ remain unchanged. Intuitively, the consumers can undo the change of the firms' financial policy.

Proof We check the conditions of the equilibrium Definition 6.5:

(i) We need to show that $(\overset{*}{x}{}^i, \hat{z}^i)$ solves the utility maximization problem given $\mathbb{B}^i(\overset{*}{q}, \omega^i, \bar{\delta}^i, A, \hat{D})$ for all i.

Consider two financial policies ξ and $\hat{\xi}$ with their corresponding dividend matrices $D := D(Y, \xi)$ and $\hat{D} := D(Y, \hat{\xi})$, assuming that firms maximize profits: $D = \sum_{j=1}^J \left(y^j + \begin{pmatrix} -q' \\ A \end{pmatrix} \xi^j\right)$ and analogously for $\hat{D}$.

Let $(x^i, z^i) \in \mathbb{B}^i(\overset{*}{q}, \omega^i, \bar{\delta}^i, A, D)$ and $(\hat{x}^i, \hat{z}^i) \in \mathbb{B}^i(\overset{*}{q}, \omega^i, \bar{\delta}^i, A, \hat{D})$ be two elements of the budget sets for which the budget constraints are binding:

$$\begin{aligned} x^i &= \omega^i + \begin{pmatrix} -q' \\ A \end{pmatrix} z^i + D\bar{\delta}^i \\ &= \omega^i + \begin{pmatrix} -q' \\ A \end{pmatrix} z^i + \sum_{j=1}^J \left(y^j + \begin{pmatrix} -q' \\ A \end{pmatrix} \xi^j\right) \bar{\delta}^i_j, \\ \hat{x}^i &= \omega^i + \begin{pmatrix} -q' \\ A \end{pmatrix} \hat{z}^i + \hat{D}\bar{\delta}^i \\ &= \omega^i + \begin{pmatrix} -q' \\ A \end{pmatrix} \hat{z}^i + \sum_{j=1}^J \left(y^j + \begin{pmatrix} -q' \\ A \end{pmatrix} \hat{\xi}^j\right) \bar{\delta}^i_j. \end{aligned}$$

Now the task is: given $(z^j, \{\xi^j, \hat{\xi}^j\}_{j\in\mathcal{J}})$, find $\hat{z}^i$ such that $x^i = \hat{x}^i$ (to undo the portfolio). The solution is given by the portfolio stated in the theorem: $\hat{z}^i = z^i + \sum_{j=1}^J (\xi^j - \hat{\xi}^j)\bar{\delta}^i_j$. Thus, for any change of financial policy by the firms, i.e. from ξ to $\hat{\xi}$, there is a portfolio $\hat{z}^i$ for each household to keep his original level of consumption x^i. This shows that restricted to the consumption allocation the budget sets are independent of the financial policy D, respectively $\hat{D}$.

(ii) Note first that in equilibrium it must hold that $\pi^{j\prime}\begin{pmatrix} -\overset{*}{q}{}' \\ A \end{pmatrix} = 0$ for each j. Otherwise firms can make infinite gains on the bond market. With binding budget constraints of the firms this yields:

$$\pi^{j\prime}\overset{*}{d}{}^j = \pi^{j\prime}\left(\overset{*}{y}{}^j + \begin{pmatrix} -\overset{*}{q}{}' \\ A \end{pmatrix}\overset{*}{\xi}{}^j\right) = \pi^{j\prime}\overset{*}{y}{}^j,$$

Thus, the objective function – a firm's profit – only depends on the production decision. As in (i), we see that the firms' budget sets restricted to the dividends are independent of the financial policy.

(iii) Achievability is obvious, since $\overset{*}{x}{}^i$, ω^i and $\overset{*}{y}{}^j$ have not changed for any $i \in \mathcal{J}$, resp. any $j \in \mathcal{J}$.

(iv) Market clearing works with choosing $\hat{z}^i$ according to the theorem. Compute

$$\begin{aligned}\sum_{i=1}^I \hat{z}^i + \sum_{j=1}^J \hat{\xi}^j &= \sum_{i=1}^I z^i + \sum_{j=1}^J (\xi^j - \hat{\xi}^j) + \sum_{j=1}^J \hat{\xi}^j \\ &= \sum_{i=1}^I z^i + \sum_{j=1}^J \xi^j = 0,\end{aligned}$$

where the last statement holds by the condition (iii) of FME. □

The following paragraph shows that this result carries over to the case of incorporated companies.

6.2.2 The Modigliani-Miller Theorem with Incorporated Companies

Apart from the activity on the bond market, firms can now also buy each other's stock. We have already described the resulting maximization problems for economic agents and firms at the beginning of the chapter. Based on the formulation there, we can adapt the requirements of an equilibrium using the definitions from above.

Definition 6.7 (Financial Market Equilibrium with Incorporated Companies) A FME for a financial economy $\mathcal{E}_F^P$ with incorporated companies consists of allocations $(\overset{*}{\phi}_{\mathcal{X}}, \overset{*}{\phi}_{\mathcal{Z}}, \overset{*}{\phi}_{\mathcal{H}}, \overset{*}{\phi}_{\mathcal{Y}}, \overset{*}{\phi}_{\mathcal{P}}, \overset{*}{\phi}_{\mathcal{T}})$ such that

(i) $(\overset{*}{x}^i, \overset{*}{z}^i, \overset{*}{\delta}^i) \in \mathbb{B}^i(\overset{*}{q}, \overset{*}{p}, \omega^i, \bar{\delta}^i, A, \overset{*}{D})$ and $\overset{*}{x}^i \in \arg\max U^i(x^i)$ for all $i \in \mathcal{I}$,
(ii) $(\overset{*}{d}^j, \overset{*}{y}^j, \overset{*}{\xi}^j, \overset{*}{\tau}^j) \in \mathbb{W}^j(\overset{*}{q}, \overset{*}{p}, A, \overset{*}{D})$ and $\overset{*}{d}^j \in \arg\max \pi^{j\prime} d^j$ for all $j \in \mathcal{J}$,
(iii) $\sum_{i\in\mathcal{I}} \overset{*}{x}^i \le \sum_{i\in\mathcal{I}} \omega^i + \sum_{j\in\mathcal{J}} \overset{*}{y}^j$,
(iv) $\sum_{i\in\mathcal{I}} \overset{*}{z}^i + \sum_{j\in\mathcal{J}} \overset{*}{\xi}^j = 0$, and $\sum_{i\in\mathcal{I}} \overset{*}{\delta}^i + \sum_{j\in\mathcal{J}} \overset{*}{\tau}^j = \mathbb{1}$,
where π^j satisfies the NAC for all j.

We have $\hat{d} \in < Y, A >$, as can be seen from (6.7).

The Modigliani-Miller theorem also holds in this setting:

Theorem 6.8 (MMT with incorporated companies) *Let* $(\overset{*}{\phi}_{\mathcal{X}}, \overset{*}{\phi}_{\mathcal{Z}}, \overset{*}{\phi}_{\mathcal{H}}, \overset{*}{\phi}_{\mathcal{Y}}, \overset{*}{\phi}_{\mathcal{P}}, \overset{*}{\phi}_{\mathcal{T}})$ *and* $(\overset{*}{q}, \overset{*}{p}, \{\pi^j\}_{j\in\mathcal{J}})$ *be a FME with* $(I - \overset{*}{\tau})$ *invertible, and let* $(\hat{\phi}_{\mathcal{P}}, \hat{\phi}_{\mathcal{T}})$ *be any financial policy allocation of the firms with* $(I - \hat{\tau})$ *invertible. Then there are* $(\hat{\phi}_{\mathcal{Z}}, \hat{\phi}_{\mathcal{H}})$ *such that*

$$(\overset{*}{\phi}_{\mathcal{X}}, \hat{\phi}_{\mathcal{Z}}, \hat{\phi}_{\mathcal{H}}, \overset{*}{\phi}_{\mathcal{Y}}, \hat{\phi}_{\mathcal{P}}, \hat{\phi}_{\mathcal{T}}),\ (\overset{*}{q}, \overset{*}{p}, \{\pi^j\}_{j\in\mathcal{J}})$$

is an FME.

So even if firms are active on the bond market and on the stock market, in equilibrium, the financial policy of the firms can be undone by the portfolio decisions of the consumers. This suggests the interpretation that the financial policy of the firms is irrelevant to the stock prices $\overset{*}{p}$. This interpretation is directly implied if $(\hat{\phi}_{\mathcal{Z}}, \hat{\phi}_{\mathcal{H}})$ are unique equilibrium allocations for $(\hat{\phi}_{\mathcal{P}}, \hat{\phi}_{\mathcal{T}})$. The proof of this version of the MMT is omitted. It follows the same logic as the proof above.[8] Neither the consumers' budget set changes nor the companies' objectives.

6.2.3 When Does the Modigliani-Miller Theorem Not Hold?

The statement that the financial policy of the firm is irrelevant hides a remarkable gap between theory and practice. It can be proven mathematically. However, the model seems to capture major points of the dynamics of reality only inadequately. For example, the news that a company would cut the dividends of its shares often leads to a very real stock price drop. Thus, there must be conditions, which are

[8]The undo portfolio turns out to be: $\hat{z}^i = \overset{*}{z}^i + (\overset{*}{\xi} - \hat{\xi})(I - \overset{*}{\tau})^{-1}(\overset{*}{d}^i + \bar{d}^i)$.

necessary for the Modigliani-Miller theorem (MMT), but need not be satisfied in reality. Let us discuss some of them:

- Perfect competition: If the markets have short sales limitations, i.e. $(\delta^i + \bar{\delta}^i) \geq 0$ or $z^i \geq 0$, then there might not be an undo portfolio.
- Equal access to financial markets: There are financial markets that are not open to private investors.
- Exclusion of bankruptcy: Considering the possibility of bankruptcy, dividends of a firm are not guaranteed, i.e. $d_1^k = \max\{0, y_1^k + A\xi^k + D_1\tau^k\}$, see [Hel81] for details.
- Assets: Consider options with the following payoff structure $A^k := \max\{0, d_1^j - K\}$, where K is the purchase price. As in the case of bankruptcy, there are nontrivial effects on the consumers' insurance possibilities.
- Tax neutrality: Unequal tax treatment of dividends is an issue.
- Symmetric information of all market participants: Consider the case where consumers do not know the firms' production policies. Instead, they have to infer the production y^j from the financial policy. Hence the latter has a nontrivial announcement effect, as we discuss in Chap. 7.

We wish to point out that complete markets is not one of the necessary conditions for the MMT. The key point in this case is merely that $d \in \text{span}\,\{Y, A\}$, i.e. the firm's dividend policy does not change the market's insurance possibilities, that is, its span. Otherwise, the bond generated by the firm's dividends could influence all prices and allocations, requiring a complete recalculation of the equilibrium.

However, if the dividend does not change the market's span, the economic agents can undo them even in an incomplete market. See [Got95] for more details on the MMT in incomplete markets.

6.3 Firm's Decision Rules

In this section we have a closer look at the decision-making process in firms. First, we question why the shareholders, who are consumers in the end of the day, would maximize profits. Then, we analyze whether shareholders with different preferences would agree on a common production plan.

6.3.1 Fisher Separation Theorem

The separation of management and ownership leads to an important problem: should a firm, from the point of view of its owners, attempt to maximize profit? Maybe it should only do so during a certain period, or try to maximize turnover instead? Perhaps the management should strive for the best product quality that is technically feasible? The owners might not even agree on a common goal. This topic is a main focus of the theory of the firm, and accordingly there is a lot of literature on it. Many interesting aspects arise, such as incentive problems. Nevertheless, we only consider the question if firm owners would agree on the goal of profit maximization.

6.3.1.1 A Simple Market with One Firm

As a first step we consider a situation where there is no uncertainty and only one firm. A single security with positive payoff A and price q is traded. Let p be the firm's market price and d its dividend at $t = 1$, while there are no dividends at $t = 0$. Furthermore, the household initially has a positive share $\bar{\delta}^i$ in the firm. This considerably simplifies the budget restriction (6.4), yielding the following maximization problem for the agent:

$$\max_{x^i \in \mathbb{R}^2_{\geq 0}, \delta^i, z^i} U(x^i) \quad \text{such that} \quad x_0^i + qz^i + p\delta^i \leq \omega_0^i + p\bar{\delta}^i \quad \text{and} \quad x_1^i \leq \omega_1^i + z^i A + \delta^i d. \tag{6.8}$$

Thus at $t = 0$, an household can consume and buy securities and shares in the firm. At $t = 1$ it gains the earnings from its initial share, security payoffs and the firm's dividends. Suppose the agent would like to consume less at $t = 0$ and more at $t = 1$. This results in a utility change. If Δx_0 and Δx_1 are very small, we can approximate the utility change by the corresponding first derivatives:

$$t = 0: \quad U(x_0 - \Delta x_0, x_1) - U(x_0, x_1) \approx \frac{\partial U(x)}{\partial x_0} \Delta x_0$$

$$t = 1: \quad U(x_0, x_1 + \Delta x_1) - U(x_0, x_1) \approx \frac{\partial U(x)}{\partial x_1} \Delta x_1.$$

There is only one way to transfer consumption between the periods: buying shares or securities. Setting aside Δx_0 consumption units at $t = 0$, one can use the free resources to buy $\frac{\Delta x_0}{q}$ units of securities. This yields a payoff of $\frac{\Delta x_0}{q} A$ at $t = 1$ to the extent of which one may finance additional consumption: $\Delta x_1 \leq \frac{\Delta x_0}{q} A$. If the agent has strictly monotonic preferences, we even have $\Delta x_1 = \frac{\Delta x_0}{q} A$.

At the optimum, this transfer of consumption may not change the net utility, i.e.:

$$\frac{\partial U(x)}{\partial x_0} \Delta x_0 = \frac{\partial U(x)}{\partial x_1} \Delta x_1.$$

Substituting $\Delta x_1 = \frac{\Delta x_0}{q} A$, we get

$$\frac{\partial U(x)}{\partial x_0} \Delta x_0 = \frac{\partial U(x)}{\partial x_1} \frac{\Delta x_0}{q} A \quad \Leftrightarrow \quad MRS_{01} = \frac{A}{q}. \tag{6.9}$$

Since there is no uncertainty about the state realized at $t = 1$, we may interpret every security as riskless. Thus, it will suffice to introduce an arbitrary security with positive payoff. Of course, this also holds for a bond. We know that the price q of a bond with payoff $A = 1$ is determined by $q = \frac{1}{1+r_f}$, where r_f denotes the riskless rate of interest. In this case the right-hand side of (6.9) becomes $MRS_{01} = 1 + r_f$. At the optimum an agent will determine his allocations such that his valuations of consumption at $t = 0$ and $t = 1$ agree with the market's.

Buying shares instead of securities leads to a similar result. Forgoing consumption at $t = 0$ allows one to buy $\frac{\Delta x_0}{p}$ shares to the firm, which in turn yield a profit of $\frac{\Delta x_0}{p} d$ consumption units at $t = 1$. Again, at the optimum this transfer of consumption may not change the net utility, and we get:

$$\frac{\partial U(x)}{\partial x_0} \Delta x_0 = \frac{\partial U(x)}{\partial x_1} \frac{\Delta x_0}{p} d \quad \Leftrightarrow \quad MRS_{01} = \frac{d}{p}.$$

Now we have both $MRS_{01} = 1 + r_f$ and $MRS_{01} = \frac{d}{p}$, so $1 + r_f = \frac{d}{p}$, and thus $p = \frac{d}{1+r_f}$.

We can divide the income at $t = 1$ by $1 + r_f$ in (6.8) to convert it to a present income, then sum up both budget restrictions. Some rewriting then leads to

$$x_0^i - \omega_0^i + \frac{x_1^i - \omega_1^i}{1+r_f} + z^i \left(q - \frac{A}{1+r_f} \right) \leq \bar{\delta}^i p + \delta^i \left(\frac{d}{1+r_f} - p \right). \tag{6.10}$$

The left-hand expression describes the value of the net swapped amounts in $t = 0$ consumption units. The right-hand side denotes the value of the firm share in consumption units. Therefore, the net transfer may not exceed the income generated by the firm share. If we assume that the economic agents have strictly monotonic preferences, (6.10) is satisfied with equality since more consumption means more utility. If we substitute $q = \frac{A}{1+r_f}$ and $p = \frac{d}{1+r_f}$, the bound simplifies to

$$x_0^i - \omega_0^i + \frac{x_1^i - \omega_1^i}{1+r_f} = \bar{\delta}^i \frac{d}{1+r_f}.$$

Hence the possible consumption increases in proportion to the dividend d. We have also considered the case $S = 1$, so the demand for a higher dividend can be translated to a demand for a high profit or net present value (NPV). Clearly, the shareholders will strive for maximum profit, i.e., carry out the project with the highest cash-flow.

This result also has a nice graphical justification. In Fig. 6.1 the product technology set is shown in grey. The households, assuming they own and control the firms, will choose a production point that maximizes their available income. In the figure this is $(\overset{*}{y}_0, \overset{*}{y}_1)$, which lies on the budget line I_3. The result is independent of the respective preferences, except that they are required to be strictly monotonic.[9] The

[9]Naturally the preferences play a crucial role in the price formation of r_f.

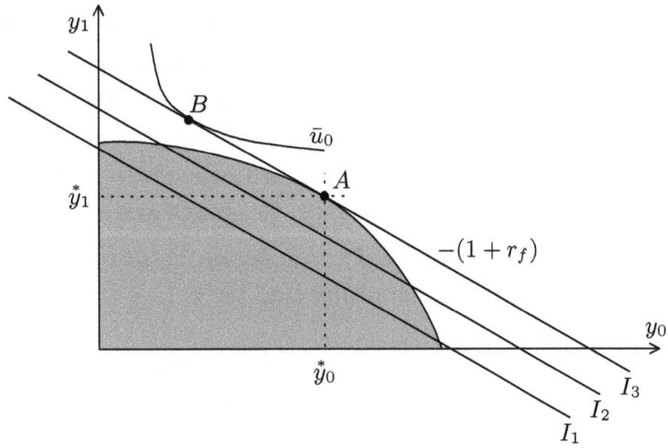

Fig. 6.1 Firms maximize their profit by choosing the line of constant profits tangential to their production technology (I_3), point A; households maximize their utility by looking for the budget line tangential to their production technology (I_3), then obtaining their optimal consumption through trading

households can achieve optimum consumption by trading their production $(\overset{*}{y}_0, \overset{*}{y}_1)$ on the market for securities.

But what happens if the households are not in direct control of the firms? If the firms maximize their profit, they will carry out the maximum achievable line of constant profits. If we take I_1, I_2, I_3 as lines of constant profits, a firm will clearly choose $(\overset{*}{y}_0, \overset{*}{y}_1)$, too. The associated profit will be paid out to the households, which in turn realize their optimal consumption allocation. In the second step, we now extend our considerations to include multiple firms.

6.3.1.2 Fisher Separation with Multiple Firms

An analogous maximization problem results for households and shareholders: an economic agent tries to maximize its temporal utility. To that end it can buy shares and securities within its resources. With multiple firms the budget restriction looks a bit more general, but is still a special case of (6.4):

$$\max_{\substack{x^i \in \mathbb{R}^2_{\geq 0} \\ \delta^i \in \mathbb{R}^J \\ z^i \in \mathbb{R}}} U(x^i) \quad \text{such that} \quad x^i_0 + qz^i + \sum_{j \in \mathcal{J}} p^j \delta^i_j \leq \omega^i_0 + \sum_{j \in \mathcal{J}} p^j \bar{\delta}^i_j$$
$$\text{and} \quad x^i_1 \leq \omega^i_1 + z^i A + \sum_{j \in \mathcal{J}} d^j \delta^i_j. \tag{6.11}$$

Similarly to the first part, we obtain the relation of prices of securities and shares. Accordingly, the budget restrictions in (6.11) can be written concisely:

$$x^i_0 - \omega^i_0 + \frac{x^i_1 - \omega^i_1}{1 + r_f} = \sum_{j \in \mathcal{J}} \bar{\delta}^i_j \frac{d^j}{1 + r_f}.$$

Introducing more firms has not changed the result: a shareholder with strictly monotonic utility function again wants the firm to maximize its profit. What can we say about the optimal production decision if the firm's owners have differing

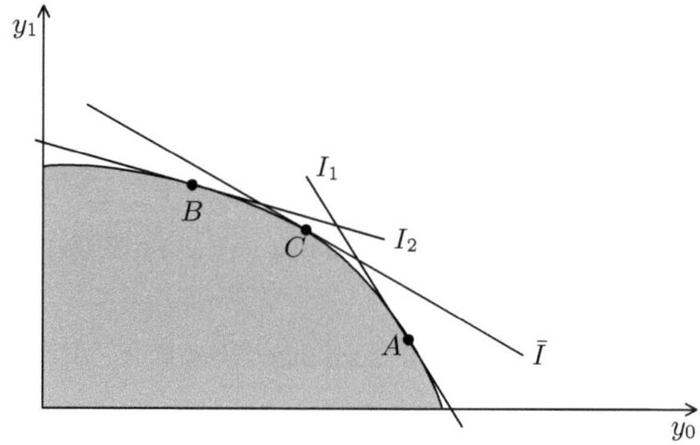

Fig. 6.2 Incomplete markets: optimal production plans for two rates of interest and their weighted average

valuations of the state prices? Up to now we did not consider any uncertainty, i.e. there was only a single future state. Varying valuations are thus just differences in the personal interest rates, as shown in Fig. 6.2. The two types of shareholders are illustrated as the slope of I_1 and I_2, showing very high and very low rates of interest, respectively. Accordingly, the former would choose production point A, while the latter would pick B. We compute an average rate of interest, weighted by the corresponding shares, which is shown as the slope of $\bar{I}$. In the next section we will show that everybody would agree on the resulting production point C. It remains crucial that the firm's production possibilities do not alter the span of the market, i.e., the set of all achievable consumption streams. This is ensured if, i.e., $< Y > \subset < A >$. This could change all relative prices, causing a utility loss for some households which would therefore vote against such a decision.

6.3.2 The Theorem of Drèze

In Sect. 6.2 the sole requirement for the firm's decision rules was that it weights its future profits by the elements of a vector $\pi^j \in \mathbb{R}^{S+1}$. If the markets are complete (rank $A = S$), this uniquely determines the firm's target function. In the case of incomplete markets, it does not have to be unique. Basically one can work with any vector that does not offer an arbitrage opportunity, i.e.,

$$\pi^{j\prime} \begin{pmatrix} -q' \\ A \end{pmatrix} = 0 \quad \text{and} \quad \pi^{j\prime} \begin{pmatrix} (Y_0 - p)' \\ Y_1 \end{pmatrix} = 0.$$

In the case of complete markets, every shareholder would agree with the uniquely determined criterion π^N. This is apparent from the first-order conditions of the utility maximization problem: the normalized utility gradients of all consumers matches the normalized target function vector π^N. In the case of incomplete markets, this unanimity between the shareholders is lost. How should one reach an agreement?

To simplify the problem, let firms be barred from running a financial policy, i.e., $\xi^j = 0$. It immediately follows that $D = Y$. The Modigliani-Miller theorem showed that in our model, nobody will want to object to this limitation. Furthermore, we will limit ourselves to the non-incorporated companies model. So we have the following underlying maximization problems:

$$\max_{\substack{x^i \in \mathcal{X}^i \\ z^i \in \mathcal{Z}^i}} U^i(x) \quad \text{such that} \quad (x^i, z^i) \in \mathbb{B}^i(q, \omega^i, \bar{\delta}^i, A, Y) \tag{6.12}$$

$$\max_{y^j \in \mathcal{Y}^j} \pi^{j\prime} y^j \quad \text{such that} \quad d^j \leq y^j.$$

We now look into firm j's general meeting of shareholders GM^j, *after* the consumers have defined their portfolios z^i. We also assume that the production policies of all other firms are available. The question is then: what can GM^j improve, given the decisions $\{\bar{z}^i\}_{i \in \mathcal{I}}, \{\bar{y}^l\}_{l \neq j}$. We will assume that GM^j only takes into consideration the consumers that own shares of firm j. Hence, let

$$\mathcal{I}_j := \{i \in \mathcal{I} \mid \bar{\delta}^i_j > 0\}, \quad \text{for all } j.$$

How does a consumer i value changes of the production policy $\bar{y}^j$? The obvious criterion is the increase in consumer's utility with the new production policy $\hat{y}^j$:

$$U^i(\bar{x}^i + (\hat{y}^j - \bar{y}^j)\bar{\delta}^i_j) U^i(\bar{x}^i).$$

Remark 6.9 The x^i implied by a given $(\bar{z}, \bar{Y})$ is

$$\bar{x}^i := \omega^i + \begin{pmatrix} -q' \\ A \end{pmatrix} \bar{z}^i + \bar{Y}\bar{\delta}^i.$$

We now give the following unanimity criterion:

Definition 6.10 (Pareto-efficient with respect to GM^j) The production policy $\bar{y}^j$ of firm j is Pareto-efficient with respect to GM^j, if $\bar{y}^j \in \mathcal{Y}^j$, and there is no $\hat{y}^j \in \mathcal{Y}^j$ with

$$U^i(\bar{x}^i + (\hat{y}^j - \bar{y}^j)\bar{\delta}^i_j) \geq U^i(\bar{x}^i) \quad \text{for all } i \in \mathcal{I}_j, \text{ and}$$

$$U^i(\bar{x}^i + (\hat{y}^j - \bar{y}^j)\bar{\delta}^i_j) > U^i(\bar{x}^i) \quad \text{for at least one } i \in \mathcal{I}_j.$$

According to this criterion, an existing production policy is only discarded if all shareholders agree. In general many production policies may fulfill the criterion, so that a manager may not know which one to choose. We therefore allow shareholders to change each other's minds by means of transfer payments: let ρ^i_j be the net side payment made by shareholder i in the voting process. Then he values the decision of GM^j according to the criterion:

$$U^i(\bar{x}^i + (\hat{y}^j - \bar{y}^j)\bar{\delta}^i_j - \rho^i_j e_1)U^i(\bar{x}^i).$$

Thus, we define

Definition 6.11 (Pareto-efficiency with side payments with respect to GM^j) The production policy $\bar{y}^j$ of firm j is Pareto-efficient with side payments w.r.t. GM^j, if $\bar{y}^j \in \mathcal{Y}^j$ and there are no $\hat{y}^j \in \mathcal{Y}^j$ and $\{\rho^i_j\}_{i\in\mathcal{I}_j}$ with $\sum_{i\in\mathcal{I}_j} \rho^i_j \geq 0$, such that $U^i(\bar{x}^i + (\hat{y}^j - \bar{y}^j)\bar{\delta}^i_j - \rho^i_j e_1) \geq U^i(\bar{x}^i)$ for all $i \in \mathcal{I}_j$, and $U^i(\bar{x}^i + (\hat{y}^j - \bar{y}^j)\bar{\delta}^i_j - \rho^i_j e_1) > U^i(\bar{x}^i)$ for at least one $i \in \mathcal{I}_j$.

The decision-finding mechanism described in this definition is rather difficult to implement: one would have to iterate through all production policies and transfer payments to find a better production policy. In the following we will describe an equivalent, direct mechanism. To that end, let $\pi^{N,i}(x^i) \in \mathbb{R}^{S+1}_{\gg 0}$ be his MRS between consumption in state s and present consumption:

$$\pi^{N,i}_s(x^i) := \frac{\partial_{x^i_s} U^i(x^i)}{\partial_{x^i_0} U^i(x^i)}, \qquad s \in \mathcal{S}.$$

Now imagine the consumers just tell the manager their vectors $\pi^{N,i}(x^i)$. He in turn chooses the production policy such that it maximizes the function $\pi^{j\prime} y^j$, where $\pi^j := \sum_i \delta^i_j \pi^{N,i}$ is treated as given by the firm.[10] Therefore, the discounting vector of firm j is the mean discounting vector of the consumers, weighted by their shares. In particular, this decision rule leads to the usual profit maximization rule if the markets are complete. The equivalence of the two mechanisms is shown by the Theorem of Drèze.

Theorem 6.12 (Theorem of Drèze) *The production policy $y^j \in \mathcal{Y}^j$ is Pareto-efficient with side payments with respect to GM^j if and only if for a given $\pi^{N,i}$, $y^j \in \arg\max_{y^j \in \mathcal{Y}^j} \pi^{N,j\prime} y^j$, where $\pi^{N,j} = \sum_i \bar{\delta}^i_j \pi^{N,i}$.*

Proof

1. *Pareto-efficiency $\Rightarrow$ Drèze criterion:*
   Consider the production plan $\bar{y}^j \in \mathcal{Y}^j$ of firm j. Suppose the Drèze criterion does not hold for $\bar{y}^j$, i.e. there exists $\hat{y}^j \in \mathcal{Y}^j$ with $\sum_i \delta^i_j \pi^i(\bar{x}^i)(\hat{y}^j - \bar{y}^j) > 0$.
   Let $\rho^i_j := \delta^i_j \pi^i(\bar{x}^i)(\hat{y}^j - \bar{y}^j) - \epsilon$, for some $\epsilon > 0$, be the net side payment of i in firm j. If ϵ is small enough, then $\sum_{i\in\mathcal{I}_j} \rho^i_j \geq 0$.

[10] We assume the MRS are honestly communicated; otherwise one would have to find incentive compatible survey mechanisms.

Let $\xi^i := \delta^i_j(\hat{y}^j - \bar{y}^j) - \rho^i_j e_1$ for $i \in \mathcal{I}_j$. Then $\pi^i(\bar{x}^i)\xi^i = \delta^i_j \pi^i(\bar{x}^i)(\hat{y}^j - \bar{y}^j) - \rho^i_j = \epsilon > 0$, for all $i \in \mathcal{I}_j$ (note that $\pi^i_{0,0} = 1$). The utility functions are monotonic and continuous, so for every $i \in \mathcal{I}_j$ there exists an $\alpha^i \in (0, 1]$ such that $U^i(\bar{x}^i + \alpha^i \xi^i) > U^i(\bar{x}^i)$. By quasi-concavity of U^i we have for all $0 < \alpha \leq \min_{i \in \mathcal{I}_j}\{\alpha^i\}$: $U^i(\bar{x}^i + \alpha\xi^i) > U^i(\bar{x}^i)$ for all $i \in \mathcal{I}_j$. $\alpha\xi^i$ can be written as $\alpha\xi^i = \delta^i_j(\tilde{y}^j - \bar{y}^j) - \alpha\rho^i_j e_1$, where $\tilde{y}^j := \bar{y}^j + \alpha(\hat{y}^j - \bar{y}^j)$. Because $\tilde{y}^j = \alpha\hat{y}^j + (1-\alpha)\bar{y}^j$, $\hat{y}^j, \bar{y}^j \in \mathcal{Y}^j$ and $\mathcal{Y}^j$ is concave, it follows that $\tilde{y}^j \in \mathcal{Y}^j$. Now $\sum_i \rho^i_j \geq 0$ and $U^i(\bar{x}^i + (\tilde{y}^j - \bar{y}^j)\delta^i_j - \alpha\rho^i_j e_1) > U^i(\bar{x}^i)$ for all $i \in \mathcal{I}_j$, therefore the Pareto criterion with side payments does not hold, which is a contradiction.

2. *Drèze criterion* $\Rightarrow$ *Pareto-efficiency with side payments:*
Consider the production plan $\bar{y}^j \in \mathcal{Y}^j$. Suppose the Pareto criterion with side payments does not hold, i.e. there are $(\hat{y}^j, \rho_j) \in \mathcal{Y}^j \times \mathbb{R}^{\mathcal{I}_j}$ with $\sum_{i \in \mathcal{I}_j} \rho^i_j \geq 0$ such that $U^i(\bar{x}^i + (\hat{y}^j - \bar{y}^j)\delta^i_j - \rho^i_j e_1) > U^i(\bar{x}^i)$ for all $i \in \mathcal{I}_j$. As in the first part, for every $i \in \mathcal{I}_j$ there exists an $\alpha^i \in (0, 1]$ such that $U^i(\bar{x}^i + \alpha^i\xi^i) > U^i(\bar{x}^i)$, hence $\pi^i(\bar{x}^i)\left((\hat{y}^j - \bar{y}^j)\delta^i_j - \rho^i_j e_1\right) > 0$ for all $i \in \mathcal{I}_j$. Since $\pi^i_{0,0}(\bar{x}^i) = 1$ and $\sum_i \rho^i_j \geq 0$, this implies $\sum_{i \in \mathbb{I}_j} \delta^i_j \pi^i(\bar{x}^i)\hat{y}^i > \sum_{i \in \mathcal{I}_j} \delta^i_j \pi^i(\bar{x}^i)\bar{y}^j$, i.e., the Drèze criterion does not hold. □

6.4 Summary

In this section we have shown that the firm can be included into our financial market economy. In the absence of frictions like taxes or bankruptcy we could even show that the firm's financial policy is irrelevant since it can be "undone" by the households. Moreover, we showed that voting according to the shares that households hold lead to Pareto-efficient allocations. This latter result is not true if the shares of the firms are traded on a stock market. In that case the best objective function of the firm is still unknown (see [MQ02] for further results).

Information Asymmetries on Financial Markets* 7

All of the books in the world contain no more information than is broadcast as video in a single large American city in a single year – Not all bits have equal value.

CARL SAGAN

So far we have assumed common knowledge about the (state-contingent) pay-offs of assets. Imagine now that some agents know the payoffs better than others. Then – besides intertemporal substitution, risk sharing and betting on the occurrence of the states of the world – a seller of an asset might want to sell it because he knows it has very low pay-offs. Anticipating this no agent would buy at a price allowing the seller to make a profit and ultimately no transaction is made. In the subprime mortgage crisis this aspect of asset markets became overwhelming so that asset markets broke down completely.

This chapter of our book shows how to model asymmetric information. It shows the effects of asymmetric information on market prices, banking and insurance contracts. A formal way allowing these aspects to be integrated into the previous model is to let agents' beliefs on the occurrence of the states of the world depend on their market observations: prices and transactions. In this sense asymmetric information is not an alternative to what we learned so far but a generalization.

Let us start with an introductory example on the topic of information that illustrates some of the fundamental ideas in a somehow not so serious way (translated from [nzz00]):

> In Gelosia, wives are not lenient towards unfaithful husbands. One morning, the queen summoned all women: "Word has reached me that at least one of our husbands has been unfaithful. If one of you discovers that your husband has cheated, you must kill him come midnight on the same day you found out."
>
> Gelosian women love to gossip, so if one of their husbands was unfaithful, the entire country would know by next morning. Only his wife would be kept in the dark out of respect. For a long time after the queen's speech, nothing has happened. Suddenly, 39 days

T. Hens, M.O. Rieger, *Financial Economics*, Springer Texts in Business and Economics, DOI 10.1007/978-3-662-49688-6_7

later, all 40 women resort to the knife and send their husbands to heaven come in a country-wide massacre.

These horrible events are an example of how insight grows through a series of conclusions. The queen mentioned "at least one" cheater. If there had only been a single one, his wife would immediately have known: since she had not heard of any affair, she had to be the betrayed one.

With two casanovas, there would have been two dead husbands one day *after* the address. The lack of news of a killing on the first night would have told their wives that they were unfaithful, since there now had to be at least two affairs – otherwise someone would have been killed on the first night already – and they had only heard of a single one. This can be formulated mathematically: no execution at midnight on the n-th day means that at least $n + 1$ husbands were unfaithful.

Thus, on the morning of the 40th day, all 40 women knew that at least 40 husbands had to have been unfaithful. Since there were only 40 women, who each had heard of only 39 affairs, they all concluded that their husband had to be the 40th culprit.

7.1 Information Revealed by Prices

The idea of the following example is due to Akerlof [Ake70][1]: If a vendor is too willing to lower prices, potential customers will think the quality of his products is rather low. This might lead to the vendor not being able to sell at all. This phenomenon is referred to as the "market for lemons" where "lemon" is a colloquial term for an inferior quality product. A typical example for this are used cars. We model the effect as follows:

Assume there is a product in two different quality levels, denoted by H (high) and L (low), where $H > L$. Let μ and $1 - \mu$ be the commonly known proportion of good and bad products, respectively. Let q be the price of the good product. The seller knows the quality Q of his product, where Q can be H or L; his utility of the sale is then $V(Q) = q - Q$. If $V(Q) < 0$, he will not make the deal. In particular, no products will be sold at all if the market price q is less than L. The buyer on the other hand does not know the quality. However, he has expectations, called *beliefs*. Let β be the buyer's belief that the product is of high quality H. Let the buyer's utility be the expected quality on the basis of his beliefs minus the price: $U(q, \beta) := \beta H + (1 - \beta)L - q$. Again, the buyer will not go through with the deal if $U(q, \beta)$ is negative.

One easily checks that $\overset{*}{\beta} = 0$ and $\overset{*}{q} = L$ forms an equilibrium in the sense that, given the belief and price, neither buyer nor seller can strictly increase their utility by deviating, i.e., not selling or buying. Only low quality products are traded; there is no market for high quality products. $\overset{*}{\beta} = 0$ and $\overset{*}{q} = L$ is even the only equilibrium: Assuming the buyers are not systematically mistaken, their belief has to be the average quality on the market. If $\overset{*}{q} = L$, only bad products are on sale,

[1] More information on information revealed by prices is revealed in [GH90].

so $\overset{*}{\beta} = 0$. If $\overset{*}{q} = H$, all qualities are offered, so $\overset{*}{\beta} = \mu$. But then $U(H, \mu) < 0$, therefore no potential buyer will show any demand.

This effect does not only hurt the buyers, but also the sellers of high quality goods, in Akelof's words:

> The cost of dishonesty, therefore, lies not only in the amount by which the purchaser is cheated; the cost also must include the loss incurred from driving legitimate business out of existence.

Another example where information is revealed by the price is the financial market: according to the efficient market hypothesis all currently known information should be already included in the market process. A consequence of this is the No-trade-Theorem: since nobody has superior information there is no reason for speculative trade, i.e. there will only be trades for other motifs (intertemporal substitution, risk sharing and betting etc.)

However, how can prices entail all information if nobody trades based on information?

This information paradox first appeared in a publication by Grossmann and Stiglitz [GS80]. It describes more generally the consequences of a classical freerider problem for information efficiency of financial markets. Suppose Fama's efficient market hypothesis holds, i.e., all prices correctly reflect the information. However, getting information or analyzing prices is not free. If all information is contained in the prices – does the model contradict its assumptions? If the prices include all information, there is no incentive to retrieve more information, i.e., nobody will pay for it. Thus not all information is contained in the prices. Conversely, assume the prices hold no information. Then it is profitable to pay for information, hence prices contain information. Therefore, the two states where the prices contain no, or all, information, can never be reached.

The information paradox can only be solved if the prices partially contain information, which contradicts the efficient market hypothesis. The (empirical) question is then whether although not completely true the efficient market hypothesis is a good enough approximation of reality or not.

One counterexample to the efficient market hypothesis being a good approximation of reality would be a successful chart analysis. It contradicts the statements of classical capital market theory, according to which no systematic information on future prices can be gained from the present price and past prices. It could be argued that chart analysis would only be used to uncover all relevant information on the market but then it should already be "priced in" according to the efficient market hypothesis. If one moves away from the strong assumptions of classical capital market theory, and takes into account various phenomena that can be observed on the financial markets, one concludes that chart analysis could in fact be used profitably. It would then aim to timely discover moods and associated trends. The field of behavioral finance deals with the description and analysis of such moods and phenomena.

7.2 Information Revealed by Trade

Not only prices can reveal information, also trades, dividends and other financial transactions can be informative. As an amusing example we want to mention that performance of companies is significantly reduced if their CEO has just bought a new house. In fact his trade (of buying the house) reveals often bad information about his company since he might have sold company options or stocks to finance his new house and he will more likely do this when he deems the stock prices high – maybe even too high. The firm 2iQ has used this idea to construct a "Directors' Confidence Index".

Let us now consider a more serious example. We want to examine the role of dividends as information revealing signals. This role has been neglected in our derivation of the Modigliani-Miller Theorem (Chap. 6). The main observation is that in the case of asymmetric information on the results of a firm's operations, dividends may indicate profitability and thus distinguish a firm from less profitable ones. Consider a scenario with two firms $i = g, b$, for *good* and *bad*, respectively. However investors do not know the type of the firm. At the end of the period, there are two possible outcomes (states) each: H_i and L_i, for high and low profits, respectively. Let $H_g > H_b$ and $L_g > L_b$. On the market, future profits are rated via risk-adjusted probabilities $\pi_i^\star$. Next, a firm can announce a dividend d_i, where we assume that if $d_i > L_i$ it will go bankrupt: a firm can never distribute a dividend higher than its worst case revenues. Since $L_g > L_b$, g can differ from b by paying $d_g = L_g$. The bad firm b cannot guarantee such a dividend because it would go bankrupt. A situation like this is called a *separating equilibrium*: it distinguishes the types of firms. The opposite, where the firms are indistinguishable, is called *pooling equilibrium*. In the latter case the prices of the two firms must be equal. In a risk-neutral economy, the price of firm i in a separating equilibrium is

$$q_i = \frac{1}{1 + r_f}(\pi_i^\star H_i + (1 - \pi_i^\star)L_i),$$

whereas in a pooling equilibrium it is

$$q = \frac{1}{2}\frac{1}{1 + r_f}\sum_{i=g,b}(\pi_i^\star H_i + (1 - \pi_i^\star)L_i).$$

The result of this very simple and intuitive example may explain why stock prices rise after a payout of dividends has been announced.

I is true that general trades on markets can reveal information Therefore, when asked why they bought a certain asset, many investors reply they did so only because other relevant groups (competitors, traders, etc.) had bought as well. This behavior is called *herding*. It is a common phenomenon on markets and is consistently validated in practice. The dynamics leading to it can be explained by the announcement effect that specific agents' actions have.

Assume that the members of a group of risk-neutral[2] investors are to decide, one after the other, whether to buy a security or not. If they don't buy, they get \$10 at the end. If they do, they get \$20 or nothing with equal probability. At the beginning, the two alternatives have the same expected value, thus all members are neutral between buying or not buying. Now assume every player gets a binary signal that is not observable by, and independent from, the other players. Let the probability p that the signal is positive, conditioned on a payment of \$20, be strictly greater than 0.5. Thus the first player observes his personal signal and has to estimate the probability of getting \$20 at the end of the game, i.e., $\mathcal{P}\,(\text{payoff} = 20 \mid \text{signal} = H)$. If this probability is more than 0.5, buying the security pays off.

The player's idea of the probability distribution is called *belief*. Furthermore, *prior* denotes the belief before observing the signal; likewise *posterior* the belief afterwards.

To describe the formation of beliefs and dependencies between them, we use Bayes' rule. At its core, it mathematically describes how beliefs should be adapted to new findings.[3] Because $p > 0.5$, the first player will decide to buy if he got the signal H, and not to buy otherwise. Thus, the other players can guess his signal from his actions. Next, the second player receives a signal H or L. If it is H and the first player bought, he will buy too. Similarly, if he got L and the first player did not buy, neither will the second. If the signals do not agree, suppose the player randomly decides for either choice with probability 0.5.

The third player again faces several possibilities: both players before him bought, or neither one did, or their choices did not agree. The first two cases are especially interesting. If the third player gets the signal H, and both players before him bought, of course so will he; analogously for the opposite case. However, if his signal contradicts the behavior of his precursors, he should also follow the first two players: Assume he sees HH. Then the first player must have got H. The probability that the second player got H as well is $2/3$, since if he got L he would have chosen L with $1/2$ probability. Given this, it is more likely that H is correct even if the signal of player 3 is low.

The longer the chain of agreeing preceding decisions, the more weight they will gain compared to the player's own signal. Hence the players base their decisions not only on their own signal, but also on their guess what the previous agents' signals could have been. This finally leads to the situation where the actions of others are the main criterion for one's own decisions. This (rational!) model for herding is called "information cascade" [BHW98].

---

[2]Risk-neutrality is just a simplification. The statements easily carry over to a setting with risk-averse agents.

[3]In fact Bayes' rule states that posterior = conditional likelihood · prior/likelihood, in symbols $\mathcal{P}(R = r|e) = \mathcal{P}(e|R = r) \cdot \mathcal{P}(R = r)/\mathcal{P}(e)$, where $\mathcal{P}(R = r|e)$ denotes the probability that a random variable R takes the value r, given the signal e.

7.3 Moral Hazard

Suppose a risk-neutral firm has a choice between two projects $i = a, b$ that carry a risk. Let each project yield X_i with probability p_i, and 0 with probability $1-p_i$. Thus the expected profit of a project is p_iX_i. We assume $p_aX_a > p_bX_b$ and $X_b > X_a$. To realize a project, the firm needs an investment I that must be financed with outside capital. The investment involves a repayment $R > I$. Both projects cover the cost R in the positive case. In the negative case ($X_i = 0$), no repayment is made. Hence, the lender or bank bears a contingency risk for the credit. The firm gains an expected profit of $U(R, i) = p_i(X_i - R)$. The expected profit of the bank is $\Pi(R, i) = p_iR - I$.

In the case of symmetric information (both partners know which project will be realized and can prove this in court) they will agree on project a, since it has a higher expected profit. The bank will completely claim the profit for itself, i.e., set $R = X_a$.[4]

Consider a situation with asymmetric information. Assume the bank has no way of discovering or proving which project was chosen. This creates an incentive: the firm can claim it will implement project b, which limits the bank's demand for the entire profit. For a given R it will then compare the profits of the two projects and choose the one with higher gains. Using our assumptions, we can determine a threshold $\hat{R}$ for R, past which project b will be favored over a. The bank's profit also depends on $\hat{R}$. If the bank picks R between 0 and $\hat{R}$, it knows the firm will choose a, thus the bank's profit is $p_aR - I$. Conversely, if R lies between $\hat{R}$ and X_b, project b will be implemented for a profit of $p_bR - I$. The bank chooses the repayment such that its profit becomes maximal. Therefore, $R = \hat{R}$ if $p_a\hat{R} > p_bX_b$, else $R = X_b$. In the first case, it does not manage to appropriate the complete profit. Thus the firm has an *information benefit*, i.e., $U(\hat{R}, i) = p_a(X_a - \hat{R}) > 0$, resulting because the bank cannot observe or legally enforce the choice of project.

A different, very concrete example for moral hazard problems can be seen as an origin for the recent financial crisis: banks gave out loans to "subprime" investors who wanted to buy houses. The risk of the loans was then partially sold to other investors (usually investment banks and hedge funds) in form, e.g. of the now infamous CDOs (for credit default options or collateralized debt obligation). Now since only the bank had information on the actual quality of the loans this led to a moral hazard problem: the banks were interested in increasing the number of loans, even if their default probability was high, since their risk was mostly sold off to investment banks and hedge funds. The latter ones did not take this effect sufficiently into account. Herding of course also played a role as more and more investment banks entered the subprime market.

[4]The bank can force its own conditions on the firm, because it is the only potential financier but not obliged to grant any credit. It is only limited by the firm's willingness to participate. Therefore it needs to choose conditions such that the firm is still interested.

7.4 Adverse Selection

In a market equilibrium prices will be set such that a (somehow) optimal allocation is reached. There are cases, however, where it is not possible to accomplish this. This phenomenon is called "adverse selection". We illustrate it with the following example[5]:

Let there be two types of firms, g and b. They differ in their expected profits. We again assume two states occurring with specific probabilities for each firm. Let H_i and L_i, $i = b, g$, be the corresponding payoffs. Let in particular $L_b := 0$ and $L_g > 0$, $p_g := \mathcal{P}(\text{payoff} = H_g) > \mathcal{P}(\text{payoff} = H_b) =: p_b$, and $p_g H_g < p_b H_b$. Each firm needs funds I to realize its operations. To cover these costs, it has to obtain a credit. However, we assume the bank may not distinguish the two types, and can only offer credits at the same conditions to everyone. The credit can only be paid back fully if the corresponding state of high returns is realized. Otherwise, the bank gets the remaining returns, which of course are lower than the repayment R previously agreed upon. By our assumptions, this means that R in the bad state is L_g or 0. The profit of firm g and b is $U(g, R) = p_g(H_g - R)$ and $U(b, R) = p_b(H_b - R)$, respectively. The bank makes a profit $\Pi(g, R) = p_g R + (1 - p_g)L_g - I$ for type g, or $\Pi(b, R) = p_b R - I$ for type b. One now easily sees that any R accepted by g will also be accepted by b.

The bank cannot design a simple credit agreement that will only be accepted by g. On the contrary, if R rises past a certain point, only b will accept the terms, i.e., the quality of debtors declines with rising interest rates.

Another well-known example for adverse selection is a model by Rotschild and Stiglitz [RS76]. They study a competitive insurance market and model distortions caused by information asymmetries:

A risk-neutral insurance firm is confronted with a demand for a specific insurance. The group of buyers consists of two types: g with low risk and b with relatively high risk. Let p_g and p_b denote the corresponding probabilities, where $p_b > p_g$. Both types incur a loss of L in the event of damage. Furthermore, both types have the same wealth W and are risk-averse expected utility maximizers. Finally, assume there are n_g good type and n_b bad type agents.

First, consider the case in which the insurance knows which type it insures. Since it operates in a competitive market, it does not make a positive profit and demands a premium q_i, $i = g, b$, equal to the expected damage of the corresponding type, i.e., $q_g = p_g L$ and $q_s = p_b L$. The wealth of type i becomes $W - d_i q_i - L + d_i L$ in the event of damage, or $W - d_i q_i$ otherwise, where d_i denotes the fraction of loss insured against. Both types maximize their expected utility. Since they are risk-averse, they will insure the entire loss, i.e., at optimum $d_g = d_b = 1$.

If the insurance cannot verifiably distinguish between the types, there are three possible cases. Either it succeeds at making the clients reveal their type by using different contracts, or they cannot be separated through contracts alone, meaning

[5] Compare [GH85] for further information.

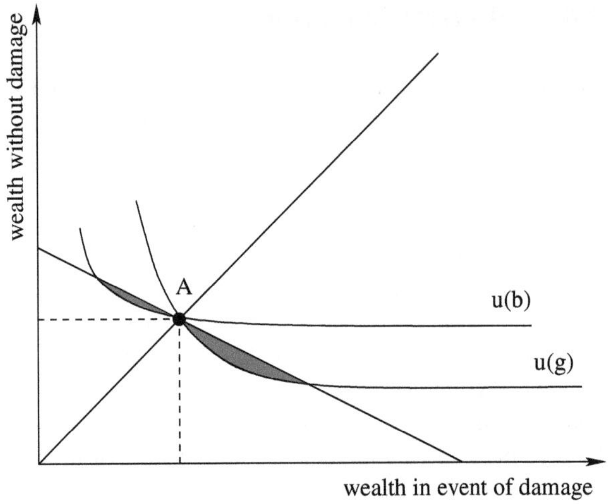

Fig. 7.1 For both types there are contracts (in *grey*) with higher utility, thus competition destabilizes A

both types get the same contract. Finally we need to check if the corresponding solution is stable, i.e., no insurance firm would deviate. Otherwise the market might collapse completely, so that nobody would be insured.

Assume the insurance firm offers a single contract for all types. This is called *pooling*. For efficiency reasons it has to lie on the 45° line. Insurance firms do not gain a profit, but they should not operate at deficit either. Thus the single premium for all insurants must be

$$q = L\frac{p_g + p_s}{n_g + n_s} =: \bar{\rho}.$$

The corresponding contract A is shown in Fig. 7.1. Unfortunately this contract is not stable in a competitive market, since by moving along a line of slope $-1/\bar{\rho}$ an insurer can motivate the good types to deviate: the bad types would stay with the firm offering A, but the good types would switch. With a premium of $\bar{\rho}L$, the insurer with the bad types would operate at a loss, while the insurer with the good types would gain a profit. Therefore, a pooling contract is not stable.

Finally, consider a contract separated by type. In Fig. 7.2, the good types get contract B, and the bad types get contract A. These correspond to the optimum for each type. However, the b type now has an incentive to obtain contract B by pretending to be of type g. To prevent this, only contracts on or below the indifference line $\bar{\mu}_s^A$ for b are possible. Denote the best contract on this line from g's point of view by C. To check the stability of such a variety of contracts, we need to find out if any participant benefits by choosing or offering another contract. D is a possibility for an efficient pooling contract. In D, the b types are certainly better off. Such a contract corresponds to the premium equal to the expected loss, i.e., $q = L\frac{p_g+p_s}{n_g+n_s} =: \bar{\rho}$. If the fraction of g types in the economy is very large, a pooling contract can also be advantageous compared to C, because D lies in the

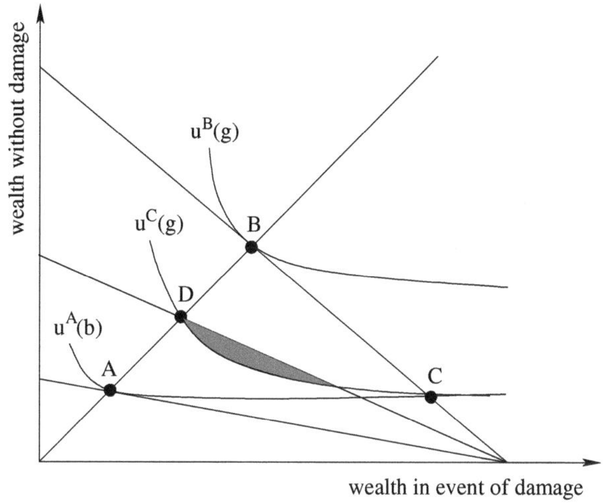

Fig. 7.2 Is the proportion of g-types large enough, it can be advantageous for them to choose D instead of C

improvement set of the g type. As we have already seen, such a contract cannot be stable: there is no equilibrium that insures both types equally. However, if the fraction of good types is low, it is possible that they would prefer C over a pooling contract. In that case, there is an equilibrium where the good types are insured by C and the bad types by A.

7.5 Summary

Financial transactions are always related to information. We have seen in this chapter first how this information can be revealed by prices. In this framework, we have discussed the market for lemons, the information paradox and chart analysis. Information can also be revealed by trades. Dividends, for instance, can work as a signal about the potential of a firm. This role has been neglected in the Modigliani-Miller Theorem. Information that is revealed by trades can also lead to rational herding through an "information cascade".

Typically, information plays a central role when few agents are involved and a market view (where the single investors do not have decisive impact) cannot be applied. In this situation we have to apply methods from game theory. In moral hazard problems, e.g., the superior information of one agent allows him to extract an information benefit. We have seen that this is related to some of the problems which caused the subprime crisis. The last class of problems we have introduced are adverse selection problems (like the Rothschild-Stiglitz model). Here it is not possible to separate different types of investors (e.g. good and bad debtors) by their actions, although it would be beneficial to some (or all) agents if that were possible. A lack of information here can even lead to a market breakdown, i.e. a situation where a specific financial good (usually an insurance) cannot be offered.

Time-Continuous Model* 8

Πάντα χωρεῖ καὶ οὐδὲν μένει.
(All is flux and nothing stays still.)
HERAKLITE, *as quoted by* PLATON.

Trading on a stock market is obviously a discrete process, as it consists of single transactions performed at distinct times. There are, however, so many transactions in such a high frequency that it is for many applications better to model them in a time-continuous setting, i.e., to assume that they take place at *all* times. In this chapter we will provide a short introduction to time-continuous models. An important difference will be that prices are exogenously given. In particular, we will derive the famous Black-Scholes model for asset pricing as it has been introduced by Fischer Black and Myron Scholes [BS73] and by Robert C. Merton [Mer73]. In 1997, the Nobel prize has been awarded for this work. The importance of this model can not be overemphasized, as the Royal Swedish Academy put it:

> Their innovative work ... has provided us with completely new ways of dealing with financial risk, both in theory and in practice. Their method has contributed substantially to the rapid growth of markets for derivatives in the last two decades.

In fact, their formula is probably the most used formula on banks and stock exchanges worldwide even today. In this chapter we will derive this celebrated formula, but we will not reduce all to one equation. There is a whole theory behind this result and we will see how this theory can be used to solve many more problems. We will also see what limitations the classical theory still has and sketch some ways how to overcome them.

In one chapter one can only give a very brief introduction into the rich field of mathematical finance. A classical reference for further studies is the book by Duffie [Duf96], a lighter source is [KK01]. We also refer the reader to the book by

T. Hens, M.O. Rieger, *Financial Economics*, Springer Texts in Business and Economics, DOI 10.1007/978-3-662-49688-6_8

Karatzas and Shreve [KS98]. All of those books provide many more references than we can.

8.1 A Rough Path to the Black-Scholes Formula

Before we start to develop the basic ideas of mathematical finance systematically, we would like to give a short overview on the derivation of the Black-Scholes formula. To do this in a concise way, we have to "walk over some dead bodies", i.e., be a little ruthless regarding mathematical precision. The concepts we use in this section will all be introduced rigorously in the sequel. For the moment, we ask you to trust us that the thin mathematical ice on which we walk is in fact stable enough for a proof. If you are afraid of drowning, just continue with the next section, and you will be safe and fine. A solid derivation of the Black-Scholes formula is given there.

The first assumption that we make to derive the Black-Scholes formula for the price of a derivative based on an asset is an assumption on the price $S(t)$ of this underlying asset. According to the information hypothesis we assume that the price of the asset changes when new information reaches the market. The information is assumed to be random, more precisely its influence on the return of the asset is assumed to be normally distributed. Moreover, there is some fundamental increase in the value of the asset which is predictable, but overlaid with the randomness of the information-driven price movements.

We can write this in the following form:

$$dS(t) = \mu S(t)\,dt + \sigma S(t)\,dB(t). \tag{8.1}$$

where μ and σ are mean and standard deviation of the asset price and $B(t)$ is a *Brownian motion*, which is (roughly) a continuous random process that has zero mean, is always independent of its past evolution and generates normally distributed returns.

The second fundamental assumption that we will make is familiar to us from our studies of time-discrete models (see Chaps. 3, 4, and 5), namely the no-arbitrage principle: we assume that there is no arbitrage opportunity. What this means precisely in the time-continuous setting will be explained in the next section. For the moment we just state that there is no trading strategy that yields riskless excess returns over the risk-free asset.

We want to price an option on the asset S. We denote the value of this option at time t by $V(S, t)$. It depends obviously on the time t and the price of the asset S.[1] At

[1]With this innocent looking assumption we have excluded many derivatives that are *path dependent*, i.e., their value does not only depend on S and t, but also on the previous prices of the underlying asset. We will come back to this later when we talk about numerical methods for the computation of asset prices.

maturity the value of the option is known. Our goal is to derive a partial differential equation for the function V and to solve this using the boundary conditions imposed particularly by our knowledge of the price at maturity.

We start by computing $\mathrm{d}V$, the incremental change of V using a formula (called Itō. formula) which we derive as Lemma (8.6). We obtain

$$\mathrm{d}V(S,t) = \left(\mu S(t)\frac{\partial V(S,t)}{\partial S} + \frac{\partial V(S,t)}{\partial t} + \frac{1}{2}\sigma^2 S(t)^2 \frac{\partial^2 V(S,t)}{\partial S^2}\right)\mathrm{d}t$$
$$+\sigma S(t)\frac{\partial V(S,t)}{\partial S}\,\mathrm{d}B(t). \tag{8.2}$$

There are now several possibilities to proceed. The probably easiest one (that makes you wonder how to find it, though) is to define a trading strategy that will turn out to replicate exactly the option. This trading strategy is defined as follows: at every time t hold one option and $-\partial V(S,t)/\partial S$ stocks. The strategy is called *delta hedge strategy*. The value of this delta hedge portfolio at time t is $V(S,t) - S(t)\partial V(S,t)/\partial S$. The incremental profit (or loss) dR that we make by following this strategy is then

$$\mathrm{d}R(t) = \mathrm{d}V - \frac{\partial V(S,t)}{\partial S}\,\mathrm{d}S.$$

We insert (8.2) and (8.1) into this formula and get

$$\mathrm{d}R(t) = \left(\frac{\partial V(S,t)}{\partial t} + \frac{1}{2}\sigma^2 S(t)^2 \frac{\partial^2 V(S,t)}{\partial S(t)^2}\right)\mathrm{d}t.$$

We notice that we have lost the stochastic terms (the ones with B) here, in other words, the incremental returns of the delta hedge portfolio are not stochastic, but deterministic and follow the above formula. But if these returns are deterministic, they are risk-free, and this implies that they should not be different from the returns of the risk free asset, since otherwise we would violate the No-arbitrage Principle. Therefore, we can equal $\mathrm{d}R$ with the incremental return of the risk-free asset of the same amount (which is $V(S,t) - S(t)\partial V(S,t)\partial S$), i.e.

$$r\left(V(S,t) - S(t)\frac{\partial V(S,t)}{\partial S(t)}\right)\mathrm{d}t = \left(\frac{\partial V(S,t)}{\partial t} + \frac{1}{2}\sigma^2 S(t)^2 \frac{\partial^2 V(S,t)}{\partial S(t)^2}\right)\mathrm{d}t.$$

Dividing by $\mathrm{d}t$, we obtain the Black-Scholes equation, a partial differential equation in the two variables S and t:

$$\frac{\partial V(S,t)}{\partial t} + \frac{1}{2}\sigma^2 S(t)^2 \frac{\partial^2 V(S,t)}{\partial S(t)^2} + rS(t)\frac{\partial V(S,t)}{\partial S} - rV(S,t) = 0. \tag{8.3}$$

Partial differential equations (PDEs) typically have infinitely many solutions, so this alone would not be of much use. However, we have boundary condition, i.e., constraints that apply on the boundary of the set where S and t are defined.

What are our boundary conditions? This depends on the option we want to price. Let us price a simple call option with exercise price K. This option allows at maturity T to buy a share for the price K. Our first boundary condition is now $V(0,t) = 0$ for all t, since the stochastic process (8.1) is constant zero if it is zero at any point in time, but a call option would not be exercised if the underlying asset is worthless, thus the option is worthless as well. On the other hand, if the asset price is extremely high, the value of the option is close to the value of the asset which gives the second boundary condition (actually, it is an asymptotic condition, i.e., the boundary is "at infinity"): $V(S,t)/S \to 1$ as $S \to \infty$. Finally, we have the condition for the value of the option at maturity which is $V(S,T) = S - K$ if $S > K$ and zero otherwise. In summary we have the following boundary value problem:

$$\begin{cases} \frac{\partial V(S,t)}{\partial t} + \frac{1}{2}\sigma^2 S(t)^2 \frac{\partial^2 V(S,t)}{\partial S(t)^2} + rS(t)\frac{\partial V(S,t)}{\partial S} - rV(S,t) = 0, \\ V(0,t) = 0 \quad \text{for all } t, \\ V(S,t)/S \to 1 \quad \text{as } S \to \infty, \text{ for all } t, \\ V(S,T) = \max(S-K, 0) \quad \text{for all } S. \end{cases}$$

This problem can in fact be solved in the following way: apply a suitable variable transformation to bring equation (8.3) to the form of a diffusion equation. Then use a standard solution ansatz and finally transfer back to the original variables.[2] The final result is:

$$V(S,t) = S\phi(d_1) - Ke^{-r(T-t)}\phi(d_2),$$

where ϕ is the normal cumulative distribution function, i.e.,

$$\phi(x) := \int_{-\infty}^{x} \frac{1}{\sigma\sqrt{2\pi}} \exp\left(-\frac{(y-\mu)^2}{2\sigma^2}\right) dy,$$

and the auxiliary variables d_1 and d_2 are given by

$$d_1 = \frac{\ln(S/K) + (r + \sigma^2/2)(T-t)}{\sigma\sqrt{T-t}},$$

$$d_2 = d_1 - \sigma\sqrt{T-t}.$$

[2] An *ansatz* is a specific functional form which we assume the solution to have in order to compute it. This assumption is a posteriori justified if we obtain a solution that is indeed of the assumed form. In the appendix more ideas on how to solve PDEs and further references are given.

If we want to study more complicated (but still path-independent) options, the corresponding boundary conditions will be more complicated, and we cannot expect to get such a neat solution. However, a numerical solution of PDEs of the form (8.3) is technically not very difficult, and it is therefore possible to price a wide range of options with this approach.

At the end of this section let us point out a couple of difficulties that we still have to address, in order to convince you that there is still something important coming in the remaining part of this chapter.

First, there is the Itō formula that we have used, but neither stated in generality, nor proved or at least motivated. Then there are a couple of concepts that we applied without really thinking too much about it. Using mathematics only led by intuition, but not by a correct foundation, can lead to wrong conclusions, and in finance wrong conclusions can be very very expensive, so maybe it is worth looking at the details. There are other reasons: besides the Black-Scholes formula, we can get more results, for instance to price path-dependent options (at least numerically). Moreover, some of the fundamental assumptions in the Black-Scholes model are arguable at best: why should stock prices follow a normal distribution? We have actually already seen that this is not really the case (as we have mentioned at first in Sect. 3.4.2). This will motivate us to study less handy, but more realistic models for stock prices. We will also introduce methods to check the predictions of the Black-Scholes model empirically. This will again enable us to improve the model substantially.

8.2 Brownian Motion and Itō Processes

We consider a state space Ω with a probability measure p.[3] We define a *process* as follows:

Definition 8.1 (Process[4]) A *process* X is a measurable function $X\colon \Omega \times [0, \infty) \to \mathbb{R}$. We call $X(t) := X(\cdot, t)$ the value of X at time t.

Processes will be used to describe the ups and downs of assets, in that they assign probabilities to states at any given time. One of the central problems in asset pricing is to find realistic and at the same time mathematically manageable classes of processes. Historically, this idea goes back to the year 1900 and Louis Bachelier's seminal and unfortunately long forgotten work [Bac00]. He already applied Brownian motion as underlying process, which is even today still the most popular process among practitioners. We will see later, in Sect. 8.8, that there are nowadays alternatives that model actual behavior of asset prices much better, but for now we will study Brownian motion:

[3]More precisely, we study a probability space $(\Omega, \mathcal{F}, p)$, where $\mathcal{F}$ is a σ-algebra and p is a probability measure on Ω with respect to $\mathcal{F}$, see Sect. A.4 for details.

[4]This and the following definitions can also be adapted to discrete-time problems.

Definition 8.2 (Brownian motion) A *standard Brownian motion* is a process B defined by the properties:

(a) $B(0) = 0$ a.s.
(b) For any times t_0, t_1 with $t_0 < t_1$, the difference $B(t_1) - B(t_0)$ is normally distributed with mean zero and variance $t_1 - t_0$.
(c) For any times $0 \leq t_0 < t_1 < t_2 < \cdots < t_n < \infty$, the random variables $B(t_0), B(t_1) - B(t_0), \ldots, B(t_n) - B(t_{n-1})$ are independently distributed.
(d) For each $w \in \Omega$, the *sample path* $t \mapsto B(w, t)$ is continuous.

The idea of Brownian motion is that at any given time the change of a quantity is completely random and following a normal distribution. Brownian motion has first been described by the botanists Jan Ingenhousz in 1785 and Robert Brown in 1827 who noticed random movements in small particles under the microscope. The mathematical theory was first developed by Thorvald Thiele in 1880, but Bachelier independently re-invented it in his work.

When describing financial markets by processes, we are particularly interested in *adapted processes*, i.e., processes that "cannot see into the future": in other words, only past events should be of interest to our model. To define this notion properly, we need the mathematical tool of the so-called "filtrations" which describe the information available at a given point in time:

Definition 8.3 (Filtration) A filtration of a measurable space Ω with σ-algebra $\mathcal{F}$ is a family of σ-algebras $\{\mathcal{F}(t)\}_{t\in(0,\infty)}$ such that

(i) $\mathcal{F}(t) \subset \mathcal{F}$ for all t,
(ii) $\mathcal{F}(t_1) \subset \mathcal{F}(t_2)$ for all $t_1 \leq t_2$.

The filtration will in a certain way reflect how much we know at a given time t: condition (ii) ensures that the knowledge can only increase, never decrease. We can now define what we mean with an "adapted process":

Definition 8.4 (adapted process) A process X is called *adapted* to the filtration $\mathcal{F}(t)$ of Ω if $X(t): \Omega \to \mathbb{R}$ is a $\mathcal{F}(t)$-measurable function for each $t \in [0, \infty)$.

We have now a mathematically precise, although maybe slightly abstract model describing the price fluctuations on a financial market. In the next step we want to trade on this market.

A *trading strategy* is described by a process that prescribes in every state w and at any time t the assets a person should hold.[5] For simplicity, we will deal with only one asset, thus we a are looking for a function $\theta: \Omega \times [0, \infty) \to \mathbb{R}$, the *trading strategy*. As example take the fixed-mix strategy that keeps the value fraction of a

[5]We have seen such a trading strategy already in Sect. 8.3: the delta hedge strategy.

risky asset constant: here θ would be defined as $\theta(w,t) = c/w$ with c constant if w denotes the price of the risky asset.

If along a sample path θ is constant, it is relatively easy to compute the total return between time t_0 and t_1 as $\theta B((t_1) - B(t_0))$. This allows us to compute the gain also if θ is piecewise constant. For general θ we need to assume that $\int_0^T \theta(t)^2\,\mathrm{d}t < \infty$ a.s., then we can define the total gain, $\int_0^T \theta(t)\,\mathrm{d}B(t)$, by approximating θ by sequences θ_n that are piecewise constant along the sample path. This method called "stochastic integration" needs in fact much more mathematical background then it seems and the interested reader is referred to [Kar88]. The general idea, however, is similar to the definition of the usual integral of a function via approximation with Riemann sums, as we know it from calculus. The stochastic integral $\int_0^T \theta(t)\,\mathrm{d}B(t)$ also shares the same fundamental properties with the standard integral, in particular it is linear, i.e., for trading strategies θ, ρ and real numbers a, b we have

$$\int_0^T a\theta(t) + b\rho(t)\,\mathrm{d}B(t) = a\int_0^T \theta(t)\,\mathrm{d}B(t) + b\int_0^T \rho(t)\,\mathrm{d}B(t).$$

Brownian motion has a mean value of zero, real assets, however, have usually average returns larger than zero. Moreover, we want to study markets with more than one asset, and the returns of the assets will differ. Therefore, we need to go a step further and define a process that adds a "drift" to the Brownian motion, i.e., an additional directed price movement. Such processes are called *Itō processes*.

Definition 8.5 (Itō process) Let B be a Brownian motion, $x \in \mathbb{R}$, $\sigma \in L^2$, i.e., σ is an adapted process with $\int_0^T \sigma(t)^2\,\mathrm{d}t < \infty$ a.s. for all t, and $\mu \in L^1$, i.e., μ is an adapted process with $\int_0^T |\mu(t)|\,\mathrm{d}t < \infty$ a.s. for all t. Then the Itō process S is defined for $t \in [0,\infty)$ as

$$S(t) = S_0 + \int_0^t \mu(s)\,\mathrm{d}s + \int_0^t \sigma(s)\,\mathrm{d}B(s).$$

Informally and shorter, we denote $dS(t) = \mu(t)\,dt + \sigma(t)\,dB(t)$, $S(0) = S_0$

Under some additional technical conditions one can prove that σ and μ are nothing else than the rate of change of mean and variance of S, precisely we have a.s.:

$$\frac{\mathrm{d}}{\mathrm{d}r}E_t(S(r))|_{r=t} = \mu_t, \quad \frac{\mathrm{d}}{\mathrm{d}r}\operatorname{var}_t(S(r))|_{r=t} = \sigma_t^2.$$

Consequently, μ is called the *drift process* and σ the *diffusion process* of S.

Again, we can define a stochastic integral computing the total gain for a trading strategy θ, but this time θ has additionally to satisfy certain additional regularity conditions, see [Duf96, Chap. 5C] for details.

If we imagine $S(t)$ to describe the price of an asset, then it would be interesting to study the process defined by a given function of this price (e.g., the payoff of a derivative based on this asset). That this can be done relatively easily is the merit of the result by Kiyoshi Itō, the Itō formula, that can be stated (in its easiest form) as follows:

Lemma 8.6 (Itō formula) *Let S be an Itō process and $f: \mathbb{R}^2 \to \mathbb{R}$ twice continuously differentiable, then the process $Y(t) := f(S(t), t)$ is an Itō process satisfying*

$$\mathrm{d}Y(t) = \left(\frac{\partial f(S(t),t)}{\partial S}\mu(t) + \frac{\partial f(S(t),t)}{\partial t} + \frac{1}{2}\frac{\partial^2 f(S(t),t)}{\partial S^2}\sigma(t)^2 \right) \mathrm{d}t$$
$$+ \frac{\partial f(S(t),t)}{\partial S}\sigma(t)\,\mathrm{d}B(t).$$

The proof of this result can be found, e.g., in [Duf96]. The intuition to it is as follows: if we expand the function $f: \mathbb{R}^2 \to \mathbb{R}$ into a Taylor series around the point $(S(t), t)$, we obtain

$$f(a,b) = f(S(t),t) + \frac{\partial f(S(t),t)}{\partial S}(a - S(t)) + \frac{\partial f(S(t),t)}{\partial t}(b - t)$$
$$+ \frac{1}{2}\frac{\partial^2 f(S(t),t)}{\partial S^2}(a - S(t))^2 + \mathcal{O}((a - S(t))^3, (b - t)^2). \tag{8.4}$$

A Brownian motion has the property that $\mathrm{d}B(t)^2$ is of order $\mathrm{d}t$, in other words, the variance is the square root of the time increase. Therefore, the higher order terms in (8.4) vanish when we approximate the derivative of f by taking the limit $a \to S(t)$, $b \to t$ and we obtain Itō's formula. (This is of course a mere intuition and should not fool us into assuming the proof would be nothing more than a straightforward expansion of this.)

8.3 A Rigorous Path to the Black-Scholes Formula

In the following we will see how the results of the previous section can be used to derive the Black-Scholes formula – this time in a rigorous way, using the tools developed in the previous section.

8.3.1 Derivation of the Black-Scholes Formula for Call Options

We describe an underlying asset S (a stock) by a *geometric Brownian motion with drift*, i.e.

$$dS(t) = \mu S(t)\,dt + \sigma S(t)\,dB(t),$$

where $S(0) = S_0$ is given. Such a process is often called *log-normal*, since $\log(S(t))$ is normally distributed for every t. The process σ is called the *volatility*.

We consider a second asset, a bond, with fixed interest rate that hence follows the price process

$$\beta(t) = \beta_0 e^{rt}, \tag{8.5}$$

where $\beta_0 > 0$ is its initial price and r is the fixed interest rate. This gives rise to the Itō process

$$d\beta(t) = r\beta(t)\,dt. \tag{8.6}$$

We can describe (8.6) as a differential equation with solution (8.5).

We call a *trading strategy* with a portfolio that contains $a(t)$ shares of stocks and $b(t)$ shares of bonds at time t (with $a, b \in L^2$) *self-financing* if for all t:

$$a(t)S(t) + b(t)\beta(t) = a(0)S_0 + b(0)\beta_0 + \int_0^t a(s)\,dS(s) + \int_0^t b(s)\,d\beta(s).$$

This condition ensures that at each time the current value of the portfolio corresponds to the initial value plus accumulated gains and losses.

We want to price an option based on the stock S. Let us consider as an example again a European call option, i.e., the right to buy at maturity T the stock at a given price K. The payoff of this call option at maturity is therefore $\max(S(T) - K, 0)$, as we have seen in Sect. 8.1.

We want to find a self-financing strategy (a, b) that replicates the payoff structure at maturity, i.e.

$$a(T)S(T) + b(T)\beta(T) = \max(S(T) - K, 0).$$

Given the existence of such a strategy, the price of the option at time t has to be $a(t)S(t) + b(t)\beta(t)$, since otherwise we would have an arbitrage opportunity. Thus, once the trading strategy has been determined, the pricing of the option is done.

Let us assume first that the price of the option at time t equals some function $V(S(t), t)$ and that V is twice differentiable (we will see later that this is the case).

We can apply the Itō formula for V and obtain:

$$\mathrm{d}V(S,t) = \left(\mu S(t)\frac{\partial V(S,t)}{\partial S} + \frac{\partial V(S,t)}{\partial t} + \frac{1}{2}\sigma^2 S(t)^2 \frac{\partial^2 V(S,t)}{\partial S^2}\right)\mathrm{d}t$$
$$+\,\sigma S(t)\frac{\partial V(S,t)}{\partial S}\,\mathrm{d}B(t). \tag{8.7}$$

We have seen this formula in Sect. 8.1, but this time a rigorous derivation led us here. Our goal is now to directly derive a hedging strategy. To this aim we assume the existence of this self-financing trading strategy (a, b) with

$$\mathrm{d}V(S,t) = a(t)dS(t) + b(t)d\beta(t). \tag{8.8}$$

Inserting the expressions for $S(t)$ and $\beta(t)$, we obtain

$$\mathrm{d}V(S,t) = (a(t)\mu S(t) + b(t)\beta(t)r)\,\mathrm{d}t + a(t)\sigma S(t)\,\mathrm{d}B(t). \tag{8.9}$$

Can we construct a and b from (8.9) and (8.7)? We can: all we have to do is to match the coefficients in the two expressions, i.e., equal the terms with $\mathrm{d}t$ and with $\mathrm{d}B(t)$ separately. Let us do this for $\mathrm{d}B(t)$ and we get

$$\sigma S(t)\frac{\partial V(S,t)}{\partial S} = a(t)\sigma S(t). \tag{8.10}$$

From this equation we obtain

$$a(t) = \frac{\partial V(S,t)}{\partial S},$$

which we insert into (8.8) to get

$$\mathrm{d}V(S,t) = \frac{\partial V(S,t)}{\partial S}dS(t) + b(t)\,\mathrm{d}\beta(t).$$

Solving this for $b(t)$ we obtain

$$b(t) = \frac{1}{\beta(t)}\left(V(S,t) - \frac{\partial V(S,t)}{\partial S}S(t)\right).$$

Now let us match the coefficients of $\mathrm{d}t$ in (8.9) and (8.7):

$$-rV(S,t) + \frac{\partial V(S,t)}{\partial t} + rS(t)\frac{\partial V(S,t)}{\partial S} - \frac{1}{2}\sigma^2\frac{\partial^2 V(S,t)}{\partial S^2} = 0.$$

Thus, we arrive again at the Black-Scholes equation. Considering the boundary condition at $t = T$, namely that $V(S, T) = \max(0, S - K)$, we can check that the following theorem gives a solution to this equation:

Theorem 8.7 (Black-Scholes formula) *The value of a European call option with strike K, maturity T and underlying asset S (described by a geometric Brownian motion with drift μ and volatility σ) is given by*

$$V(S,t) = S\phi(d_1) - K\mathrm{e}^{-r(T-t)}\phi(d_2),$$

where ϕ is the normal cumulative distribution function, i.e.,

$$\phi(x) := \int_{-\infty}^{x} \frac{1}{\sigma\sqrt{2\pi}} \exp\left(-\frac{(y-\mu)^2}{2\sigma^2}\right) \mathrm{d}y,$$

and the auxiliary variables d_1 and d_2 are given by

$$d_1 = \frac{\ln(S/K) + (r + \sigma^2/2)(T-t)}{\sigma\sqrt{T-t}},$$

$$d_2 = d_1 - \sigma\sqrt{T-t}.$$

To be rigorous, we need to double check that the assumptions made about the existence of the hedging strategy are in fact satisfied, i.e., we need to prove that a different price would indeed allow for arbitrage strategies. For simplicity, we prove this for $t = 0$, but the proof carries over to all times $t \leq T$.

Suppose that the price of the option at time zero is larger than $V(S_0, 0)$. Consider the trading strategy $(-1, a, b)$ in option, stock and bond with $a(t)$ and $b(t)$ as given above. We have $a(T)S(T) + b(T)\beta(T) = \max(S(T) - K, 0)$ which is the value of the option at time T, thus we have made a riskless profit (an arbitrage), namely the difference between the price for which we sold the option and $V(S_0, 0)$.

In the same way, if the price of the option at time zero is smaller than $V(S_0, 0)$, the strategy $(1, -a, -b)$ is an arbitrage. In other words, the no-arbitrage condition implies indeed that the price is $V(S_0, 0)$.

The Black-Scholes model can be easily extended to higher dimensional situations, i.e., to $N \geq 1$ underlying assets. To model this, we define a vector $S(t) = (S_1(t), \ldots, S_N(t))$ and a D-dimensional Brownian motion $B(t) = (B_1(t), \ldots, B_D(t))$. Moreover, we define a *drift vector* $\mu \in \mathbb{R}^N$ and a *volatility matrix* $\sigma \in \mathbb{R}^{D\times N}$. Then the asset process for the n-th asset can be written as

$$\mathrm{d}S_n(t) = \mu_n S_n(t)\,\mathrm{d}t + S_n(t)\sigma\,\mathrm{d}B(t).$$

We will encounter this model again briefly in Sect. 8.6. For details on such multi-dimensional processes we refer to [Duf96] or other textbooks on mathematical finance.

8.3.2 Put-Call Parity

Can we use the result for call options derived above to price put options? A put option with strike K and maturity T on an underlying S gives the right to sell a share of S at time T for the price K. The payoff of a put option at maturity is therefore $\max(K - S, 0)$.

The put-call parity will provide us with a way to price put options, when we know the price for a call option and vice versa, provided that both have the same underlying, the same strike and the same maturity. To derive this parity, we consider the following two portfolios:

- One put option and one share.
- One call option and K bonds that pay each 1 at maturity.

Computing the payoff of these portfolios at maturity, we notice that both pay K if $S \leq K$ and S if $S \geq K$, therefore both portfolios have the same value – also at times $t < T$. (Otherwise we would have a natural arbitrage opportunity.)

Let us denote the value of the put option at time t by $P(t)$, the value of the call option by $C(t)$ and the value of stock and bond by $S(t)$ and $R(t)$, respectively. Then the following relationship holds:

$$C(t) + KR(t) = P(t) + S(t).$$

This relation is called *call-put parity*.

If the bond pays a constant interest rate r, then $R(t) = \mathrm{e}^{-r(T-t)}$, thus the value of a put option is

$$P(t) = C(t) - S(t) + K\mathrm{e}^{-r(T-t)}.$$

Similarly, we could compute the value of a call option from the value of a put option, always provided we know $S(t)$ and r.

8.4 Exotic Options and the Monte Carlo Method

The options we have priced so far are often referred to as "plain vanilla options". They are straightforward in that they have only one underlying and their value at maturity only depends on the value of their underlying at maturity. However, there is an abundance of other options nowadays that do not satisfy this assumption. Let us list some examples of these *exotic options*:

8.4.1 Barrier Option

There are different variants of barrier options. A down-and-out call, e.g., has the same payoff as a call, *provided* the price of the underlying never falls below a certain "barrier" level. Otherwise the option pays back zero at maturity.

8.4.2 Asian Option

This is essentially a call option where the strike is given by the *average* of the price over time.

8.4.3 Fixed-Strike Average

This is a call option where instead of the price of the underlying, its average over times is used to compute the final payoff.

8.4.4 Variance Swap

The payoff of this option is determined by the difference between observed variance (i.e., the square of the volatility) and a predefined value.

8.4.5 Rainbow Option

Here not only one, but several underlying assets are used. The payoff of this option is determined by the average of these underlyings. However, as in the case of the barrier option, there is no payoff if a predefined barrier is hit. In the case of a rainbow option, however, it is sufficient if *one* of the underlyings falls below this barrier level at some point.

This (by no means complete) list of exotic options demonstrates not only the creativity of issuers, but also the need for more advanced methods for pricing of such options. It is important to notice that these exotic options are not all rare. In fact, some of them (in particular barrier options and rainbow options) are used to construct structured financial products for the retail market which are enjoying a huge popularity in recent years, particularly in Europe and East Asia.[6] A typical structured product is the following "reverse convertible": at maturity (after one year) this product yields the average return of three selected stocks, but not more than 10 %. There is a capital protection, i.e., if this average is below the starting price at maturity, the starting price is paid instead, but this capital protection is only valid if

[6]Restrictive laws seem to hinder their success in the US.

none of the three stocks fell below of 70 % of its starting price during the one year period. We see immediately that we need several options to hedge such a structured product, in particular a rainbow option.

In fact, variants of this product are among the most popular structured products on the market. They are particularly popular with retail customers. For reasons for this popularity and a theoretical analysis of structured products see [Rie10, Rie12, HR14]. Fundamental for the understanding of these products is the study of the probability that a barrier is reached. For theoretical work on this, see [SR09].

How can we price exotic options? For many of these options there are by now sophisticated methods to obtain pricing formulas using similar methods as for the Black-Scholes formula. Some examples for this can be found, e.g., in [KK01, Chap. 4.1]. There are, however, cases where it is either not possible to follow this route or options are so new that there are simply no mathematical results available yet. In such situations numerical approximation methods are used extensively. Moreover, they have the additional advantage that exchanging the underlying stochastic process with a more sophisticated and realistic variant than the geometric Brownian motion is usually much simpler.[7] In the following we will sketch just one of the many methods to price options numerically, the *Monte Carlo method.*

The key idea for this method is to use risk-free probabilities, also called equivalent martingale measure (see Sect. 4.2.2) and to take the discounted expected value over the payoff of the option. This expected value can be computed by simulating n independent random paths for the underlying(s) and computing the value of the option in each of these cases.

How can we implement this idea? The first difficulty we encounter is that the price of the underlying follows a continuous stochastic process, but we can compute only finitely many values. Thus we need to discretize the process. In the case of a geometric Brownian motion this means that we use N discrete time steps $t = 0, T/N, 2T/N, \ldots,$ and generate for each time step an independent random number $y(t)$ that follows a standard normal distribution. The approximation for a Brownian motion $B(t)$ is then for $n = 1, \ldots, N$ given by

$$B(nT/N) = B((n-1)T/N) + \sqrt{T/N}y(nT/N).$$

Between these points, we can interpolate $B(t)$, e.g., by piecewise affine functions.

Having constructed a realization of the Brownian process $B(t)$, we can compute the price $S(t)$ and – given its path – the payoff of the option at maturity.

If we have several underlyings, the path of each of them can be constructed in the same way. However, if they are not independent (which will usually be the case for stocks or indices as underlyings) we need to construct a process that respects the correlations between them, which adds more difficulties.

The method relies heavily on a good source of random numbers. This is occasionally a crucial issue, since computers do not actually provide real random

[7]The need for studying other processes will be discussed in Sect. 8.8.

numbers, but only numbers that look random, but are actually the result of a computation. These pseudo-random numbers are nowadays, however, quite reliable, as long as they are implemented correctly. A drawback of the method is certainly its relatively large computational cost, since n as well as N have to be large to yield a good approximation. On the other hand the method is universal and can be applied to any option and with quite complicated price processes. More information on the Monte Carlo method can be found in [Rub81, Gla03].

8.5 Connections to the Multi-period Model

As mentioned in Chap. 4, we can price any redundant asset with priced assets by the absence of arbitrage as stated in FTAP. Thus, we can easily come up with two different methods to price any derivative asset, whose payoffs we can replicate with existing assets. One is by forming a hedge portfolio and the other is based on using risk-neutral probabilities.

As in the Black-Scholes formula we first assume that the underlying asset price process is described by a binomial model with *up* and *down* moves as we have done in Sect. 4.2. In the absence of arbitrage, both pricing methods result in the formula for basic two period model of European option:

$$C_0 = \frac{1}{R_f}\left(\frac{R_f - d}{u - d}C_1(u) + \frac{u - R_f}{u - d}C_1(d)\right), \tag{8.11}$$

where C_1 shows the final payoff of the option. In the case of call option, the final payoff of the option is equal to $(S_1 - K)^+$, so in the state of *up* move $C_1(u) = (Su - K)^+$ and in the state of *down* move $C_1(d) = (Sd - K)^+$. Moreover, risk-neutral probabilities are defined for *up* move and *down* move by

$$\pi^\star = \frac{R_f - d}{u - d}, \qquad 1 - \pi^\star = \frac{u - R_f}{u - d},$$

respectively. In the multi-period discrete time settings (Chap. 5), by induction, the pricing formula can be expressed with risk-neutral probabilities:

$$C_0 = \frac{1}{R_f^n}\left[\sum_{j=0}^{n}\frac{n!}{j!(n-j)!}(\pi^\star)^j(1-\pi^\star)^{n-j}(Su^jd^{n-j} - K)^+\right]. \tag{8.12}$$

To eliminate the positivity condition in the pricing equation in the formula, we can define a variable m such that it denotes the minimum number of up moves for the stock price to satisfy the positivity of the final payoff function as follows:

$$m = \min\left\{j, j \in \{0, 1, \ldots, n\} : (Su^jd^{n-j} - K) \geq 0\right\}. \tag{8.13}$$

Thus, we can conclude for all $j \geq m$ option expires in-the-money and otherwise it expires out-of-money. Thus we have the relation for m:

$$m > \frac{\ln\left[K/(Sd^n)\right]}{\ln(u/d)}.$$

Then, if we use m in the pricing equation (8.12), we eliminate the positivity condition and we have

$$C_0 = \frac{1}{R_f^n}\left[\sum_{j=m}^{j} \frac{n!}{j!(n-j)!}(\pi^\star)^j(1-\pi^\star)^{n-j}(Su^jd^{n-j}-K)\right],$$

which yields

$$C_0 = S\sum_{j=m}^{n} \frac{n!(\pi^\star)^j(1-\pi^\star)^{n-j}}{j!(n-j)!}\frac{u^jd^{n-j}}{R_f^n} - KR_f^{-n}\sum_{j=m}^{n}\frac{n!(\pi^\star)^j(1-\pi^\star)^{n-j}}{j!(n-j)!}.$$

Here, we can define a new binomial probability

$$\pi^{\star\star} = \pi^\star\frac{u}{R_f} \quad \text{and} \quad 1-\pi^{\star\star} = (1-\pi^\star)\frac{d}{R_f},$$

since $\pi^\star$ is the risk-neutral probability. Then the pricing formula reduces to

$$C_0 = S\sum_{j=m}^{n}\frac{n!}{j!(n-j)!}(\pi^{\star\star})^j(1-\pi^{\star\star})^{n-j} - KR_f^{-n}\sum_{j=m}^{n}\frac{n!(\pi^\star)^j(1-\pi^\star)^{n-j}}{j!(n-j)!}.$$

Here, one can notice that the summations in the pricing equation correspond to the probabilities of binomially distributed random variables with parameters $(n, \pi^{\star\star})$ and $(n, \pi^\star)$ respectively, that take values larger or equal than m, which is defined as in equation (8.13).

We find it useful to remind the binomial distribution. A random variable distributed binomially with parameters (n, p), as n shows the number of trials or repetitions and p shows the probability of success of an event for each trial, has a probability distribution function

$$\mathbb{P}(X \leq x) = \sum_{j=0}^{x}\frac{n!}{j!(n-j)!}(p)^j(1-p)^{n-j}$$

and

$$\mathbb{P}(X \geq x) = 1-\mathbb{P}(X \leq x) = \sum_{j=x}^{n}\frac{n!}{j!(n-j)!}(p)^j(1-p)^{n-j}.$$

In the pricing equation, we have probabilities $\pi^\star$ and $\pi^{\star\star}$ as in the binomial success probability. We could interpret them as the adjusted risk-neutral probabilities of the underlying asset prices up move at each node given that the option expires in-the-money. Thus we can express the probabilities as $\mathbb{Q}(m; n, \pi^{\star\star})$ and $\mathbb{Q}(m; n, \pi^\star)$ and so the pricing equation has a very similar formula to the Black-Scholes formula in continuous time setting as follows:

$$C_0 = S\mathbb{Q}(m; n, \pi^{\star\star}) - KR_f^{-n}\mathbb{Q}(m; n, \pi^\star).$$

We have the same analogy with the Black-Scholes formula expressed with current asset price and discounted strike price with the expire in-the-money probabilities $\mathbb{Q}(m; n, \pi^\star)$ and $\mathbb{Q}(m; n, \pi^{\star\star})$. These probabilities come from binomial distributions unlike the standard Black-Scholes model. From multiperiod to continuous time, we take the limit of the time step number to infinity by keeping the time to maturity T finite. We define time-to-maturity $T = \Delta tn$ and we take $n \to \infty$. When we take the limit the discount term has an exponential form with the instantaneous risk-free rate r, i.e. $e^r = R_f$. Then, under continuous time we can explicitly state the up move probabilities:

$$\pi^\star = \frac{e^{rT} - d}{u - d}, \qquad \pi^{\star\star} = ue^{-rT}.$$

For the convergence, analogously, we would expect that as $n \to \infty$, the expire in-the-money probabilities $\mathbb{Q}(m; n, \pi^\star)$ and $\mathbb{Q}(m; n, \pi^{\star\star})$ would converge to the standard normal distribution with the log-normal prices. Thus, we remind a very nice result of theorem called de Moivre Laplace limit theorem, which is a special case of the central limit theorem. It states that since the binomial random variable is actually a sum of Bernoulli random variables, as $n \to \infty$, the binomial distribution converges to the normal distribution. For the proof of the theorem and the details one can refer to Grimmett et al. [GS01]. Thus, we have

$$\mathbb{Q}(m; n, p) \to \int_m^\infty N(x)\,dx, \qquad \text{as } n \to \infty,$$

where X is a binomial random variable and $N(\cdot)$ denotes the normal probability density function (see Sect. A.2). By standardizing the normal distribution, we have

$$y = \frac{x - \mathbb{E}(X)}{\sqrt{\text{var}(X)}}, \qquad z = \frac{m - \mathbb{E}(X)}{\sqrt{\text{var}(X)}}.$$

Then, we have

$$\mathbb{Q}(m; n, p) \to \int_z^\infty N(y)\,dy =: \phi(-z), \tag{8.14}$$

where the function $\phi(\cdot)$ represents the cumulative distribution of the normal distribution.

From this point, the only task left is to show that $-z$ actually corresponds to d_1 (in the Black Scholes formula) under the probability $\pi^{\star\star}$ and d_2 under the probability $\pi^\star$. To show this, we first express the asset price in terms of time to maturity, where $\Delta t = T/n$:

$$S_T = Su^x d^{n-x}.$$

We take the logarithm to get

$$\ln\left(\frac{S_T}{S}\right) = x \ln\left(\frac{u}{d}\right) + n\ln(d).$$

Thus, we can express the expected value and the variance of the random variable X as follows:

$$\mathbb{E}[X] = \frac{\mathbb{E}[\ln(S_T/S)] - n\ln(d)}{\ln(u/d)},$$

$$\text{var}[X] = \frac{\text{var}[\ln(S_T/S)]}{\ln(u/d)^2}.$$

Hence, we need to find the expectation and the variance of the logarithmic return variable in order to compute them for the random variable X. We define the variable m again for continuous-time setting by doing a simple trick such that we can always find ε, $0 \le \varepsilon < 1$:

$$m = \frac{\ln[K/(Sd^n)]}{\ln(u/d)} + \varepsilon, \tag{8.15}$$

as $n \to \infty$. By using the relation given by (8.15) and the expectation and variance of the random variable X, we can express the value $-z$ in (8.14) as follows:

$$-z = \frac{-m + \mathbb{E}[X]}{\sqrt{\mathbb{V}[X]}} = \frac{\ln(S/K) + \mathbb{E}[\ln(S_T/S)]}{\sqrt{\mathbb{V}[\ln(S_T/S)]}} - \frac{\varepsilon \ln(u/d)}{\sqrt{\text{var}[\ln(S_T/S)]}}.$$

Moreover, we know that

$$\sqrt{\text{var}[\ln(S_T/S)]} = \ln(u/d)\sqrt{\text{var}[X]} = \ln(u/d)\sqrt{n\pi^\star(1-\pi^\star)}.$$

by binomial distribution. Thus, we have

$$-z = \frac{\ln(S/K) + \mathbb{E}[\ln(S_T/S)]}{\sqrt{\text{var}[\ln(S_T/S)]}},$$

as $n \to \infty$. One can show that the variance does not change under equivalent probability measures (this follows from Girsanov's Theorem, see [KS98]). We need some specifications for up move and down move and the probability:

$$u = e^{\sigma\sqrt{\Delta t}}, \quad d = e^{-\sigma\sqrt{\Delta t}}, \quad p = \frac{1}{2} + \frac{1}{2}\frac{\mu\sqrt{\Delta t}}{\sigma},$$

where p denotes the objective probability of the up move so that we can conclude with those values suggested first by Cox, Ross and Rubinstein [CRR79], the expected log return and variance can be expressed as

$$\mathbb{E}[\ln(S_T/S)] \to \mu T, \qquad \mathrm{var}[\ln(S_T/S)] \to \sigma^2 T.$$

Note that one can come up with different parameterizations to satisfy the same expressions. Then we can use the normality of $\ln(S_T/S)$ to calculate the expectation under different probabilities that will give the corresponding values of d_1 and d_2. We know that if $\ln(S_T/S)$ is normally distributed than S_T/S will be log-normally distributed and we have the following relation for log-normal distributions:

$$\ln\left(\mathbb{E}(S_T/S)\right) = \mathbb{E}\left(\ln(S_T/S)\right) + \frac{1}{2}\,\mathrm{var}\left(\ln(S_T/S)\right). \tag{8.16}$$

By using (8.16) and the values of the probabilities we find for the probability $\pi^{\star\star}$

$$\mathbb{E}[\ln(S_T/S)] = rT + \frac{\sigma^2 T}{2}$$

and for $\pi^{\star}$

$$\mathbb{E}[\ln(S_T/S)] = rT - \frac{\sigma^2 T}{2}.$$

Hence, we finally reach the original values of the Black-Scholes model. The value corresponding to probability $\pi^{\star\star}$ is

$$d_1 = -z = \frac{\ln(S/K) + \mathbb{E}[\ln(S_T/S)]}{\sqrt{\mathrm{var}[\ln(S_T/S)]}} = \frac{\ln(S/K) + rT + \sigma^2 T/2}{\sigma\sqrt{T}}$$

and the value corresponding to the probability $\pi^{\star}$ is

$$d_2 = -z = \frac{\ln(S/K) + rT + \sigma^2 T/2}{\sigma\sqrt{T}}.$$

Thus, the formula (8.11) of the binomial model reduces to

$$C_0 = S\phi(d_1) - Ke^{-rT}\phi(d_2),$$

as $n \to \infty$, where we used that

$$\lim_{n\to\infty} \left(1 + \frac{rT}{n}\right)^{-n} = \lim_{n\to\infty} e^{-rT} = R_f^{-1}.$$

8.6 Time-Continuity and the Mutual Fund Theorem

How should an investor choose his trading strategy? Portfolio selection is – besides asset pricing – one of the central problems that we want to solve in financial economics.[8]

The CAPM in Chap. 3 was the first asset pricing model that we have encountered and one of its consequences was the Two-Fund Separation Theorem (Sect. 3.1.5) that stated that every mean-variance investor in a CAPM market (i.e., in a market where everybody is a mean-variance investor) should hold the same portfolio of risky assets. Only the combination with the riskless asset is used to account for his risk attitudes.

We have criticized the mean-variance approach as a model for rational or behavioral preferences (compare Sect. 2.3.2), and so it seems that the consequences of the CAPM can only be seen as a result of using a mathematically appealing, but unfortunately overly simple decision model. This is, however, not entirely true: in fact we will show that in continuous-time trading under certain assumptions on the underlying asset process and the risk attitudes of the investor the so-called *Mutual Fund Theorem* holds, which is in a certain sense a generalization of the Two-Fund Theorem to the continuous-time setting. This theorem holds in particular not only for a mean-variance investor, but in fact for a large class of rational investors (in the sense of Expected Utility Theory).

The Mutual Fund Theorem has been proved by Merton [Mer72]. We state here a simplified variant of a version stated in the book by Karatzas and Shreve [KS98]. The (not so easy) proof of this theorem can be found there. As preparation for the theorem we need to make a couple of definitions. In particular, we need to define what we call an *optimal trading strategy* for an expected utility maximizing investor.

We assume that there are $N \geq 1$ underlying assets driven by a $D \geq 1$ dimensional geometric Brownian motion $B(t)$ (see Definition 8.2). Let $S(t) \in \mathbb{R}^N$ be the price vector of the assets and $\sigma \in \mathbb{R}^{D\times N}$ the volatility matrix.

Let us now construct the utility that we aim to maximize. First, we notice that in a time-continous framework with finite investment horizon we need to distinguish two utility functions: one that describes the utility derived from consumption during the investment time and one that describes the utility derived from final wealth. We

[8] Compare also [CV02].

denote these two utility functions by u_1 and u_2. We allow u_1 to be time-dependent, thus $u_1: \mathbb{R} \times [0, \infty) \to [-\infty, +\infty)$ and $u_2: [0, \infty) \to [-\infty, +\infty)$.[9]

Our gaol is now to formulate the optimization problem. The investment horizon is denoted by T, i.e., we invest at time $t = 0$ and sell at time $t = T$. In between, we are allowed (and need) to take money out of the investment for consumption. This consumption plus the amount of money we have at time T determines our utility.

The consumption over time is described by the function $c(t)$. Moreover, we have a trading strategy $\theta(t)$ such that we never have to face a utility of $-\infty$.[10] Then we want to maximize the expected utility

$$U := \mathbb{E}\left(\int_0^T u_1(t, c(t))\, dt + u_2(X(T))\right)$$

in consumption plan c and self-financing trading strategy θ.

Let us now assume that the utility functions u_1 and u_2 are "reasonable" in the sense that they are strictly concave (i.e., the investor is strictly risk-averse), increasing and satisfy the following technical conditions (a more precise formulation of these statements can be found in [KS98]):

Assumption 8.8 *We assume that u_1 and u_2 satisfy the following conditions*[11]*:*

(i) $I_1 := (u_1')^{-1}$ *and* $I_1 := (u_2')^{-1}$ *have polynomial growth.*
(ii) $u_1(I_1)$ *and* $u_2(I_2)$ *have polynomial growth.*
(iii) I_1 *is Hölder continuous.*
(iv) Either $\partial I_1(t, y)/\partial y$ *is strictly negative for a.e.* y *or* $\partial I_2'(y)/\partial y$ *is strictly negative a.e. (or both).*

These assumptions still leave ample room for the choice of the utility functions, but nevertheless Robert Merton proved the following result [Mer72]:

Theorem 8.9 (Mutual Fund Theorem) *Assume that the volatility of the underlying assets is given by σ and the dividend process is given by δ. Assume that the investor can invest risk free for a return of r and borrow money for a return of b.*

[9] It is of course possible that u_1 is time-independent and that both utilities are in effect the same, however, in reality this is unlikely to be the case: consider an investment problem over one year without earnings, then the consumption in that year induces probably a smaller utility than the saved money at the end of that year does, simply because the latter has to be used in all the subsequent years still to come.

[10] More precisely: let $X(t)$ denote the process of the value of the portfolio defined by the investment strategy (i.e., X depends in particular on S, c and θ), then the expected value of the intertemporal consumption, $\int_0^T \min\{0, u_1(t, c(t))\}\, dt$, has to be larger than $-\infty$ and the expected value of the final wealth, $\min\{0, u_2(X(T))\}$ has to be larger than $-\infty$, as well.

[11] For mathematical terminology compare Appendix A.

Assume that σ, δ, r and b are smooth functions of the time t.[12] *Then any agent with preferences satisfying Assumption 8.8 should hold a mutual fund containing the assets in the proportion*

$$(\sigma'(t))^{-1}\theta(t) = (\sigma(t)\sigma'(t))^{-1}(b(t) + \delta(t) - r(t)\mathbf{1})$$

plus a risk-free asset in order to maximize the expected utility U.

This generalizes in a certain way the Two-Fund Separation Theorem to a large class of preferences. It shows in particular that the Two-Fund Separation Theorem was not just a weird artifact of the mean-variance assumption, but that there is some deeper insight behind it. Even more so, the Mutual Fund Theorem does not always apply to mean-variance preferences, since we know from Sect. 2.3 that the mean-variance approach corresponds to an expected utility maximization with a quadratic utility function; a quadratic utility function, however, cannot be strictly concave and increasing, thus the Mutual Fund Theorem would not be applicable. If we assume, e.g., returns with normal distribution, then mean-variance and expected utility preferences coincide, so in this case the Mutual Fund Theorem applies.

What is more important: the Mutual Fund Theorem holds for a large class of strictly concave expected utility functions. But what about its validity in reality?

First, observations about investment decisions show a strong heterogeneity in asset allocations. One possible explanation might be that expectations of investors often are heterogeneous. In fact, many observations on financial markets can be explained by trading expectations, and the Mutual Fund Theorem ignores this aspect completely.

But still, there are situations where expectations should be homogeneous, take e.g., the case of a client advisor at a bank: his (or the bank's) expectations on the market are likely to be more correct than the ones of his clients. In fact, the clients might believe his expectations completely and just expect him to invest their money in a way that is providing them with an optimal risk profile, i.e., to optimize the client's utility function, given the advisor's expectations. Therefore, the expectations are homogeneous and the Mutual Fund Theorem should hold, i.e., the advisor should suggest the same portfolio of risky assets to all of his clients, regardless of their risk attitudes. This is certainly puzzling and quite contrary to our experience: banks structure the risky part of portfolios differently, depending on the client's risk profile. Does the Mutual Fund Theorem prove that this is suboptimal and a policy change would lead to a mutual benefit for client and bank?

[12]Up to now we have only considered the case where the risk-free rate and the volatility were constant and borrowing and investing in the risk-free asset had the same fixed interest rate. Moreover, we have not considered dividends. All of these extensions are not essential to understand this theorem, but are stated for completeness. These extensions include as special case particular the setting we have previously used where $r = b$, $\delta = 0$ and σ and r being constant in time. For details see [KS98].

Before we simply "buy" such a surprising result and recommend to everyone to follow it, some careful skepticism might still be in place. In fact, this is always a good idea, when we derive a very surprising result from an abstract model. In any case, a discrepancy between real life on financial markets and the theory is always an intriguing observation that helps us to improve at least one of them – either the model or the real life... Let us therefore reflect about the problem for a moment: what could be possible explanations for this puzzle?

Essentially, one can argue from three point of views: the economics point, the behavioral finance point and the mathematical finance point.

The classical *economist* could say that the model simply overlooks market friction, e.g., transaction costs. Considering such costs, an active continuous-time trading structure is not feasible, and it might be better to stick to some simpler (but less general) portfolio.

The advocate of *behavioral finance* could argue that investment decisions and markets are far away from being rational. Investors would therefore exhibit preferences that are not covered by the Mutual Fund Theorem, in particular they might be risk-seeking for some wealth levels, have reference points, overweight small probabilities and are – in essence – too messy to fit into the neat model of Merton. Moreover, due to these irrationalities, markets show anomalies that make them deviate from the models we have studied so far. In particular, future returns might not always be independent of past returns.

The devotee of *mathematical finance* finally would partially agree to both and start improving his models in order to capture transaction costs on the one hand and more realistic stock market processes on the other hand.

What do *we* say on this matter? Probably that all three are right: they observe weak points in the model that have to be addressed one way or the other. We have already seen, e.g., that the utility function implied by option prices is not everywhere concave (see Sect. 4.6.3). Moreover, we will collect some more evidence for the need for more complicated processes (see Sect. 8.8).

To sum up: the puzzle gives rise to improving the theory, and some directions for an improvement we will discuss in the next sections.

8.7 Market Equilibria in Continuous Time

To formulate an equilibrium model in continuous time requires more complicated tools than in discrete time, although the fundamental ideas are very similar. To attain the existence of an equilibrium, generally we need very strong assumptions. Yet, unfortunately, we might not get continuous and diffusion prices from a general equilibrium theory although this is assumed in most of the literature of finance and Arbitrage Pricing Theory. As in the multi-period equilibrium model, prices are determined by the state prices with the next period payoff structure. In complete markets, these state prices are unique, non-negative adapted processes.

In this part, we try to show briefly how the equilibrium in continuous time would look like and what possible implications for asset prices would be. For this, we adopt the general model of Duffie and Zame [DZ89]. According to this approach, the endowments are stochastic processes and each agent has time additive differentiable expected utility functions. The model has a very crucial implication that unlike the most of the general equilibrium models, under this model, we can have continuous stochastic asset prices.

As in the previous sections of this chapter, we work with a probability space $(\Omega, \mathcal{F}, \mathbb{P})$ equipped with a standard filtration $\{\mathcal{F}_t, 0 \le t \le T\}$. We assume the consumption process c_t is square integrable and the cumulative dividend process is defined as

$$dD_t^n = \mu_D(t)dt + \sigma_D(t)dB(t), \qquad \text{for all } n = 1, \dots, K,$$

such that it has a finite variance and for the $n = 0$, $D_t^0 = 0$, for all $t < T$ and for $t = T$, $D_T^0 = 1$. Thus, for the asset prices $S^0, S^1, \dots, S^K$ the gain process is expressed as $G = D + S$ from which we can express the gain process as geometric Brownian motion. With this, we can define the cumulative gains with predictable[13] squared integrable portfolio process $\theta = (\theta^0, \dots, \theta^K)$. This portfolio θ is generally defined as piecewise constant function over time intervals, which is very reasonable when we consider the transaction costs.

Definition 8.10 A feasible consumption-portfolio plan (c_t^i, θ_t^i) for agent $i = 1, \dots, I$ is defined as a pair for time t such that

$$\theta_t \cdot S_t = \int_0^t \theta_s \, dG(s) + \int_0^t p_s(\omega_s^i - c_s^i) \, ds.$$

Definition 8.11 A feasible consumption-portfolio plan (c_t^i, θ_t^i) is optimal if there is no other feasible consumption-portfolio plan $(\tilde{c}_t^i, \theta_t^i)$ such that $U^i(\tilde{c}^i) > U^i(c^i)$ for agent i.

Definition 8.12 An equilibrium for an economy

$$\mathbf{E} = \big((S, p), (c^1, \theta^1), \dots, (c^I, \theta^I)\big)$$

is a collection such that given the security price and commodity spot price processes

- For every agent i, the consumption-portfolio plan is optimal, i.e. maximizes their utility,
- Markets clear, i.e. $\sum_i \theta^i = 0$ and $\sum_i (c^i - \omega^i) = 0$.

[13] For technical definition, e.g. predictable process, see [Duf96]

For the sufficient conditions for the existence of such an equilibrium, one should refer to Duffie [Duf86]. We continue to give the equilibrium condition for the model.

Assumption 8.13 *Each agent has the following utility function U^i with the following representation:*

$$U^i(c) = \mathbb{E}\left[\int_0^T u^i(c_t, t)\,\mathrm{d}t\right],$$

where $u^i : \mathbb{R}_+ \to \mathbb{R}$ is differentiable and concave and c_t denotes the non-negative consumption process. Moreover, to avoid unbounded spot prices for consumption, $\lim_{c\downarrow 0} u_i'(c,t) = +\infty$.

Assumption 8.14 *The aggregate endowment process $\omega = \sum_{i=1}^I \omega^i$ follows a stochastic process which satisfies the following differential equation:*

$$\mathrm{d}\omega_t = \mu_\omega(t)\mathrm{d}t + \sigma_\omega(t)\mathrm{d}B(t),$$

such that the volatility term satisfies

$$\mathbb{E}\left[\int_0^T \sigma_\omega^2(t)\,\mathrm{d}t\right] < \infty.$$

Assumption 8.15 *Equivalent assumption for spanning in discrete time to show any feasible consumption plan is attainable by trading in financial markets is that the processes $M^1, M^2, \ldots, M^N$ are martingales, i.e.*

$$M_t^k = \mathbb{E}[D_t^k | \mathcal{F}_t], \quad \textit{for all } t \in [0, T].$$

The representative agent model for a given equilibrium economy $\mathbf{E} = \big((S,p), (c^1, \theta^1), \ldots, (c^I, \theta^I)\big)$ is a single agent (U_λ, ω) maximizing the following utility form:

$$U_\lambda(c) = \sup_{c^1,\ldots,c^I} \sum_i \lambda_i U^i(c^i)$$

subject to $\sum_i c^i \leq c$ for some coefficient vectors $\lambda \in \mathbb{R}_+^I$ and the equilibrium prices are the same (S,p). By the equilibrium functional form, we can express the single representative agent's utility as

$$U_\lambda(c) = \mathbf{E}\left[\int_0^T u_\lambda(c_t, t)\,\mathrm{d}t\right],$$

where

$$u_\lambda(c, t) = \sup_{c_1,\ldots,c_I) \in \mathbb{R}^I_+} \sum_i \lambda_i u^i(c^i, t)$$

subject to $\sum_i c^i \leq c$.

Theorem 8.16 *Under the Assumptions 8.13 and 8.15, on the defined economy, there exists an equilibrium with a representative agent (U_λ, ω), such that the real security prices $\hat{S}_t$ satisfy the following representative agent pricing formula for any time t:*

$$\hat{S}_t = \frac{1}{u'_\lambda(\omega_t, t)} \mathbb{E}\left[\int_t^T u'_\lambda(\omega_s, s)\, \mathrm{d}\hat{D}_s | \mathcal{F}_t\right],$$

for all $t \in [0, T)$.

For the proof one should refer to Duffie and Zame [DZ89].

The last expression is analogous to discrete time, because it has the same characteristics regarding the no-arbitrage condition of the equilibrium asset price formulation. Under some dividend and utility specifications, one can derive a C-CAPM model in a straightforward way. Moreover, the model obviously produces stochastic asset prices under some utility specifications. With the right utility choice, the Black-Scholes model can be supported by an equilibrium model. However, for the Black-Scholed model to hold one does not need to know the equilibrium allocations, the only thing is to estimate the market price of risk as long as the underlying asset price functional is given by the no-arbitrage condition. In fact, we also know that the no arbitrage condition is also a necessary condition for an equilibrium. Black-Scholes can be considered as a partial equilibrium model of prices.

8.8 Limitations of the Black-Scholes Model and Extensions

The Black-Scholes model as we have derived it relies on a list of assumptions, in particular:

1. Trading in the assets is continuous in time.
2. The price of the underlying asset follows a geometric Brownian motion with drift.
3. The market is arbitrage free.
4. There are no short-sell constraints.
5. Assets are arbitrarily divisible.

6. There are no frictions, like transaction costs or taxes.
7. There is a fixed risk-free rate for which money can be invested or borrowed.
8. There is no dividend payment.

Some of these assumptions could be relaxed relatively easily (e.g., the dividend payment or the fixed interest rate). Others are realistic approximations to reality and well-accepted (e.g., trading in continuous time or that assets are arbitrarily divisible). There are, however, some assumptions that are rather restrictive. In the following we will see that there is some empirical evidence that suggests particularly that the model of a geometric Brownian motion is not sufficient to explain all empirical facts about asset returns that we observe on the market. For more information on generalizations of Black-Scholes we refer the reader to [Duf96].

8.8.1 Volatility Smile and Other Unfriendly Effects

The geometric Brownian motion assumes that volatility is constant in time. Is this a reasonable assumption? A standard way to figure this out is to compute the *implied volatility*, i.e., the volatility that an asset should have, assuming that the price of the call option traded on this asset obeys the Black-Scholes formula. For standard assets like the S&P 500 we can compute the implied volatility for various values of strike K and maturity T, thus we obtain a function in the two variables K and T that is usually represented as a three-dimensional plot called the *(implied) volatility surface*.

If the assumptions of the Black-Scholes model were perfectly right, this surface would be flat. However, like the earth, this surface is not flat:

- At-the-money options tend to have lower implied volatility than options with a strike far away from the current price of the underlying. This curved shape reminded some researchers of a smile, thus the name *volatility smile*.
- There is also a time-dependence of the implied volatility: different maturities lead to different implied volatilities, an effect which is called *term structure of volatility*.

Both effects are not as friendly as a smile usually is, because they exhort us to improve the underlying assumptions. We could say with Huxley that "the great tragedy of science is the slaying of a beautiful hypothesis by an ugly fact". It is in particular not correct to assume that the volatility is constant in time, and that returns are normally distributed. Indeed, that stock returns have fat tails (i.e., that the probability for extreme events is larger than for a normal distribution) is a well-established empirical fact that we will explain in the next section. These fat tails can explain the volatility smile: options with strikes far away from the current price of the underlying are priced in a way that seems to imply a larger volatility, since only such a larger volatility could explain the high probabilities associated to strong price movements, given the assumption that returns are normally distributed. There is support for this explanation from the US market: before the crash in 1987 there was no smile, as if investors had not been aware of the potential for large price changes. Since then, a volatility smile can be observed on the US market as well. One could say that investors have learned their lesson...

The term structure, on the other hand, can often be explained by the expectation of news. A typical example are earning reports: stock options with maturity shortly after the earning report show higher implied volatility than options with a later maturity.

8.8.2 Not Normal: Alternatives to Normally Distributed Returns

We have already mentioned that standard assets like stocks and bonds do not usually have normally distributed returns (compare Fig. 8.1). This was already observed in the early twentieth century by various researchers [Mit15, Oli26, Mil27]. There are various ways to confirm this observation empirically. A simple method is to measure whether returns of real assets have significant skewness or excess kurtosis (compare Sect. A.2). To this aim one can use a kind of Monte Carlo method: let us assume you have the data of N past returns of an asset. First simulate N random returns under the assumption of normality, then compute skewness and excess kurtosis of this (finite) distribution, finally iterate this n times to get a large sample of approximated normal distributions, each with N data points. If the measured skewness of the data

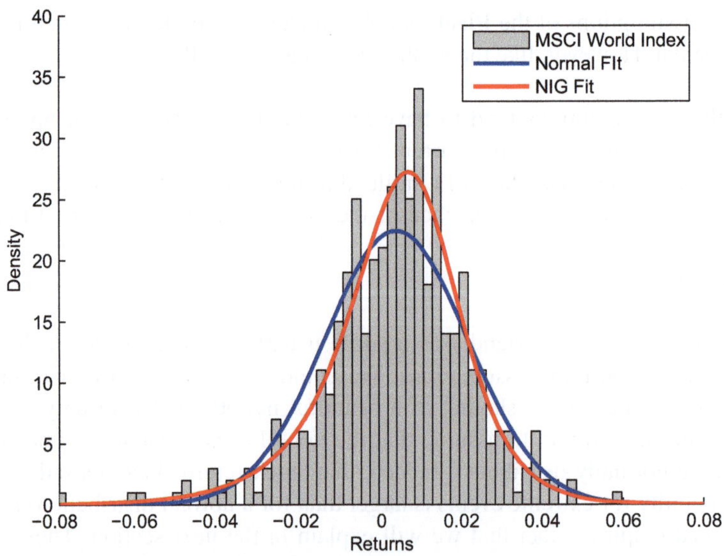

Fig. 8.1 The distribution of daily returns of the MSCI World from January 1, 1970 to April 30, 2007. The best fit with a normal distribution shows that the fat tails are underestimated. A better fit can be obtained by using NIG distributions as we explain below

sample exceeds the skewness of the vast majority of simulated distributions, then it is very unlikely that the data follows a normal distribution. More precisely, if m of n simulated values are below the data value, then the probability is approximately $p = (n - m)/n$, which is very small if $m \approx n$. The probability p is called *simulated p-value*. Similarly, excess kurtosis can be tested. This method is called *simulated p-test*. Some results for stocks, bonds and hedge funds that show that most asset classes show excess kurtosis and skewness can be found, e.g., in [RSW].

The fact that assets are not normally distributed has several important consequences: first, asset pricing based on the Black-Scholes formula needs to be corrected (remember this unfriendly volatility smile!). Second, judging investments by mean-variance which would be correct under the assumption of normally distributed returns is also not correct, since it underestimates the risk of large losses (fat tails in the distribution!).

But how could one improve the model? There are in fact many possible ways to do this. In this section we will outline some of them. In order to apply them for option pricing, we need also a stochastic process that generates non-normal distributions that allow for skewness and fat tails. This will be possible in the general framework of Lévy processes, see Sect. 8.8.3.

As an example for a class of very versatile distributions we introduce the *normal inverse Gaussian distributions* (short: *NIG*) that has been introduced to financial applications by Barndorff-Nielsen [BN97].[14] They are specified by four parameters that roughly correspond to the first four moments. In other words, they allow to model distributions with skewness and excess kurtosis, as we encounter them in the data. A NIG distribution is defined by the following probability density function

$$NIG(x; \alpha, \beta, \mu, \delta) := \frac{\alpha\delta}{\pi} e^{\delta\sqrt{\alpha^2-\beta^2}+\beta(x-\mu)} \frac{K_1(\alpha\sqrt{\delta^2 + (x-\mu)^2})}{\sqrt{\delta^2 + (x-\mu)^2}},$$

where $x, \mu \in \mathbb{R}, 0 \leq \delta, 0 \leq |\beta| \leq \alpha$ and K_1 is the modified Bessel function of the third kind with index 1 (see [BN97] and [Sch08]). The mean, the variance, the skewness and excess kurtosis[15] of $X \sim NIG(\alpha, \beta, \mu, \delta)$ are given by

$$\begin{aligned}
\mathbb{E}(x) &= \mu + \frac{\chi\delta}{(1-\chi^2)^{1/2}}, \\
\mathrm{var}(x) &= \frac{\delta}{\alpha(1-\chi^2)^{3/2}},
\end{aligned}$$

[14]Originally, NIG distributions have been used in physics, more precisely in the modeling of turbulence and sand grain distributions. Only years later they made it into finance.

[15]Excess kurtosis refers to the amount of kurtosis that exceeds that of the normal distribution.

$$S(x) = \frac{3\,\chi}{(\delta\alpha)^{1/2}(1-\chi^2)^{1/4}},$$

$$K(x) = 3\frac{4\,\chi^2+1}{\delta\alpha(1-\chi^2)^{1/2}},$$

where $\chi = \beta/\alpha$.

While the normal distribution has zero skewness and a kurtosis equal to three, we see that a $NIG(\alpha, \beta, \mu, \delta)$ distributed variable has parameter-dependent moments, which are interacting with one another. The four parameters $\alpha, \beta, \mu, \delta$ have natural interpretations relating to the overall shape of the density function: the parameter α controls the steepness of the density, in the sense that the steepness increases monotonically with an increasing α. This also has implications for the tail behavior: large values of α imply light tails, while smaller values of α imply heavier tails as illustrated in Fig. 8.2. The parameter β is a skewness parameter, in the sense that $\beta < 0$ implies a density skew to the left and $\beta > 0$ implies a density skew to the right, i.e., the skewness of the density increases as β increases. In the symmetric case where the parameter β is equal to 0, the density is symmetric around μ. Figure 8.3 shows the dependency on β. Finally, the parameter δ is akin to the standard deviation σ of the normal distribution and represents a measure of the spread of returns.

NIG distributions have a nice property: when we combine two independent NIG distributions $NIG(\alpha_1, \beta_1, \mu_1, \sigma_1)$ and $NIG(\alpha_2, \beta_2, \mu_2, \sigma_2)$, the result is again a NIG distribution, *provided* that $\alpha_1 = \alpha_2$ and $\beta_1 = \beta_2$.

As an example for a fit with NIG, we consider the daily returns of the S&P 500 from January 4, 1988 to May 4, 2007.[16] Using a likelihood estimate gives the parameters $\alpha = 183.7$, $\beta = -8.0$, $\mu = 0.000$ and $\sigma = 0.0033$ for the NIG distribution. This corresponds to a mean of 0.02 %, a variance of 0.002 %,

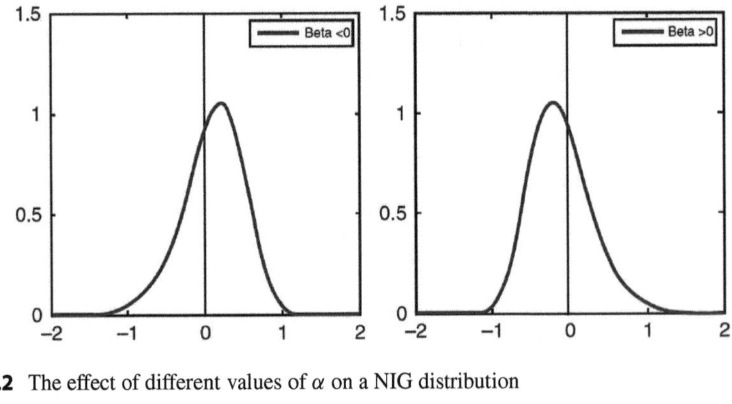

Fig. 8.2 The effect of different values of α on a NIG distribution

[16]Other data gives very similar results. For illustration we concentrate on one particular case. See [RSW] for details.

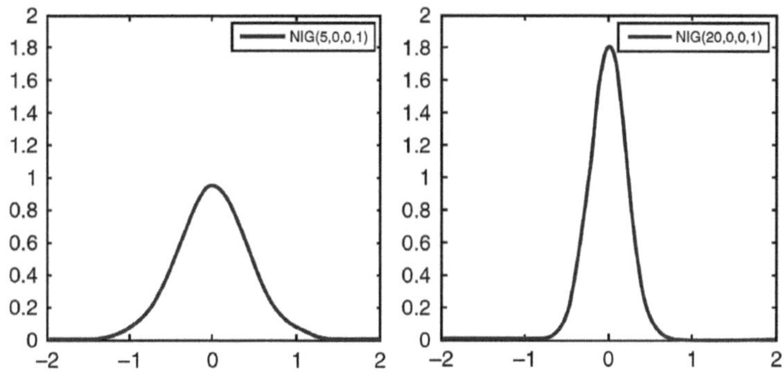

Fig. 8.3 The effect of different values of β on a NIG distribution

a skewness of -0.17 and an excess kurtosis of 5.03. It is interesting to notice that an likelihood estimate with only the two parameters of the normal distribution gives a much larger variance of 0.42 %. In other words: most of the variance that we observe when approximating returns by a normal distribution is quite likely only an artifact of higher moments. The likelihood of the estimate using NIG increases by about 2 %, so besides theoretical reasons in favor of a distribution capable of modeling skewness and fat tails, there is also a quantitative gain in the approximation accuracy. On the other hand, there is of course a cost in dealing with a more complicated model.

What other approaches are there to replace normal distributions? There are in fact several other models, maybe the best-known is the *Lévy skew alpha-stable distribution*, named after the French mathematician Paul Lévy who invented them in 1925 [Lév25] and introduced to finance by the French-American mathematician Benoît Mandelbrot in 1963 [Man63], who later gained popular fame for his "Mandelbrot set" and scientific fame for a number of important contributions to fractal geometry. The central difference to the NIG distributions is that Lévy skew alpha stable distributions (LSASD) do usually not have finite variance. This property might frighten us a little, after all we have spent substantial time with mean-variance approaches, where the assumption of a finite variance was as natural as it was crucial. But maybe we want to risk the revolution and forego the finiteness of the variance if we can gain something for it in exchange. In fact, there is not everything bad about LSASD: in particular, combining two independent LSASDs yields again a LSASD. (This property is meant when we talk about "stability".)

How are LSASDs defined? Like the NIG distrubution, the LSASD depends on four parameters. They are denoted by α, β, c and μ. The distribution is defined as the *Fourier transformation* (see Sect. A.5) of a characteristic function φ, i.e.

$$f(x;\alpha,\beta,c,\mu) = \frac{1}{2\pi}\int_{-\infty}^{+\infty} \varphi(t)e^{-itx}\,dt.$$

The characteristic function is defined as

$$\varphi(t) = e^{it\mu - |ct|^{\alpha}\,(1-\beta\,\mathrm{sign}(t)\Phi(\alpha,t))},$$

where sign(t) gives the sign of t (i.e., $+1$ or -1) and $\Phi(\alpha, t)$ is given by

$$\Phi(\alpha,t) := \begin{cases} \tan(\pi\alpha/2) & ,\text{for } \alpha \neq 1, \\ -(2/\pi)\log|t| & ,\text{for } \alpha = 1. \end{cases}$$

The parameters can be interpreted similar to the NIG distribution: β is a measure of asymmetry, where the distribution is symmetric around the shift parameter μ if $\beta = 0$. The parameter c is a scale factor that describes the width of the distribution and α specifies the asymptotic behavior of the distribution. (Like in the case of the NIG distribution there are other parameterizations in the literature.)

It can be proved that the variance of the LSASD is infinite if $\alpha < 2$ and that several prominent distributions are special cases: for $\alpha = 2$ we obtain the normal (Gaussian) distribution, for $\alpha = 1$ and $\beta = 0$ we obtain the *Cauchy distribution* and for $\alpha = 1/2$ and $\beta = 1$ the *Lévy distribution*. Finally, for $c \to 0$ or $\alpha \to 0$ we obtain the Dirac distribution, i.e., a certain outcome at μ (compare Sect. A.4).

The most notable property of Lévy distributions is that the sum of arbitrary random variables with tails following the power-law $|x|^{-(\alpha+1)}$ with $\alpha > 0$ (and hence with infinite variance) tend to the stable Lévy distribution $f(\alpha, 0, c, 0)$.

Empirical values for α that describe actual price movements of stocks and commodities quite well have been found already in [Man63] as $\alpha \approx 1.7$. Applying this model to asset pricing is, however, a non-trivial task. The foundation for this will be laid in the next section, where we show how processes can be constructed that generate non-normal return distributions.

8.8.3 Jumping Up and Down: Lévy Processes

We have seen in the last section that there are better models for the return distribution of assets than the classical log-normal distribution. In this section we present a generalized class of stochastic processes, the Lévy processes, that allow for such forms of outcome distributions, but also includes the classical Brownian motion. Our exposition follows the introductory text by Jan Kallsen [Kal06] and the book by Jean Bertoin [Ber98].

The key idea of Lévy processes X is to generalize the notion of linear functions in time to stochastic processes – in the sense that linear functions are characterized by constant increments in time and Lévy processes are characterized by constant random distributions of their increments. More precisely, we assume that for every t the difference $X_{t+\delta} - X_t$ follows the same probability distribution. If we remind ourselves on Definition 8.2 of the Brownian motion, we see that this is a part of condition (b). To complete the definition of Lévy processes we need to assume more, namely that the process starts in zero (condition (a) of Definition 8.2) and that increments are independent (condition (d)). All together we define:

Definition 8.17 (Lévy process) A *Lévy process* is a process X defined by the properties:

(a) $X(0) = 0$ a.s.
(b) For any times t_0, t_1 with $t_0 < t_1$, the difference $X(t_1) - X(t_0)$ follows a fixed distribution.
(c) For any times $0 \leq t_0 < t_1 < t_2 < \cdots < t_n < \infty$, the random variables $X(t_0), X(t_1) - X(t_0), \ldots, X(t_n) - X(t_{n-1})$ are independently distributed.

We see that Brownian motions are special cases of Lévy processes that are additionally continuous and have normally distributed increments.

How can we characterize Lévy processes? A central tool we will use is the Fourier transformation (see Sect. A.5). We start with considering small time step increments of the process X, i.e., the difference between $X(t + \Delta t)$ and $X(t)$. These differences all follow a fixed distribution (according to condition (b)), thus $X(t + \Delta t) - X(t) \stackrel{d}{=} X(\Delta t) - X(0) \stackrel{d}{=} X(\Delta t)$, where $\stackrel{d}{=}$ means that they coincide as distributions. We now consider the Forier transform of X. Due to the independence property (c) we obtain

$$\hat{X}(t) = (\hat{X}(\Delta t))^{t/\Delta t}.$$

If $\hat{X}(\Delta t)$ is nowhere zero, we can write it as the exponential of some function $\psi: \mathbb{R} \to \mathbb{C}$, thus obtaining

$$\hat{X}(t) = \exp(\psi t),$$

in other words, we could say that the logarithm of $\hat{X}$ depends linearly on t.

Let us take a closer look on ψ. First we notice that

$$\mathbb{E}(X(t)) = \frac{\mathbb{E}(X(\Delta t))}{\Delta t} t,$$

thus $\mathbb{E}(X(\Delta t))$ has to be of order Δt as $\Delta t \to 0$. Second, we consider the variance

$$\operatorname{var}(X(t)) = \frac{\operatorname{var}(X(\Delta t))}{\Delta t} t,$$

which shows that also the variance has to be of order Δt.

There are now three important cases how we can choose X such that expected value and variance of its increments are of order Δt:

1. X can be deterministic, i.e., $X(\Delta t) = b\,\Delta t$ for some $b \in \mathbb{R}$. The characteristic function $\hat{X}(t)$ then becomes $\hat{X}(t)(u) = \exp(iub\,\Delta t)$, where $i^2 = -1$ (see Sect. A.5).
2. X can be non-deterministic and continuous: if Q is the distribution of $X(\Delta t)/\sqrt{\Delta t}$, then $\hat{X}(\Delta t)(u) = \hat{Q}(u\sqrt{\Delta t})$. If we assume for simplicity that the expected value of $X(\Delta t)/\sqrt{\Delta t}$ is zero, then a Taylor expansion of $\hat{Q}$ yields that $\hat{X}(\Delta t)(u)$ is of order $\exp(-\frac{1}{2}cu^2\Delta t)$ as $\Delta t \to 0$, where $c := \operatorname{var}(Q)$.
3. X can be non-deterministic and discontinuous: take $\lambda > 0$ and let X change with a probability $\lambda\Delta t$ according to the distribution Q. (With the probability $1 - \lambda\Delta t$ the process remains constant.) The characteristic function is in this case $\hat{X}(\Delta t) = (1 - \lambda\Delta t) + \lambda\Delta t\hat{Q}$. Using the Taylor expansion for the exponential function, we can write this for $\Delta t \to 0$ as $\hat{X}(\Delta t) = \exp(\lambda(\hat{Q}(u) - 1)\Delta t)$.

A general Lévy process can now be written as a sum of these three components – plus some remainder term that is discussed in [Kal06] and [Ber98]. This yields to the general Lévy-Chintschin Formula, where we set

$$h(x) := \begin{cases} x, & |x| \leq 1, \\ 0, & |x| > 1 \end{cases}$$

and the Lévy measure $F := \lambda Q$:

Lemma 8.18 (Lévy-Chintschin Formula) *Let X be a Lévy process, then its characteristic function $\hat{X}$ can be described as*

$$\hat{X}(t)(u) = \exp\left(\left(iub(h) - \frac{1}{2}u^2c + \int (e^{iux} - 1 - iuh(x))F(dx)\right)t\right).$$

The variance of a Lévy process has therefore two main sources: the jumps which are described by the measure F (third term in the formula) and by a continuous movement described by the parameter c (second part).

If we want to design a process with a specific distribution (e.g., a NIG distribution), we can choose F as this distribution and the above formula will yield a Lévy process for this distribution. We just need to apply the formula to obtain $\hat{X}$ and then the Fourier transformation to get X. Lévy processes are therefore an extremely flexible and useful class of processes for modeling financial data. Sometimes they are, however, too general and complex to obtain results, e.g., on asset pricing, without further assumptions.

One of many technically simpler subclasses of Lévy processes which has recently caught attention are *stable Lévy processes with exponential decay*, as introduced by Boyarchenko and Levendorskiĭ [BL02]. This class of processes encompasses in particular Brownian motion, NIG processes, hyperbolic processes etc., so many important models are covered by this generalization. On the other hand, the class is small enough in order to use analytical methods for asset pricing. Unlike in the setting of the classical Black-Scholes formula they often do not lead to a PDE, but instead to an equation involving *pseudo differential operators* (see Sect. A.5 for a rough intuition). For details on asset pricing in this general framework and further generalizations we refer the reader also to [BL02] and [FS07].

We see that departing from the simple world of Brownian motions makes things harder and requires sophisticated mathematical tools, but, as Douglas Adams put it:

> It is a mistake to think you can solve any major problems just with potatoes.

8.8.4 Drifting Away: Heston and GARCH Models

When considering Brownian motion as a model for stock price movements, we assume constant volatility. We have seen in Sect. 8.8.1 that the volatility implied by options depends on their maturity. This could be explained by a time-varying volatility which is in fact supported by the data. Figure 8.4 shows a volatility index starting from 1990 up to 2007. It is obvious that the volatility is not just random, but that there are periods with high and periods with low volatility. This phenomenon is called *volatility drift*.

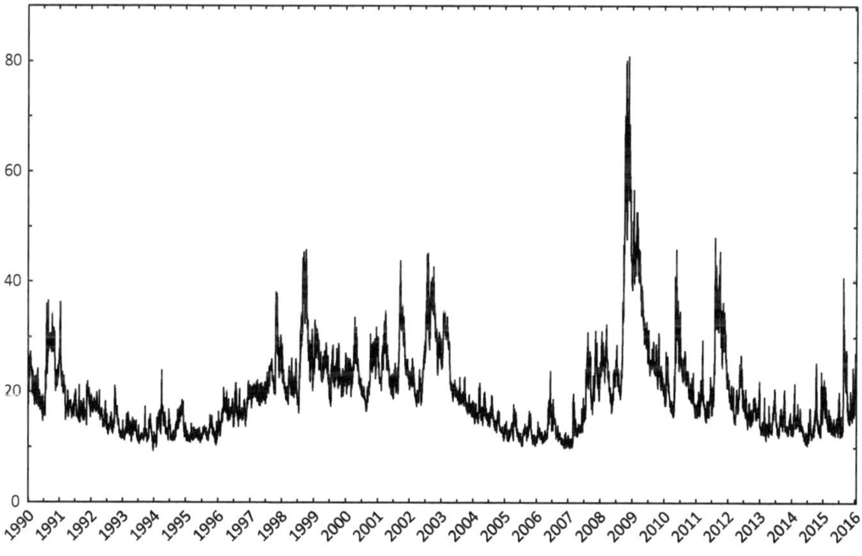

Fig. 8.4 Volatility index (VIX) from 1990 to 2016

If we want to take this effect into account, we need to describe the volatility itself by a random process. Therefore we need two stochastic differential equations:

$$dS(t) = \mu S(t)\,\mathrm{d}t + \sqrt{\sigma(t)}S(t)\,\mathrm{d}B_1(t),$$
$$d\sigma(t) = \alpha(\sigma(t))\,\mathrm{d}t + \beta(\sigma(t))\,\mathrm{d}B_2(t),$$

where α and β are given and B_1, B_2 are both Brownian motions that may correlate with each other with correlation $\rho \in [-1, +1]$.

A typical feature (a "stylized fact") that can be observed for the volatility is that it tends to revert to the mean, i.e., there is a long-term average ω such that the volatility eventually returns to this value. Another stylized fact is that the fluctuation of the volatility tends to be larger when the volatility is large. Both observations lead to a class of standard models with $\alpha(\sigma(t)) = \theta(\omega - \sigma(t))$ (where $\theta > 0$ is a constant) and $\beta(\sigma(t)) = \xi\sigma(t)^{\gamma}$ (where $\xi > 0$ and $\gamma > 0$ are also constants). While θ describes how strongly the process tends to return to its mean, ξ is the (constant part of the) volatility of σ. Finally, γ is an exponent that describes how strong the volatility of σ increases when σ increases, i.e., how strongly "the volatility makes the volatility become volatile".

There are three frequently used models that fall into this framework:

- The *Heston model* assumes $\gamma = 1/2$. The variance process is in this case called a *CIR process*, named after its inventors John C. Cox, Jonathan E. Ingersoll and Stephen A. Ross [CIR85].
- The *Generalized Autoregressive Conditional Heteroskedasticity model*[17] (short: *GARCH*) assumes $\gamma = 1$.
- The $3/2$ *model* assumes $\gamma = 3/2$.

It is difficult to decide which of these models is most appropriate. If we use the data on the volatility index and measure how large the standard deviation of its daily changes is (where we always collect 50 data points with similar volatility to compute the standard deviation) we obtain Fig. 8.5 that gives a best-fit exponent of $\gamma \approx 0.6$. This simple approach is of course not reliable enough to decide which process is best in this case, but the data shows clearly that there is a strong positive correlation between the volatility σ and its standard deviation as proposed by all three models.

There are recent approaches to base this (and other) stylized facts on models of interacting agents on financial markets, see the survey article by Lux [Lux09] for details and further references.

There is one more interesting feature about the volatility that is not captured by the above models: volatility tends to be lower in bull markets, and higher in bear markets. This is certainly counterintuitive, since the risk-return tradeoff should

[17] A vector of random variables is *heteroskedastic* if the random variables have different variances.

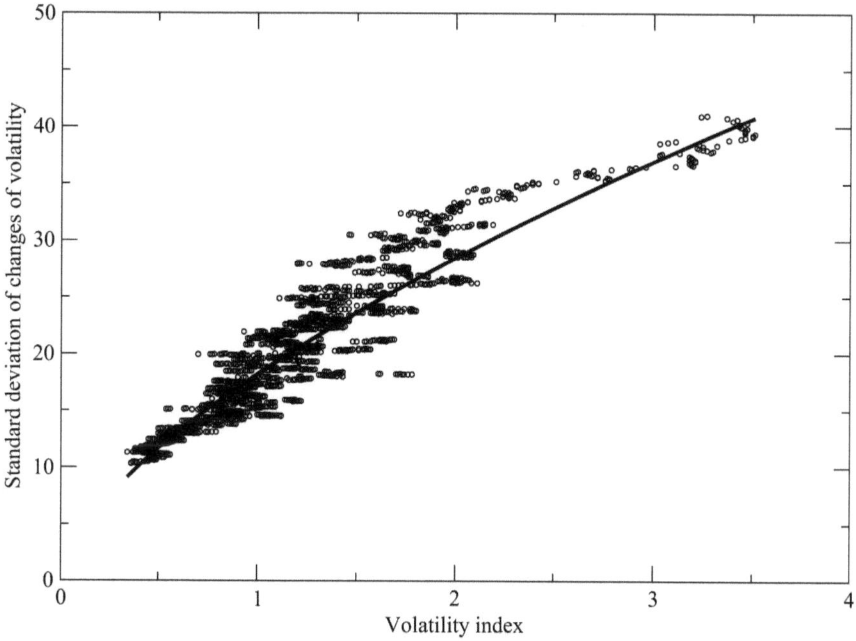

Fig. 8.5 Standard deviation of the volatility index as a function of the volatility index. For each day the standard deviation of the daily volatility changes of the 50 days with a volatility closest to the volatility of the original day is computed and plotted with respect to the average volatility of these 50 days

reward a high volatility with larger returns, but the opposite is the case. The effect (sometimes called "leverage effect") can be observed in many markets, compare Fig. 8.6. An analysis for monthly returns of the S&P 100 yields an interesting pattern: the correlation is strong in losses, but weak in gains (compare Fig. 8.7).

To model this volatility asymmetry one can use more sophisticated stochastic processes, e.g. the APARCH process.

Attempts for an explanation of this effect can be found, e.g. in [BHS01] and [HS09], but there is still no universally accepted model that explains why volatility and stock prices correlate negatively. Recent empirical research where this asymmetry has been measured in a large number of countries found that the asymmetry seems to be particularly high when there are many private investors on the market. Assuming that private investors are more prone to behavioral biases than institutional investors, this might point to behavioral factors as one cause for volatility asymmetry [TR10].

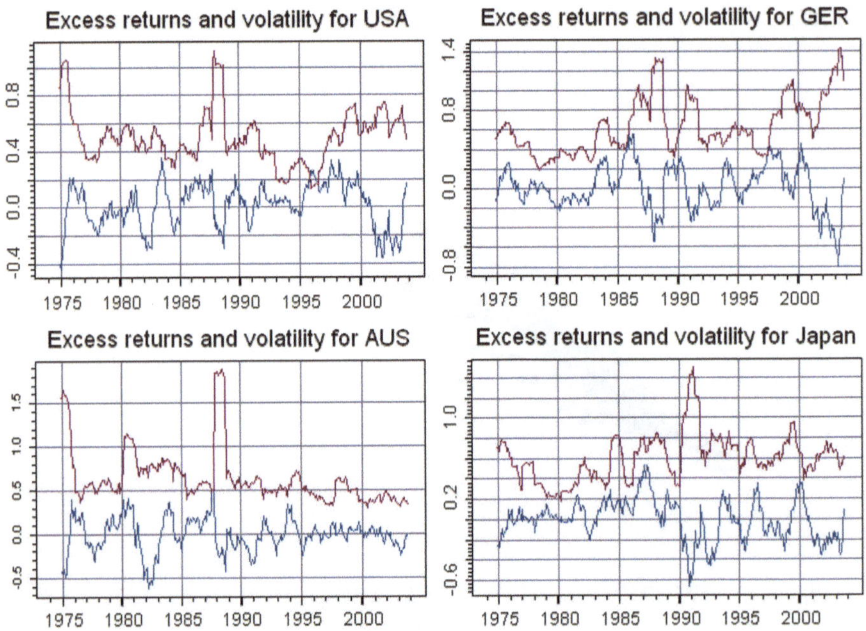

Fig. 8.6 Volatility stock market returns in the USA, Germany, Australia and Japan (compare [Pap04]). The strong negative correlation is in most cases evident

8.9 Summary

Asset pricing is one of the central topics in finance and fundamental for the understanding of many other areas. Moreover, it is one of the most important applications of finance to the "real world". Probably in no other area so many mathematicians and other theorists are employed at major banks and solve very practical problems. Its fundamental ideas are to consider trading on a market as a time-continuous process, and to generalize the idea of replication (hedging) of options into this setting.

Based on these ideas, we have seen a heuristic and a more solid route to the historically most important asset pricing model, the Black-Scholes formula (Sects. 8.1, 8.2, and 8.3). The mathematical apparatus to derive this formula was already sophisticated (stochastic processes and stochastic integration, the Itō formula etc.). The resulting formula for call and put options, however, is relatively simple and hence readily applied to practical option pricing problems.

There are many different options ("exotic options") that have been introduced in Sect. 8.4 and we have seen some numerical methods for the approximation of their prices (Sect. 8.4).

The Mutual-Fund Theorem (Sect. 8.6) was a very general extension of the Two-Fund Separation Theorem in the time-continuous case. It says that every investor

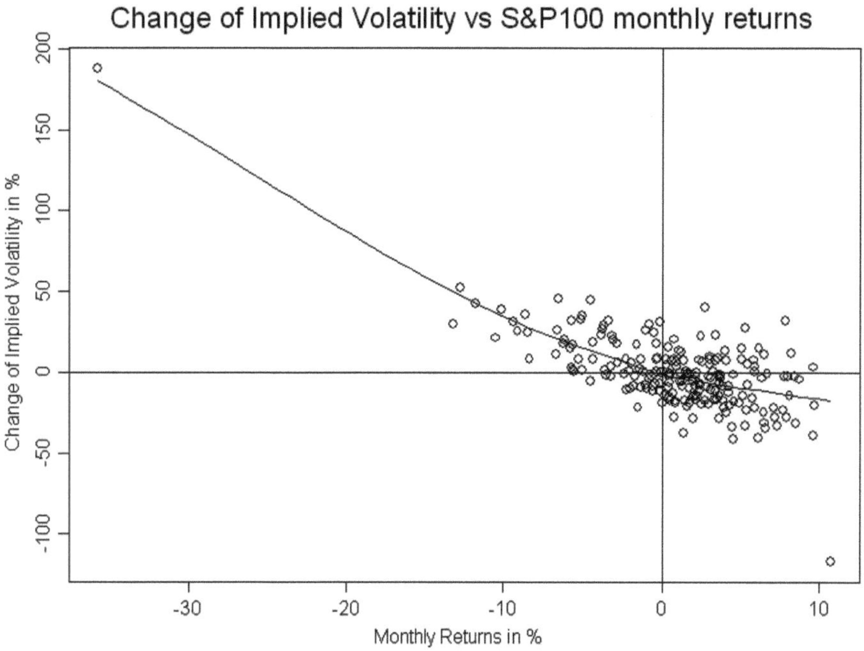

Fig. 8.7 A closer look on the volatility index versus the S&P 100 monthly returns (compare [Pap04]). We observe a strong negative correlation in losses, but a not so clear pattern in gains

should hold the same mutual fund as risky assets – regardless of his precise risk attitudes. We have discussed several reasons (heterogeneous beliefs, behavioral biases, more complicated underlying processes) why the result nevertheless might not hold in practical applications.

In Sect. 8.8, we have studied various ways to relax the restrictive and not always realistic assumptions of the Black-Scholes model, since we encountered effects on stock markets (volatility smile, term structure) that could not readily be understood with the Black-Scholes model. In particular, we have studied alternative return distributions, like NIG, and more general processes than the Brownian motion, in particular Lévy processes and processes with volatility drift.

We have encountered in this last chapter a lot of phenomena that are not completely understood. Regardless of all the progress made in recent years, there are certainly more open problems than well-established models, and what is a standard method today, might become outdated tomorrow. But all of this just demonstrates the old wisdom of research: *as we enlarge the island of knowledge, we increase the coast of questions.*

Mathematics

A

How can it be that mathematics, being after all a product of human thought independent of experience, is so admirably adapted to the objects of reality?
ALBERT EINSTEIN

A.1 Linear Algebra

We recommend the reader who is not familiar with the following notions to take a look into a standard text book on linear algebra, our favorite one is [Jän94]. The brief reminders at the start of this section might help to get used to our notation.

A.1.1 Vectors

A *vector* x is, for the purpose of this text book, simply a tuple of N real numbers, i.e., $x = (x_1, \dots, x_N)$. The space of all such vectors is denoted by $\mathbb{R}^N$. Why do we need this definition for arbitrary N given that we only live in a three-dimensional world? Well, vectors are not necessarily points in the real space, but can denote other things, e.g., returns of assets, and the number of assets can be arbitrarily large, thus we need in fact to consider vectors in this generality.

For two vectors $x, y \in \mathbb{R}^N$, we say that

$$\begin{aligned} x \geq y &\iff x_i \geq y_i \text{ for all } i = 1, \dots, N, \\ x > y &\iff x_i \geq y_i \text{ for all } i \text{ and } x_i > y_i \text{ at least for one } i, \\ x \gg y &\iff x_i > y_i \text{ for all } i. \end{aligned}$$

T. Hens, M.O. Rieger, *Financial Economics*, Springer Texts in Business and Economics, DOI 10.1007/978-3-662-49688-6

We define the (standard) *scalar product* of x and y in $\mathbb{R}^N$ by

$$x \cdot y := \sum_{1=1}^{N} x_i y_i = x_1 y_1 + x_2 y_2 + \cdots + x_N y_N.$$

The scalar product is zero if the vectors are orthogonal. The (Euclidean) *norm* of a vector $x \in \mathbb{R}^N$ is defined by $|x| := \sqrt{x_1^2 + \cdots + x_N^2}$.

A.1.2 Matrices

A *matrix* $A \in \mathbb{R}^{M\times N}$ is (for our purposes) nothing else than a rectangular box of real numbers a_{ij}, $i = 1, \ldots, M, j = 1, \ldots N$. We write $A = (a_{ij})$ or

$$A = \begin{pmatrix} a_{11} & \cdots & a_{1N} \\ \vdots & \ddots & \vdots \\ a_{M1} & \cdots & a_{MN} \end{pmatrix}.$$

The *transposed* matrix A^T (sometimes denoted as A') is defined by "flipping" the indices, i.e., $A^T = (a_{ji})_{i=1,\ldots,M,\, j=1,\ldots N}$. Every vector in $\mathbb{R}^N$ can be understood as a matrix in $\mathbb{R}^{1\times N}$ or $\mathbb{R}^{N\times 1}$. Two matrices $A, B \in \mathbb{R}^{M\times N}$ are added by adding their respective entries, i.e., $(A+B)_{ij} := a_{ij} + b_{ij}$. We can multiply two matrices $A = (a_{ij})$ and $B = (b_{kl})$ if $A \in \mathbb{R}^{M\times N}$ and $B \in \mathbb{R}^{N\times S}$, where the product AB is defined as c_{mn} with

$$c_{mn} := a_{m1} b_{1n} + \cdots + a_{mN} b_{Nn}.$$

It is important to notice that AB is in general not the same as BA (if defined at all). We define the *identity matrix* by

$$Id := \begin{pmatrix} 1 & & 0 \\ & \ddots & \\ 0 & & 1 \end{pmatrix} \in \mathbb{R}^{N\times N}.$$

We define A^{-1} as the *inverse matrix*, i.e., $A^{-1}A = Id$. For instance, if

$$A = \begin{pmatrix} 1 & 1 \\ 0 & 1 \end{pmatrix}, \quad \text{then} \quad A^{-1} = \begin{pmatrix} 1 & -1 \\ 0 & 1 \end{pmatrix},$$

which you can easily check by multiplying A and A^{-1}.

It is sometimes useful to construct a matrix in $\mathbb{R}^{N\times N}$ out of a vector in $\mathbb{R}^N$ by taking the vector entries as diagonal elements of the matrix and setting all other entries zero. This operation is denoted by the operator Λ, sometimes also written as diag. More precisely, we define for $x \in \mathbb{R}^N$:

$$\Lambda x := \operatorname{diag} x := \begin{pmatrix} x_1 & 0 & 0 & 0 \\ 0 & x_2 & & \\ \vdots & & \ddots & \vdots \\ 0 & & \dots & x_N \end{pmatrix}.$$

A.1.3 Linear Maps

Matrices can be used to describe *linear maps*, i.e., maps F from a vector space X (in our case mostly $\mathbb{R}^M$) to a vector space Y (in our case mostly $\mathbb{R}^N$) satisfying the following *linearity conditions*:

(i) $F(x + y) = F(x) + F(y)$ for all $x, y \in \mathbb{R}^M$,
(ii) $F(\lambda x) = \lambda F(x)$ for all $x \in \mathbb{R}^M$ and for all $\lambda \in \mathbb{R}$.

Every linear map F from $\mathbb{R}^M$ to $\mathbb{R}^N$ can be represented by a matrix A in $\mathbb{R}^{M\times N}$ such that $F(x) = Ax$, where x is considered to be in $\mathbb{R}^{N\times 1}$. The composition of two linear maps F and G can be expressed as the product of their corresponding matrices A and B, i.e., $G(F(x)) = (G \circ F)(x) = ABx$.

A special case of linear maps are the *linear functionals*. These are linear maps from a vector space X to $\mathbb{R}$. (A simple example for a linear functional on $\mathbb{R}^3$ is the distance from the x_1-x_2-plane.) An important fact is that linear functionals can always be expressed as a scalar product with a fixed vector. (In the above example, this vector is $(0, 0, 1)$.) This is guaranteed by the Riesz Representation Theorem, named after the Austro-Hungarian mathematician Frigyes Riesz:

Theorem A.1 (Riesz Representation Theorem) *Every linear functional $F: X \to \mathbb{R}$ on a Hilbert space (i.e., a complete vector space with a scalar product) X can be represented as scalar product with a vector $v \in X$, i.e., $F(x) = \langle x, v\rangle$ for all $x \in X$.*

A.1.4 Subspaces, Dimension and Hyperplanes

The *span* of vectors $x_1, x_2, \dots, x_N$ is defined as

$$\operatorname{span}\{x_1, \dots, x_N\} := \{x = \lambda_1 x_1 + \dots + \lambda_N x_N \mid \lambda_1, \dots, \lambda_N \in \mathbb{R}\}.$$

In $\mathbb{R}^3$, for instance, the span of the vectors $(1, 0, 0)$ and $(0, 1, 0)$ contains all vectors of the form $(a, b, 0)$ with a and b arbitrary real numbers, but, e.g., not the vector $(1, 1, 1)$.

We call a set in $\mathbb{R}^N$ which can be represented as a span of vectors in $\mathbb{R}^N$ a *subspace*. In $\mathbb{R}^3$, subspaces are either lines or planes through zero – or $\mathbb{R}^3$ itself.

Let d be the smallest number of vectors that are needed to represent a subspace Z of $\mathbb{R}^N$ as their span. It is easy to see that $1 \leq d \leq N$. We call the number d the *dimension* of the subspace, in short $\dim Z$.

We call an $N-1$ dimensional subspace of $\mathbb{R}^N$ a *hyperplane*. (In the case $N = 3$, such a hyperplane is simply a plane.)

We say, a vector $x \in \mathbb{R}^N$ is *normal* on the hyperplane Z if $x \cdot z = 0$ for all $z \in Z$. The geometric intuition (in $\mathbb{R}^3$) for this is that x stands upright on the plane Z, forming right angles with Z.

Sometimes we want to consider "shifted" subspaces, e.g., planes that do not go through the zero point. We call such sets *affine subspaces*. They are defined as sets A for which there exists a vector $x = (x_1, \ldots, x_N)$ such that $A - x := \{(y_1 - x_1, \ldots, y_N - x_N) \mid (y_1, \ldots, y_N) \in A\}$ is a subspace. Analogously, we can define *affine hyperplanes*.

A.1.5 Convex Sets and the Separation Theorem

We call a set $K \in \mathbb{R}^N$ *convex* if for all $x, y \in K$ the entire straight line segment between x and y is in K. Formally: if $x, y \in K$ then for all $\lambda \in [0, 1]$ we have $\lambda x + (1 - \lambda y) \in K$.

A set K in $\mathbb{R}^N$ is called *closed* if every sequence in K that converges has a limit in K (and not outside K). A set K in $\mathbb{R}^N$ is called *bounded* if it fits for some $r > 0$ into a ball $B_r := \{x \in \mathbb{R}^N \mid |x| < r\}$. A set K in $\mathbb{R}^N$ is called *compact* if it is closed and bounded.

We need the following theorem, e.g., in the proof of the First Asset Pricing Theorem:

Theorem A.2 (Separation Theorem) *Let $K, M \subset \mathbb{R}^N$ convex sets with $K \cap M = \emptyset$. If K is compact and M is closed then there exists some $x \in \mathbb{R}^N$ such that*

$$\sup_{y \in M} x \cdot y < \inf_{y \in K} x \cdot y.$$

The Theorem is called "Separation Theorem", since it says, in geometrical terms, that every pair of disjoint, compact, convex sets can be separated by a hyperplane. (The normal vector on the hyperplane is x.)

There are infinite-dimensional generalizations of this theorem. Things are, however, more complicated since compactness cannot be so easily defined in infinite dimensions.

A.2 Basic Notions of Statistics

We give a brief summary of what we assume the reader knows in statistics, mainly to familiarize with our notation.

Elementary statistics studies properties of data samples x_i with $i = 1, \ldots, N$, which form a vector $x = (x_1, \ldots, x_N) \in \mathbb{R}^N$. We can generalize this by considering the x_i as events with certain probabilities p_i, where $p_i \geq 0$ and $\sum_{i=1}^N p_i = 1$ (in the simplest case $p_i := 1/N$), or even more generally by considering general probability measures p on $\mathbb{R}$ (compare Sect. A.4). In the following, we recall all definitions in the language of probability measures and in the language of random variables.

A.2.1 Mean and Expected Value

The *(arithmetic) mean* (or average) μ or $\bar{x}$ of a vector $x \in \mathbb{R}^N$ is defined as

$$\bar{x} := \frac{1}{N} \sum_{i=1}^{N} x_i.$$

For a general probability measure p the *mean* (or *expected value*) μ or $\mathbb{E}(p)$ is defined as the mean squared deviation

$$\mathbb{E}(p) := \int_{\mathbb{R}} x \, \mathrm{d}p(x),$$

and for a real-valued random variable X on a probability space Ω with probability measure p (i.e., X is a map from the "state space" Ω to $\mathbb{R}$), the mean is given by

$$\mathbb{E}(X) := \int_{\Omega} X \, \mathrm{d}p.$$

A.2.2 Variance

The *variance* σ^2 or $\mathrm{var}(x)$ of a vector $x \in \mathbb{R}^N$ is defined as

$$\mathrm{var}(x) := \frac{1}{n} \sum_{i=1}^{N} (x_i - \bar{x})^2.$$

For a general probability measure p the *variance* $\mathrm{var}(p)$ becomes

$$\mathrm{var}(p) := \int_{\mathbb{R}} (x - \mathbb{E}(x))^2 \, \mathrm{d}p(x).$$

For a random variable X, the variance can be defined as

$$\operatorname{var}(X) := \mathbb{E}((X - \mathbb{E}(X))^2).$$

Intuitively, the variance describes how much a random variable “fluctuates”.

A.2.3 Normal Distribution

The most frequently used probability distributions are *normal distributions* (also called “Gauss distributions”), defined as $N(\mu, \sigma^2)\,\mathrm{d}x$, where

$$N(\mu, \sigma^2) := \frac{1}{\sigma\sqrt{2\pi}}\,\mathrm{e}^{-\frac{(x-\mu)^2}{2\sigma^2}},$$

and $\mu \in \mathbb{R}$ and $\sigma > 0$. We sometimes abbreviate $N(\sigma^2) := N(0, \sigma^2)$.

The mean value of a normal distribution is μ and its variance is σ^2. The importance of the normal distribution arises from the *Central Limit Theorem*. It states that the mean of a sufficiently large number of mutually independent random variables, each with finite mean and variance, will be approximately normally distributed. This explains why normal distributions appear in so diverse fields as physics, biology, psychology or economics.

A.2.4 Covariance and Correlation

The *marginals* of a probability measure T on $\mathbb{R} \times \mathbb{R}$ are the probability measures p and q defined as

$$p(A) := \int_{\mathbb{R}} \mathrm{d}T(A, y) \quad \text{and} \quad q(A) := \int_{\mathbb{R}} \mathrm{d}T(x, A)$$

for $A \subset \mathbb{R}$.

The *covariance* of a probability measure T on $\mathbb{R} \times \mathbb{R}$ with marginals p and q is defined as

$$\operatorname{cov}(p, q) := \operatorname{cov}(T) := \int_{\mathbb{R}} \int_{\mathbb{R}} (x - \mathbb{E}(p))(y - \mathbb{E}(q))\,\mathrm{d}T(x, y).$$

Using random variables X, Y, we can represent the covariance as

$$\begin{aligned}\operatorname{cov}(X, Y) &:= \mathbb{E}((X - \mathbb{E}(X))(Y - \mathbb{E}(Y)) \\ &= \mathbb{E}(X \cdot Y) - \mathbb{E}(X)\mathbb{E}(Y).\end{aligned}$$

If X and Y are statistically independent, then their covariance is zero. The converse, however, is not true: if X and Y have covariance zero, they are not necessarily independent. The covariance of a variable with itself is the variance, i.e., $\text{cov}(X, X) = \text{var}(X)$.

Intuitively, covariance is the measure of how much two random variables "follow each other". This idea leads to the following definition of the *correlation*, for a probability measure T on $\mathbb{R} \times \mathbb{R}$:

$$\text{corr}(T) := \frac{\text{cov}(T)}{\text{var}(p)\,\text{var}(q)},$$

or, in terms of random variables:

$$\text{corr}(X, Y) := \frac{\text{cov}(X, Y)}{\text{var}(X)\,\text{var}(Y)}.$$

The correlation takes values between -1 and $+1$, where $+1$ implies a linear dependence between X and Y, -1 an antilinear dependence, and values around zero no visible dependence. However, $\text{corr}(X, Y) = 0$ does not imply independence. In the language of probability measures, a correlation of $+1$ would, e.g., imply that T has a support along a line in $\mathbb{R}^2$ with positive slope.

Occasionally, we will need the *covariance matrix*: if we have two vector-valued random variables X and Y, then the covariance matrix COV is defined by

$$COV := COV(X, Y) := \mathbb{E}((X - \mathbb{E}(X))(Y - \mathbb{E}(Y)^T).$$

In this book, the elements of COV are denoted by $\text{cov}_{i,j}$.

A.2.5 Skewness and Higher Order Moments

We call

$$\mu_k := \int_{\mathbb{R}} (x - \mathbb{E}(p))^k \,\mathrm{d}p(x)$$

the *k-th central moment* of the probability measure p. For a random variable X we define it as $\mathbb{E}((X - \mathbb{E}(X))^k)$. We see immediately that the variance is the second central moment and the mean is (essentially) the first. We define the *skewness* as the *standardized third central moment*, i.e., μ_3/σ^3, where σ^2 denotes the variance.

Symmetric probability measures have zero skewness, since

$$\begin{aligned}
\int_{\mathbb{R}} (x-\mathbb{E}(p))^k \,\mathrm{d}p(x) \\
&= \int_{-\infty}^{\mathbb{E}(p)} (x-\mathbb{E}(p))^k \,\mathrm{d}p(x) + \int_{\mathbb{E}(p)}^{+\infty} (x-\mathbb{E}(p))^k \,\mathrm{d}p(x) \\
&= \int_{-\infty}^{\mathbb{E}(p)} (x-\mathbb{E}(p))^k \,\mathrm{d}p(x) - \int_{\mathbb{E}(p)}^{+\infty} (\mathbb{E}(p)-x)^k \,\mathrm{d}p(\mathbb{E}(p)-x) \\
&= 0.
\end{aligned}$$

Every normal distribution is symmetric and therefore has zero skewness.

The skewness measures how "tilted" or how "asymmetric" a distribution is. Stock market returns are typically skewed, since very large losses are more likely than very large gains, although on average the returns are positive. Derivatives are usually extremely skewed, as can be seen from their highly asymmetric payoff-diagram.

The fourth standardized central moment leads to the definition of *kurtosis*. For a random variable X it is defined as $\mathbb{E}((X-\mathbb{E}(X))^4)/\sigma^4$. Since the normal distribution has a kurtosis of 3, one defines the *excess kurtosis* as $\mathbb{E}((X - \mathbb{E}(X))^4)/\sigma^4 - 3$. Kurtosis describes how "peaked" a distribution is and how quickly its tails decay to zero. Positive excess kurtosis means that the distribution has a prominent peak and its tails are "fat", i.e., it decays slower to zero than a normal distribution. Many alternative investments (hedge funds and insurance linked securities (ILS) are prominent examples) have fat tails, meaning that extreme outcomes (usually extreme losses) are more likely than suggested by considering mean and variance alone. Negative excess kurtosis occurs, e.g., in lotteries with two outcomes (coin-flipping), where tails and peak are both zero.

A.3 Basics in Topology

Sometimes it is necessary to have a deeper understanding of notions like "convergence" or "continuity" that we frequently use. In the following we will provide a little background on this. An excellent and very pleasantly written mathematical book for further reading is [Jän95].

A.3.1 Open Sets

Let X be an arbitrary set. We can call a system $\mathcal{T}$ of subsets of X *open* if the following conditions hold:

(i) For any subset $\{O_\lambda\}_{\lambda\in\Lambda}$ of $\mathcal{T}$, the union of all O_λ is also in $\mathcal{T}$. ("Unions of open sets are open.")

(ii) For any two sets O_1, O_2 in $\mathcal{T}$, their intersection $O_1 \cap O_2$ is also in $\mathcal{T}$. ("The intersection of two open sets is open.")
(iii) The empty set $\emptyset$ and the whole set X are both in $\mathcal{T}$.

You can easily convince yourself that the standard open sets in $\mathbb{R}$ fit this definition.

All other "topological" concepts, like closed sets, compactness and continuity, can be based on the notion of open sets. Therefore it is possible to study these properties not only in $\mathbb{R}$ or $\mathbb{R}^N$, but also in very abstract sets, e.g., we can define convergence of probability measures or we can talk about a compact set of functions.

Let us give the definition of a closed set: a set $A \subset X$ is called *closed* in X if its complement, i.e., $X \setminus A$, is open.

Compactness can also be defined based on this concept. Since we deal only with compact subsets of $\mathbb{R}^N$, it is sufficient to know that in this case compact sets are closed and bounded (and vice versa). In infinite-dimensional situations, this is not true!

Finally, it is also possible to define continuity solely based on the concept of open sets:

A function $f: X \to Y$ is *continuous* if for every open set $U \subset Y$, also the inverse image $f^{-1}(U) \subset X$ is open.

As illustration for this definition think about simple examples of continuous and non-continuous functions from $\mathbb{R}$ to $\mathbb{R}$ and convince yourself that it holds for continuous functions, but not for discontinuous functions!

A set is *disconnected* if we can divide it into two disjoint open sets. As an example, take the set $X := (0,1) \cup (1,2)$ in $\mathbb{R}$. This set is disconnected, since $A := (0,1)$ and $B := (1,2)$ are disjoint open subsets with $A \cup B = X$. A set is *connected* if it is not disconnected. If $f: X \to Y$ is continuous and X is connected, then Y is also connected. If we consider subsets of $\mathbb{R}^n$, every connected set is *path-connected* (and vice versa), i.e., we can always find a continuous curve path between any two points in the set.

A.3.2 Convergence and Metrics

We can also define convergence of sequences $x_n \subset X$ via open sets:

A sequence $x_n \subset X$ *converges* to $x \in X$ if for all open sets $U \subset X$ with $x \in U$ there exists an $n_0 \in N$ such that starting from this n_0, the sequence $(x_n)_{n \geq n_0}$ lies in U. (In other words: for all $m \geq n_0$, we have $x_m \in U$.)

How does this notion correspond to the usual notion of convergence that you probably have learned ($x_n \to x$ if the distance between x_n and x converges to zero)? First, if we have a *metric* d, i.e., a way to measure the distance $d(x, y)$ between any two elements x, y in X (in a meaningful way, i.e., $d(x, x) = 0$, $d(x, y) = d(y, x)$ and $d(x, y) + d(y, z) \geq d(x, z)$, compare [Jän95] for details), then any open set can be characterized as follows:

A set $U \subset X$ is open if, for any $x \in U$, there is a small ball $B(x, \varepsilon) := \{y \in X \mid \mathrm{d}(x, y) < \varepsilon\}$ which is entirely contained in U.

The usual notion of convergence (that $x_n \to x$ if $d(x_n, x) \to 0$) corresponds to the definition based on open sets.

On the other hand, we can define, e.g., closed sets based on a definition of convergence: A is closed if the limit of any converging sequence $x_n \subset A$ is itself in A, in other words $x_n \subset A$ and $x_n \to x$ implies $x \in A$.

Similarly, if X and Y both have a metric, then a function $f\colon X \to Y$ is continuous at $x \in X$ if $x_n \to x$ implies $f(x_n) \to f(x)$.

A very useful result on convergence is the Theorem of Bolzano and Weierstrass. We present a simplified version:

Theorem A.3 (Bolzano-Weierstrass Theorem) *Let x_n be a sequence of vectors on a bounded and closed set $S \subset \mathbb{R}^k$, then there is a subsequence of the x_n that converges to a limit $x \in S$.*

A.4 How to Use Probability Measures

In this section we set the notion of probability measures on a mathematically solid foundation. This will enable us to formulate many problems in a more general setting and to see that lotteries with finitely many outcomes and lotteries with continuous outcome distributions are only two special instances of general lotteries. Many results can then derived much simpler. There is, however, some more severe mathematics involved, although we will skip a lot of more subtle details. Therefore we need to apologize to the specialists for pretending that things are easier than they actually are, and to the non-specialist for presenting things that are quite theoretical. Nevertheless we hope that both can forgive us and appreciate the goal of bridging the gulf between measure theory and applications in finance.

We first need to define what we want to "measure" with a probability measure. Given a set of all possible states, a *state space* Ω, we want to assign probabilities to subsets of this state space, i.e., to *events*. As example you may think of the state space as the set of all possible returns of an asset $\mathbb{R}$, and as events like "the asset yields a return larger than x" which would correspond to the subset (x, ∞) of $\mathbb{R}$.

Unfortunately, for measure theoretical reasons it is generally not possible to assign meaningful probabilities to *all* subsets of a state space,[1] therefore we have to restrict ourself to subsets that are in a so-called "σ-algebra" (sometimes also called "tribe"). This point is not essential for most of our applications, but it helps to know the properties these subsets satisfy:

Definition A.4 (σ-algebra) A collection $\mathcal{F}$ of subsets of Ω is called σ-algebra if the following three conditions are satisfied:

[1]If you want to know what can happen when you allow for arbitrary subsets, take a look on the creepy Banach-Tarski Paradox, see e.g. [Fre88]: you could decompose a massive ball into five pieces, put them together again and – the new (still massive) ball is twice as large as before!

(i) If a set B is in $\mathcal{F}$, then its complement $\Omega \setminus B$ is also in $\mathcal{F}$. (This condition ensures that if an event B has a certain probability, then its complement, i.e., the event that B does *not* happen, has also a well-defined probability.)
(iii) If $B_1, B_2, \ldots$ are in $\mathcal{F}$, then also their union $B_1 \cup B_2 \cup \ldots$ is in $\mathcal{F}$. (This condition ensures that separate events that have a probability, can be summarized to one event with a certain probability.)
(iii) The empty set $\emptyset$ and the whole state space Ω are both in $\mathcal{F}$. (This condition simply ensures that the event "nothing happens" and the event "something happens" both have a well-defined probability.)

The mathematically less inclined reader is again ensured that all events he usually encounters are in a σ-algebra and hence have a mathematically well-defined probability. It is enough to remember that there is a potential mathematically problem hidden that fortunately has been solved in a clever way by the definition of σ-algebras.

After this technical prelude we can now define probability measures:

Definition A.5 (Probability measure) A probability measure p on a state space Ω is a map that assigns to subsets of Ω that are in some σ-algebra $\mathcal{F}$ a number in $[0, \infty]$ and satisfies

(i) $p(\emptyset) = 0$ (the empty set has measure zero),
(ii) for pairwise disjoint sets $B_i \subset \Omega$, we have

$$p\left(\bigcup_{i=1}^{\infty} B_i\right) = \sum_{i=1}^{\infty} (B_i),$$

(iii) $p(\Omega) = 1$ (the whole state space has measure one).

In general, not only the empty set will have the probability zero (even if we deal only with a finite numbers of events). In the general case that does not mean that the event can never happen: as example think of the probability to get the number π when randomly picking a number between 0 and 10. This probability will be zero if the probability distribution is uniform, however, it is perfectly possible to get π.

If we want to restrict our analysis of a certain situation only to the "relevant" events, i.e., to the events with non-zero probability, it is handy to use the following notation:

Definition A.6 ("Almost" and null sets) Let $(\Omega, \mathcal{F}, p)$ be a probability space. We say that an event $B \subset \Omega$ happens *almost surely* (abbreviated *a.s.*) if $p(B) = 1$, and we say that a property that holds on a set $B \subset \Omega$ with $p(B) = 1$ holds for *almost every* element in Ω (abbreviated *a.e.*). We call all sets B with $p(B) = 0$ *nullsets*.

We are mostly (but not exclusively) interested in probability measures on $\mathbb{R}$. (The outcome distribution of an asset, e.g., is nothing else than a probability measure on $\mathbb{R}$.) It is possible to characterize probability measures on $\mathbb{R}$ in a handy way. This can be done by applying Lebesgue's Decomposition Theorem and the Radon-Nikodym Theorem to get the following result:

Theorem A.7 *Let p be a probability measure on $\mathbb{R}$. For $A \subset \mathbb{R}$ let $|A|$ denote the usual (Lebesgue) measure of the set A. Then there exists an integrable function $f\colon \mathbb{R} \to [0,\infty]$ and a singular measure p_s such that $p = f\,\mathrm{d}x + p_s$. Here $\mathrm{d}x$ denotes the usual (Lebesgue) measure on $\mathbb{R}$, and "singular" means that there exist disjoint sets $A, B \subset \mathbb{R}$ with $A \cup B = \mathbb{R}$ and $p_s(A) = 0$ and $|B| = 0$.*

We can now decompose the singular part even more. For this we need first the following definition:

Definition A.8 (Dirac measure) The *Dirac measure* δ_x assigns to every set $A \subset \mathbb{R}$ the value one if $x \in A$ and zero if $x \notin A$. δ_x is a probability measure on $\mathbb{R}$.

With this definition we can decompose the singular measure p_s into a linear combination of Dirac measures plus a remainder:

$$p_s = \sum_{i=1}^{\infty} \lambda_i \delta_{x_i} + p_c,$$

where $\lambda_i > 0$, $x_i \in \mathbb{R}$, and the remainder p_c is called the "Cantor-part" of the measure. In most of our applications we assume that all probability measures p satisfy $p_c = 0$.

We finally need to find a way to integrate with respect to a general probability measure. This is quite natural when we want to integrate over a step function, so let us do this first:

Definition A.9 Let f be a step function, i.e., there is an increasing sequence of $x_i \in \mathbb{R}$ such that f is constant on each interval $[x_i, x_{i+1})$ with value f_i. Let p be a probability measure on $\mathbb{R}$. Then we define

$$\int fp := \sum_i p([x_i, x_{i+1}))f_i.$$

For general integrable functions f we define $\int fp$ as limit of $\int f^n p$ where (f^n) is a sequence of step functions which approximates f.

How can we make use of this new formalism? We illustrate this with the following example: if we write down the EUT for a lottery A with finitely many outcomes $x_1, \dots, x_n$ and probabilities $p_i, \dots, p_n$ we get

$$EUT(A) = \sum_{i=1}^{n} u(x_i)p_i.$$

If a lottery B has instead a continuous outcome distribution with density f, we have to replace this definition with an integral formulation

$$EUT(B) = \int u(x)f(x)\,\mathrm{d}x.$$

But what if we have a mixture of both lotteries? With our new formalism we can express all these cases in one simple formula:

$$EUT = \int u\,\mathrm{d}p,$$

where dp denotes a probability measure. (The little "d" reminds us that we are dealing with an integration.) If, for instance, $dp = f\,\mathrm{d}x + \sum_{i=1}^{n} \lambda_i \delta_{x_i}$, a short computation gives

$$EUT = \int u\,\mathrm{d}p = \sum_{i=1}^{n} u(x_i)\lambda_i + \int u(x)f(x)\,\mathrm{d}x.$$

This new way of dealing with probability measures is not only convenient, but also allows us to discuss situations involving discrete *and* continuous lotteries, for instance when we want to approximate a continuous lottery by finite lotteries. (We do this in Sect. 2.4.5 and in Sect. 2.4.6, when we discuss continuity properties of decision theories.)

We conclude this section with a useful result, the Jensen Inequality:

Theorem A.10 (Jensen Inequality) *Let $u\colon \mathbb{R} \to \mathbb{R}$ be a concave function and let dp be a probability measure on $\mathbb{R}$. Let μ be the expected value of dp, i.e., $\mu := E[p] := \int x\,\mathrm{d}p$. Then we have*

$$u(\mu) \geq \int u\,\mathrm{d}p.$$

Proof Since u is concave, the tangent on u in μ lies nowhere below u. In other words, there is a line $g(x) = u(\mu) + a(x - \mu)$ (for some constant a), such that

$g(x) \geq u(x)$ for all $x \in \mathbb{R}$. Now we can estimate

$$\int u \, \mathrm{d}p \leq \int g \, \mathrm{d}p = \int \alpha(x - \mu) + u(\mu) \, \mathrm{d}p$$
$$= \alpha \int x \, \mathrm{d}p - \alpha\mu \int \mathrm{d}p + u(\mu) \int \mathrm{d}p$$
$$= \alpha\mu - \alpha\mu + u(\mu) = u(\mu).$$

This concludes the proof. □

Jensen's Inequality can be generalized in several ways:

- The inequality holds on a subinterval of $\mathbb{R}$ if u is concave only on this subinterval.
- It can be generalized from $\mathbb{R}$ to $\mathbb{R}^n$.
- We can obtain a strict inequality sign if u is not only concave, but *strictly* concave (as long as $\mathrm{d}p \neq \delta_\mu$).
- We obtain the inverse inequalities if u is not concave, but convex.[2]

It is a nice little exercise to prove these statements.

A.5 Calculus, Fourier Transformations and Partial Differential Equations

Let us start with some calculus facts that you should be familiar with.[3]

A function $f: \mathbb{R} \to \mathbb{R}$ is *differentiable* if

$$\frac{\mathrm{d}f(x)}{\mathrm{d}x} := \lim_{h \to 0} \frac{f(x+h) - f(x)}{h}$$

exists and is finite. We call $f'(x) := \mathrm{d}f(x)/\mathrm{d}x$ the *derivative* of f (with respect to the variable x). If f' is continuous, we call f *continuously differentiable*.

A function $f: \mathbb{R} \to \mathbb{R}$ is *Hölder continuous with Hölder exponent* α at x if there is a constant C and an exponent $\alpha \in (0, 1]$ such that $|f(x + h) - f(x)| \leq Ch^\alpha$ for all $h > 0$. If $\alpha = 1$, we say that the function is *Lipschitz continuous*. The following result is non-trivial and very useful:

Theorem A.11 (Rademacher Theorem) *Every Lipschitz continuous function is a.e. differentiable.*

[2]Indeed, this is the form in which the inequality is presented in mathematical text books. In finance we more frequently deal with concave functions.

[3]If this is all new for you, we advise you to study your favorite calculus textbook first.

Next we consider functions with more than one variable. As an example take the elevation h of a hilly area that depends on two variables, say, latitude x and longitude y. We can take *partial derivatives* of the elevation function h, e.g., we can consider its slope in the direction of x which we denote by $\partial h(x,y)/\partial x$. This means that we cut out a thin slice of the hills along the x-direction and look at the derivative of the elevation on this slice (which is just a function in one variable). Of course we can generalize this idea to arbitrarily many dimensions. An example for a vector composed of partial derivatives is the *gradient* of a function $h: (x_1, \ldots, x_n) \to h(x_1, \ldots, x_n) \in \mathbb{R}$ which is defined by

$$\nabla h(x) := \begin{pmatrix} \frac{\partial h(x)}{\partial x_1} \\ \vdots \\ \frac{\partial h(x)}{\partial x_n} \end{pmatrix},$$

where $x = (x_1, \ldots, x_n)$ and sometimes Dh is written instead of ∇h.

In the following we need to extend the real numbers $\mathbb{R}$ to *complex numbers* $\mathbb{C}$ that can be written as a linear combination of a real and an *imaginary number*, i.e., $z = x + iy \in \mathbb{C}$ with $x, y \in \mathbb{R}$ and $i^2 = -1$. For complex numbers particularly the Euler formula holds:

$$e^{i\phi} = \cos\phi + i \sin\phi.$$

We can now define a *Fourier transformation*: it maps a function $f: \mathbb{R} \to \mathbb{R}$ to its *Fourier transform* $\mathcal{F}f: \mathbb{R} \to \mathbb{C}$ and is defined by

$$(\mathcal{F}f)(\xi) := \frac{1}{\sqrt{2\pi}} \int_{-\infty}^{+\infty} e^{-i\xi t} f(t)\, dt.$$

The Fourier transformation describes in a certain way the frequency distribution of f. The most natural application is in acoustics where f describes the oscillation (e.g., of the air) and its Fourier transform corresponds to the distribution of frequencies, i.e., its sound. A sine function would lead to a Dirac distribution as Fourier transform: in other words there is only one frequency, the sound is a pure tone. Such a tone would sound very harsh and artificial, since natural tones (like the sound of a piano) are composed of many different tones, i.e., they correspond to a weighted sum of sine functions. Their Fourier transform is therefore a weighted sum of Dirac distributions. Another example is a normal distribution: their Fourier transform is again a normal distribution.

The Fourier transformation has a couple of interesting properties. We collect the most important in the following lemma:

Lemma A.12 (Properties of the Fourier transformation) *Let $\mathcal{F}$ be the Fourier transformation, then:*

(i) $\mathcal{F}$ *is a linear map.*
(ii) The inverse of the Fourier transformation is given by

$$(\mathcal{F}^{-1}f)(x) = \frac{1}{\sqrt{2\pi}} \int_{-\infty}^{+\infty} e^{ixt}\hat{f}(t)\, dt.$$

(iii) The Fourier transform of a derivative is a polynomial, more precisely:

$$\mathcal{F}\left(\frac{\partial^n}{\partial x^n} f(x)\right)(\xi) = (i\xi)^n (\mathcal{F}f)(\xi).$$

(iv) It is possible to define an appropriate space (of functions or distributions) such that $\mathcal{F}$ *is a bijective map on this space.*

Property (iii) can be summarized by saying that the Fourier transformation turns derivatives into a product – although this has nothing to do with the marketing of options... The property is often the main reason to use the Fourier transformation and could even be used as an alternative definition of differentiation. While this is certainly a complicated way for reaching this goal, it enables us to generalize the definition of derivatives: if we take the Fourier transform of a multiplication of f with, e.g., $i|\xi|^{1/2}$, we get something like a "half" derivative. If that sounds esoteric to you, then be ensured that it is not: very much to the contrary it is surprisingly useful. The concept leads to the definition of *pseudo differential operators* and they are needed, e.g., in solving certain asset pricing problems when the underlying process is of a more complicated form. We mention this point and generally use Fourier transformations when we discuss Lévy processes in Sect. 8.8.

A *partial differential equation* (short: *PDE*) is an equation that contains different partial derivatives of an unknown function and is used to determine this function. In contrast, an *ordinary differential equation* (short: *ODE*) only involves one kind of derivative (e.g., only derivatives with respect to t). As an example for a simple PDE we consider the *heat equation*[4] which is needed to solve the Black-Scholes equation (see Sect. 8):

$$\frac{\partial u(x,t)}{\partial t} = \frac{\partial^2 u(x,t)}{\partial x^2},$$

where $t \in [0, T]$ and $x \in \mathbb{R}$. (In the original physics model for heat transport, t is the time and x the space variable, whereas $u(x,t)$ is the temperature at time t and position x.) The above equation therefore means that the partial derivative of u with respect to t equals its second partial derivative with respect to x.

[4]This PDE was originally used to describe the transport of heat in a material. There are, however, various other applications for this equation, therefore it is sometimes also called, e.g., *diffusion equation*.

Typically, PDEs can only be solved uniquely when we have additional conditions. In the case above this would be an initial condition (specifying u at $t = 0$) plus some boundary conditions (e.g., specifying the behavior of u for $x \to \pm\infty$).

PDEs are a central modeling tool in all scientific disciplines that rely on sophisticated mathematical models (like physics, chemistry, biology, engineering – and some areas in finance). Therefore their analytical and numerical investigation is very important. But how can we solve such a PDE? First, we need to stress that there is no general method that works for all kinds of PDEs. Very much to the contrary, specific methods need to be developed for different situations, and there is a whole research area in mathematics dedicated to this. In the case of the simple linear heat equation given above, things look better, of course: there are in fact several methods that can be applied. In the following we sketch one particularly simple method (the *separation of variables*). Other methods that could be used are the Fourier transformation,[5] variational methods or the finite element method. The latter is the standard way for numerical computations and works for a large class of PDEs. We refer the reader to [Eva98] and [RR04] for in-depth introductions to PDEs.

The key idea of the separation of variables is to look not for all possible solutions, but only for solutions of a special form, namely $u(x,t) = a(x) \cdot b(t)$. Once we have found such a solution, we only need to prove uniqueness, and we know that the solution we have found is not *any* solution, but *the only* solution. In fact, the uniqueness proof will be omitted here, but can be found in most mathematical textbooks on PDEs.

Using the ansatz $u(x,t) = a(x) \cdot b(t)$ we can rewrite the PDE as follows:

$$a(x)b'(t) = a''(x)b(t).$$

Sorting terms (and assuming that non of them vanishes, which is another point that would have to be justified later) we obtain

$$\frac{b'(t)}{b(t)} = \frac{a''(x)}{a(x)}.$$

The central observation is now the following: the left side only depends on t and the right side only on x. Since both sides agree for all x and t, both terms have to be constant. Let us call this constant $-\lambda$, then we get two ordinary differential

[5] Here we can exploit that the Fourier transform of a derivative is a simple multiplication. Thus after taking the Fourier transform of a PDE some of the derivatives become multiplications (compare Lemma A.12). The result is typically an ODE that can be solved much easier, either analytically or numerically. Finally the solution needs to be Fourier transformed again to return to the original formulation.

equations:

$$b'(t) = -\lambda b(t),$$
$$a''(x) = -\lambda a(x).$$

The first of these equations can be solved by an exponential function:

$$b(t) = b(0)e^{-\lambda t},$$

the second can be solved by a combination of sine and cosine functions, e.g.

$$a(x) = \sin(x/\sqrt{\lambda}).$$

To determine the precise form of a and b we need to take into account the initial and boundary conditions of the heat equation: we superimpose solutions for a such that $a(0)$ corresponds to the initial condition . For instance, if the boundary condition is given by $u(0,t) = u(1,t) = 0$, then any $a_k(x) = \sin(kx/\pi)$ for $k \in \mathbb{N}$ satisfies the boundary condition, since $\sin(k\pi) = 0$ for all $k \in \mathbb{N}$. Denote $u_k(x) = a_k(x) \cdot b_k(x)$, where $b_k(x) = b(0)e^{-kt/\pi}$. Then a weighted sum of the u_k can be constructed to fit the initial condition. This sum still solves the heat equation and the boundary and initial condition, since the heat equation is linear, i.e., weighted sums of solutions are also solutions.

A.6 General Axioms for Expected Utility Theory

There are many different ways to obtain a general characterization of Expected Utility Theory for arbitrary probability measures. In the following we sketch a rather new approach by Chatterjee and Krishna [CK]. The details of this derivation can be found in their article.

Let Z be a compact metric space. Let $\mathcal{P}(Z)$ be the set of all probability measures on Z.

The Independence Axiom can essentially be stated as in the finite case:

Axiom A.13 (Independence) *Let $p, q, r \in \mathcal{P}(Z)$. Let $p \succ q$ and $\lambda \in (0, 1]$, then $\lambda p + (1 - \lambda r) \succ \lambda q + (1 - \lambda)r$.*

To state the Continuity Axiom we need to generalize the notion of continuity via the concept of open sets (compare Sect. A.3): we say that a function $f: X \to Y$ is *continuous* if, for all open sets $U \subset Y$, the set $f^{-1}(U)$ is open.

Open sets in $\mathcal{P}(Z)$ can be defined via weak-$\star$ convergence (compare Sect. 2.4.5): first, define closed sets as sets which contain the limit of any converging sequence (compare Sect. A.3). Second, define open sets as complement to these closed sets. We can do more and construct even a metric d that measures the distance between two probability measures and reflects the same convergence. Such a metric is given by the so-called "Wasserstein metric", compare, e.g., [AGS05].

The Continuity Axiom then becomes:

Axiom A.14 (Continuity) *The sets* $\{q \in \mathcal{P}(Z) \mid q \succ p\}$ *and* $\{q \in \mathcal{P}(Z) \mid p \succ q\}$ *are open.*

Theorem A.15 *Let Z be a compact metric space (e.g. a bounded and closed interval in* $\mathbb{R}$*). Let* $\succeq$ *be a preference relation, i.e., a complete and transitive relation on* $\mathcal{P}(Z)$*, satisfying the Continuity and Independence Axioms, then* $\succeq$ *can be represented by a von Neumann-Morgenstern Expected Utility function* $u: Z \to \mathbb{R}$.

To prove Theorem A.15, Chatterjee and Krishna use intermediate steps: they prove that the Independence Axiom together with the Continuity Axiom implies a new axiom, the *Translation Invariance Axiom*. Together with the Continuity Axiom, this new axiom implies the existence of a EUT function, representing $\succeq$. Translation Invariance can be stated as follows:

Axiom A.16 (Translation Invariance) *Let r be a signed measure on Z with average* $r(Z) = 0$*, in other words, let r be the difference of two probability measures on Z. Let* $p, q \in \mathcal{P}(Z)$*. Assume moreover that* $p + r, q + r \in \mathcal{P}(Z)$*. Then* $p \succeq q$ *implies* $p + r \succeq q + r$.

The intuition behind the translation invariance is that adding a signed measure r to a lottery does not change the preference relation. This means that making certain outcomes more likely, others less likely *in the same way* for p and q, does not change the original preference between p and q. This is morally the same as the Independence Axiom and mathematically at least close enough to show the equivalence of both axioms (under the condition of continuity) relatively easy.

How can we now use the Independence Axiom to construct an EUT function?

First, one can prove that the indifference sets under the preference relation are "thin". This means: for any $q \in \mathcal{P}(Z)$ and any $\varepsilon > 0$ there are $p, r \in \mathcal{P}(Z)$ which are "close" to q, i.e., $d(p, q) < \varepsilon$ and $d(q, r) < \varepsilon$, and that $p \succ q \succ r$.

Second, one can show that, for any $p \in \mathcal{P}(Z)$, the contour sets $\{q \in \mathcal{P}(Z) \mid q \succ p\}$, $\{q \in \mathcal{P}(Z) \mid q \sim p\}$ and $\{q \in \mathcal{P}(Z) \mid q \prec p\}$ are all convex.

$\mathcal{P}(Z)$ is a convex subset of a vector space. We can pick a measure $o \in \mathcal{P}(Z)$ and some $\delta_z \in \mathcal{P}(Z)$ with $\delta_z \succ o$. Let us choose moreover some $q \in \mathcal{P}(Z)$, $q \neq \delta_z$, $q \neq o$. The structure of the indifference sets derived above allows us to find a continuous affine functional $f: \mathcal{P}(Z) \to \mathbb{R}$ such that $f(o) = 0, f(\delta_z) = 1$ and such that the indifference set of q is a contour set of f, i.e., for all $p \in \mathcal{P}(Z)$ with $q \succ p$ we have $f(q) > f(p)$.[6]

[6] More precisely, we first restrict ourselves to a finite dimensional subset, such that the existence of the affine functional f can be deduced from the Separating Hyperplane Theorem (see Sect. A.1). Later one can show that f is independent of the choice of this finite dimensional subset.

Using the translation invariance, one can show that f reflects the preferences on all of $\mathcal{P}(Z)$. With a translation, we can also assume that f is not only affine, but linear.

In the final step, we define $u: Z \to \mathbb{R}$ by $u(z) := f(\delta_z)$. We have to show that this definition is correct, i.e., that

$$U(p) := \int_Z u(z)\,\mathrm{d}p(z) = f(p).$$

This is easy to see for measures with finite support: let $p = \sum_{i=1}^n p_i \delta_{z_i}$, then by linearity of f,

$$\int_Z u(z)\,\mathrm{d}p(z) = \sum_{i=1}^n p_i u(z_i) = \sum_{i=1}^n p_i f(z_i) = f\left(\sum_{i=1}^n p_i \delta_{z_i}\right) = f(p).$$

We can approximate any measure $p \in \mathcal{P}(Z)$ by measures with finite support. Since f is continuous, this proves that $U(p) = f(p)$ for all $p \in \mathcal{P}(Z)$ and thus u is an expected utility function representing the preference relation $\succeq$.

Solutions to Tests

B

师傅领进门，修行在个人

Teachers open the door. You enter it by yourself.
CHINESE PROVERB

The tests are meant to provide an immediate feedback when studying by yourself, hence we give solutions to all questions. Although some of the questions are tricky and require some thinking about the context of the chapter, the student should be able of answering most questions correctly after working through a chapter. If this is not the case, we would recommend to the reader, to study the chapter a little bit more in detail. A good result, however, can only ensure that the basic concepts have been understood and memorized.

Chapter 2

Exercise:	1	2	3	4	5	6	7	8	9	10	11	12	13	14
Answers:			×	×				×	×		×	×		
	×		×		×		×	×		×			×	×
		×		×			×				×	×	×	×
			×	×		×							×	
											×			×

T. Hens, M.O. Rieger, *Financial Economics*, Springer Texts in Business and Economics, DOI 10.1007/978-3-662-49688-6

Chapter 3

Exercise:	1	2	3	4	5	6	7	8	9	10	11	12	13	14	15	16
Answers:	×					×		×			×	×				
			×	×	×				×			×		×		×
		×					×			×		×	×			
	×															
				×											×	

Chapter 4

Exercise:	1	2	3	4	5	6	7	8	9	10
Answers:	×	×	×		×	×			×	×
		×	×		×		×	×		×
		×	×		×					
		×	×	×						×

Chapter 5

Exercise:	1	2	3	4	5	6
Answers:		×		×		
			×		×	×
	×	×	×	×		×

References

[AB03] C.D. Aliprantis and O. Burkinshaw, *Locally Solid Riesz Spaces with Applications to Economics*, Mathematical surveys and monographs, vol. 105, American Mathematical Society, 2003.

[Abd00] M. Abdeallaoui, *Parameter-Free Elicitation of Utilities and Probability Weighting Functions*, Management Science **46** (2000), 1497–1512.

[Abe89] A. Abel, *Asset Prices under Heterogeneous Beliefs: Implications for the Equity Premium Puzzle*, Rodney L. White Center for Financial Research, Working Paper No. 9–89, 1989.

[Abe90] ———, *Asset Prices under Habit Formation and Catching up with the Joneses*, The American Economic Review **80** (1990), 38–42.

[AGS05] L. Ambrosio, N. Gigli, and G. Savaré, *Gradient flows in metric spaces and in the space of probability measures*, Birkhäuser Verlag, Basel, 2005.

[Ake70] G.A. Akerlof, *The Market for "Lemons": Quality Uncertainty and the Market Mechanisms*, The Quarterly Journal of Economics **84** (1970), no. 3, 488–500.

[Arr74] K.J. Arrow, *The Use of Unbounded Utility Functions in Expected-Utility Maximization: Response*, The Quarterly Journal Of Economics **88** (1974), no. 1, 136–138.

[Aum77] R.J. Aumann, *The St. Petersburg Paradox: A Discussion of Some Recent Comments*, Journal of Economic Theory **14** (1977), no. 2, 443–445.

[Bac00] L. Bachelier, *Théorie de la Spéculation*, Annales Scientifiques de l'Ecole Normale Superieure **3** (1900), no. 17, 21–86.

[Ban81] R.W. Banz, *The Relationship between Return and Market Value of Common Stocks*, Journal of Financial Economics **9** (1981), no. 1, 3–18.

[BC97] M.H. Birnbaum and A. Chavez, *Tests of Theories of Decision Making: Violations of Branch Independence and Distribution Independence*, Organizational Behavior and Human Decision Processes **71** (1997), no. 2, 161–194.

[BE82] L.E. Blume and D. Easley, *Learning to Be Rational*, Journal of Economic Theory **26** (1982), no. 2, 340–351.

[BE92] L. Blume and D. Easley, *Evolution and Market Behavior*, Journal of Economic Theory **58** (1992), no. 1, 9–40.

[Ber38] D. Bernoulli, *Specimen Theoriae de Mensura Sortis*, Commentarii Academiae Scientiarum Imperialis Petropolitanae (Proceedings of the royal academy of science, St. Petersburg) (1738).

[Ber98] J. Bertoin, *Lévy Processes*, Cambridge University Press, 1998.

[Bew80] T. Bewley, *The optimum quantity of money*, Models of Monetary Economics (1980), 169–210.

[BH97] W.A. Brock and C.H. Hommes, *A Rational Route to Randomness*, Econometrica: Journal of the Econometric Society (1997), 1059–1095.

[BH98] ———, *Heterogeneous Beliefs and Routes to Chaos, in a Simple Asset Pricing Model*, Journal of Economic Dynamics and Control **22** (1998), 1235–1274.

T. Hens, M.O. Rieger, *Financial Economics*, Springer Texts in Business and Economics, DOI 10.1007/978-3-662-49688-6

[BHS01] N. Barberis, M. Huang, and T. Santos, *Prospect Theory and Asset Prices*, The Quarterly Journal of Economics **116** (2001), no. 1, 1–53.

[BHW98] S. Bikhchandani, D. Hirshleifer, and I. Welch, *Learning from the Behavior of Others: Conformity, Fads, and Informational Cascades*, The Journal of Economic Perspectives **12** (1998), no. 3, 151–170.

[Bir05] M.H. Birnbaum, *A Comparison of Five Models that Predict Violations of First-Order Stochastic Dominance in Risky Decision Making*, Journal of Risk and Uncertainty **31** (2005), no. 3, 263–287.

[BL02] S. Boyarchenko and Levendorskiĭ, *Barrier Options and Touch-and-Out Options under Regular Levy Processes of Exponential Type*, The Annals of Applied Probability **12** (2002), no. 4, 1261–1298.

[Bla05] P.R. Blavatskyy, *Back to the St. Petersburg Paradox?*, Management Science **51** (2005), no. 4, 677–678.

[BN97] O.E. Barndorff-Nielsen, *Normal Inverse Gaussian Distributions and Stochastic Volatility Modelling*, Scandinavian Journal of Statistics **24** (1997), no. 1, 1–13.

[BN98] M.H. Birnbaum and J.B. Navarrete, *Testing Descriptive Utility Theories: Violations of Stochastic Dominance and Cumulative Independence*, Journal of Risk and Uncertainty **17** (1998), 49–78.

[BP00] H. Bleichrodt and J.L. Pinto, *A Parameter-Free Elicitation of the Probability Weighting Function in Medical Decision Analysis*, Management science **46** (2000), 1485–1496.

[Bre79] D.T. Breeden, *An Intertemporal Asset Pricing Model with Stochastic Consumption and Investment Opportunities*, Journal of Financial Economics **7** (1979), no. 3, 265–296.

[BS73] F. Black and M. Scholes, *The Pricing of Options and Corporate Liabilities*, The Journal of Political Economy **81** (1973), no. 3, 637–654.

[BSV98] N. Barberis, A. Shleifer, and R. Vishny, *A Model of Investor Sentiment*, Journal of Financial Economics **49** (1998), no. 3, 307–343.

[BT95] S. Benartzi and R.H. Thaler, *Myopic Loss Aversion and the Equity Premium Puzzle*, The Quarterly Journal of Economics **110** (1995), no. 1, 73–92.

[BX00] M. Brennan and Y. Xia, *Stochastic Interest Rates and the Bond-Stock Mix*, European Finance Review **4** (2000), 197–210.

[CH94] C. Camerer and T.H. Ho, *Violations of the Betweenness Axiom and Nonlinearity in Probability*, Journal of Risk and Uncertainty **8** (1994), 167–196.

[CIR85] J.C. Cox, J.E. Ingersoll, and S.A. Ross, *A Theory of the Term Structure of Interest Rates*, Econometrica **53** (1985), 385–407.

[CK] K. Chatterjee and R.V. Krishina, *A Geometric Approach to Continuous Expected Utility*, Economic Letters **In press**.

[CKL96] L. Chan, J. Karceski, and J. Lakonishok, *Momentum Strategies*, Journal of Finance **51** (1996), 1681–1711.

[CMW97] N. Canner, N.G. Mankiw, and D.N. Weil, *An Asset Allocation Puzzle*, The American Economic Review **87** (1997), no. 1, 181–191.

[Coc01] J.H. Cochrane, *Asset Pricing*, Princeton University Press Princeton, NJ, 2001.

[Con82] G.M. Constantinides, *Intertemporal Asset Pricing with Heterogeneous Consumers and Without Demand Aggregation*, The Journal of Business **55** (1982), no. 2, 253–267.

[Coo64] P. Cootner, *Stock Prices: Random vs. Systematic Changes*, The random character of stock market prices (MIT Press, Cambridge, MA) (1964), 231–252.

[Cov84] T.M. Cover, *An algorithm for maximizing expected log investment return*, IEEE ToIT **IT-30** (1984), no. 2.

[CRR79] John C. Cox, Stephen A. Ross, and Mark Rubinstein, *Option pricing: A simplifeid approach*, Journal of Financial Economics **7** (1979), 229–263.

[CS09] M. Chauvet and Z. Senyuz, *A Joint Dynamic Bi-Factor Model of the Yield Curve and the Economy as a Predictor of Business Cycles.*

[CTW10] A.C.W. Chui, S. Titman, and K.C.J. Wei, *Individualism and Momentum around the World*, The Journal of Finance **65** (2010), 361–392.

[CV02] J.Y. Campbell and L. Viceira, *Strategic Asset Allocation: Portfolio Choice for Long-Term Investors*, Oxford University Press, 2002.

[DBT85] W.F.M. De Bondt and R. Thaler, *Does the Stock Market Overreact?*, The Journal of Finance **40** (1985), no. 3, 793–805.

[DDT80] E. Dierker, H. Dierker, and W. Trockel, *Continuous Mean Demand Functions Derived from Non-Convex Preferences*, Journal of Mathematical Economics **7** (1980), no. 1, 27–33.

[DGH06] E. De Giorgi and T. Hens, *Making Prospect Theory Fit for Finance*, Journal of Financial Markets and Portfolio Management **20** (2006), 339–360.

[DGHP05] E. De Giorgi, T. Hens, and T. Post, *Prospect Theory and the Size and Value Premium Puzzles*, NHH Finance & Management Science Discussion Paper No. 20/2005, 2005.

[DGHR10] E. De Giorgi, T. Hens, and M.O. Rieger, *Financial Market Equilibria With Cumulative Prospect Therory*, Journal of Mathematical Economics **46** (2010), no. 5, 633–651.

[DGP08] E. De Giorgi and T. Post, *Second Order Stochastic Dominance, Reward-Risk Portfolio Selection and the CAPM*, The Journal of Financial and Quantitative Analysis **43** (2008), no. 2, 525–546.

[DHM09] K. Detlefsen, W. Härdle, and R. Moro, *Empirical pricing kernels and investor preference*, Mathematical Methods in Economics and Finance (2009).

[DHS98] K. Daniel, D. Hirshleifer, and A. Subrahmanyam, *Investor Psychology and Security Market Under-and Overreactions*, The Journal of Finance **53** (1998), no. 6, 1839–1885.

[DLSSW90] J.B. De Long, A. Shleifer, L.H. Summers, and R.J. Waldmann, *Noise Trader Risk in Financial Markets*, The Journal of Political Economy **98** (1990), no. 4, 703–738.

[DLW91] Shleifer A. Summers L. De Long, B. and R. Waldmann, *The survival of noise traders in financial markets*, Journal of Business **64** (1991), 1–19.

[DR92] P. Dybvig and S. Ross, *The New Palgrace Dictionary of Money and Finance*, Newmann, Milgate and Eatwell, 1992.

[Duf86] D. Duffie, *Stochastic Equilibria: Existence, Spanning Number, and theNo Expected Financial Gain from Trade'Hypothesis*, Econometrica: Journal of the Econometric Society **54** (1986), 1161–1183.

[Duf88] ———, *Security Markets*, Academic Press Boston, 1988.

[Duf96] ———, *Dynamic Asset Pricing Theory*, Princeton University Press, Princeton, 1996.

[DZ89] D. Duffie and W. Zame, *The Consumption-Based Capital Asset Pricing Model*, Econometrica **57** (1989), no. 6, 1279–1297.

[EHSH06] I. Evstigneev, T. Hens, and K. Schenk-Hoppé, *Evolutionary Stable Stock Markets, Economic Theory*, Springer Verlag **27** (2006), no. 2, 449–468.

[Ell61] D. Ellsberg, *Risk, Ambiguity, and the Savage Axioms*, The Quarterly Journal of Economics **75** (1961), no. 4, 643–669.

[Eva98] L.C. Evans, *Partial Differential Equations*, American Mathematical Society, Providence, R.I., 1998.

[Fel50] W. Feller, *An Introduction to Probability Theory and its Applications*, 3 ed., vol. 1, Wiley India Pvt. Ltd., 1950.

[Fel69] M.S. Feldstein, *Mean-Variance Analysis in the Theory of the Firm under Uncertainty*, Review of Economic Studies **36** (1969), 5–12.

[FF92] E.F. Fama and K.R. French, *The Cross-Section of Expected Stock Returns*, The Journal of Finance **47** (1992), no. 2, 427–465.

[FF98] ———, *Value versus Growth: The International Evidence*, The Journal of Finance **53** (1998), no. 6, 1975–1999.

[Fre88] R.M. French, *The Banach-Tarski Theorem*, The Mathematical Intelligencer **10** (1988), no. 4, 21–28.

[FS07] W. Farkas and C. Schwab, *Anisotropic Stable Levy Copula Processes-Analysis and Numerical Pricing Methods*, Mathematical Models and Methods in Applied Sciences **17** (2007), 1405–1443.

[GH85] D. Gale and M. Hellwig, *Incentive-Compatible Debt Contracts: The One-Period Problem*, The Review of Economic Studies **52** (1985), no. 4, 647–663.

[GH90] G.M. Grossman and E. Helpman, *Trade, Innovation, and Growth*, The American Economic Review **80** (1990), no. 2, 86–91.

[GH06] A. Gerber and T. Hens, *Modelling Alpha-Oppurtunities within the CAPM*, Tech. report, SSRN working paper, 2006.

[GL90] G. Genotte and H. Leland, *Hedging and Crashes*, American Economic Review **80** (1990), 999–1021.

[Gla03] P. Glasserman, *Monte Carlo Methods in Financial Engineering*, Springer Verlag, 2003.

[Gol04] C. Gollier, *The Economics of Risk and Time*, The MIT Press, 2004.

[Got95] P. Gottardi, *An Analysis of the Conditions for the Validity of Modigliani-Miller Theorem with Incomplete Markets*, Economic Theory **5** (1995), 191–207.

[GR07] A. Gerber and K.I.M. Rohde, *Anomalies in Intertemporal Choice?*, Swiss Finance Institute Research Paper Series **07–12** (2007).

[GS80] S.J. Grossman and J.E. Stiglitz, *On the Impossibility of Informationally Efficient Markets*, The American Economic Review **70** (1980), no. 3, 393–408.

[GS01] G. Grimmett and D. Stirzaker, *Probability and Random Processes*, Oxford University Press, 2001.

[Hel81] M. Hellwig, *Bankruptcy, Limited Liability, and the Modigliani-Miller Theorem*, The American Economic Review **71** (1981), 155–170.

[Hic86] J.R. Hicks, *A Revision of Demand Theory*, Oxford University Press, 1986.

[HJ91] L.P. Hansen and R. Jagannathan, *Implications of Security Market Data for Models of Dynamic Economies*, The Journal of Political Economy **99** (1991), no. 2, 225–262.

[HK78] M. Harrison and D. Kreps, *Speculative Investor Behavior in a Stock Market with Heterogeneous Expectations*, Quaterly Journal of Economics **92** (1978), 323–336.

[HP03] T. Hens and B. Pilgrim, *General Equilibrium Foundations of Finance: Structure of Incomplete Markets Models*, Kluwer Academic Publishers, 2003.

[HR14] T. Hens and M.O. Rieger, *Can utility optimization explain the demand for structured investment products?*, Quantitative Finance **14** (2014), 673–681.

[HS99] H. Hong and J.C. Stein, *A Unified Theory of Underreaction, Momentum Trading, and Overreaction in Asset Markets*, The Journal of Finance **54** (1999), no. 6, 2143–2184.

[HS09] T. Hens and S.C. Steude, *The Leverage Effect without Leverage: An Experimental Study*, Finance Research Letters **6** (2009), no. 2, 83–94.

[HSH05] T. Hens and K.R. Schenk-Hoppé, *Evolutionary Stability of Portfolio Rules in Incomplete Markets*, Journal of Mathematical Economics **41** (2005), 43–66.

[HSH09] T. Hens and K.R. Schenk-Hoppé, *Handbook of Financial Markets: Dynamics and Evolution*, North-Holland, 2009.

[HW07a] T. Hens and M. Wang, *Hat Finance eine kulturelle Dimension?*, Finanzmärkte – Effizienz und Sicherheit (Brigitte Strebel-Aerni, ed.), Schulthess, 2007.

[HW07b] T. Hens and P. Woehrmann, *Strategic Asset Allocation and Market Timing: A Reinforcement Learning Approach*, Computational Economics **29** (2007).

[Ine05] A.M. Ineichen, *The Critique of Pure Alpha*, UBS Global Equity Research (March) (2005).

[Ing07] J. Ingersoll, *Non-Monotonicity of the Tversky-Kahneman Probability-Weighting Function: A Cautionary Note*, European Financial Management (2007).

[ISW08] Z. Ivković, C. Sialm, and S. Weisbenner, *Portfolio Concentration and Performance of Individual Investors*, Journal of Financial and Quantitative Analysis **43** (2008), no. 3, 613–656.

[Jac00] J.C. Jackwerth, *Recovering risk aversion from option prices and realized returns*, Review of Financial Studies **13** (2000), no. 2, 433.

[Jän94] K. Jänich, *Linear Algebra*, Springer Verlag, 1994.

[Jän95] ______, *Topology*, Springer Verlag, 1995.

[JB03] E.T. Jaynes and G.L. Bretthorst, *Probability Theory: the Logic of Science*, Cambridge University Press, 2003.

[JK95a] E. Jouini and H. Kallal, *Arbitrage in Securities Markets with Short Sales Constraints*, Mathematical Finance **5** (1995), 197–232.

[JK95b] ______, *Martingales and Arbitrage in Securities Markets with Transaction Costs*, Journal of Economic Theory **66** (1995), 178–197.

[JT93] N. Jegadeesh and S. Titman, *Returns to Buying Winners and Selling Losers: Implications for Stock Market Efficiency*, The Journal of Finance **48** (1993), no. 1, 65–91.

[JW96] R. Jagannathan and Z. Wang, *The Conditional CAPM and the Cross-Section of Expected Returns*, Journal of Finance **51** (1996), no. 1, 3–53.

[Kal06] J. Kallsen, *A Didactic Note on Affine Stochastic Volatility Models*, from stochastic calculus to mathematical finance (2006), 343–368.

[Kar78] U.S. Karmarkar, *Subjectively Weighted Utility: A Descriptive Extension of the Expected Utility Model*, Organizational Behavior and Human Performance **21** (1978), 61–72.

[Kar88] I. Karatzas, *On the Pricing of American Options*, Applied Mathematics and Optimization **17** (1988), no. 1, 37–60.

[KDS99] D. Kahneman, E. Diener, and N. Schwarz, *Well-Being: The Foundations of Hedonic Psychology*, Russell Sage Foundation, 1999.

[Key88] J.M. Keynes, *Memorandum for the Estates Committee, Kings College, Cambridge*, The Collected Writings of John Maynard Keynes, 1988.

[KK01] R. Korn and E. Korn, *Option Pricing and Portfolio Optimization*, Graduate Studies in Mathematics, vol. 31, American Mathematical Society, 2001.

[Kni21] F.H. Knight, *Risk, Uncertainty and Profit*, Houghton Mifflin Company, 1921.

[KR95] G. Keren and P. Roelofsma, *Immediacy and Certainty in Intertemporal Choice*, Organizational Behavior and Human Decision Processes **63** (1995), no. 3, 287–297.

[KS98] I. Karatzas and S. Shreve, *Methods of Mathematical Finance*, Springer, 1998.

[KT79] D. Kahneman and A. Tversky, *Prospect Theory: An Analysis of Decision Under Risk*, Econometrica **47** (1979), no. 2, 263–291.

[KW01] M. Kilka and M. Weber, *What Determines the Shape of the Proability Weighting Function Under Uncertainty?*, Management Science **47** (2001), no. 12, 1712–1726.

[LAP99] B. LeBaron, W.B. Arthur, and R. Palmer, *Time Series Properties of an Artificial Stock Market*, Journal of Economic Dynamics and Control **23** (1999), 1487–1516.

[Len04] Y. Lengwiler, *A Monetary Policy Simulation Game*, The Journal of Economic Education **35** (2004), no. 2, 175–183.

[Lév25] P. Lévy, *Calcul des Probabilities*, Guthier-Villars, Paris, 1925.

[Lév05] H. Lévy, *Stochastic Dominance: Investment Decision Making Under Uncertainty*, Springer Verlag, 2005.

[LJ78] R.E. Lucas Jr, *Asset Prices in an Exchange Economy*, Econometrica **46** (1978), no. 6, 1429–1445.

[LL03] M. Lévy and H. Lévy, *Prospect Theory: Much Ado About Nothing?*, Management Science **48** (2003), no. 10, 1334–1349.

[LS00] C. Lee and B. Swaminathan, *Price Momentum and Trading Volume*, Journal of Finance **55** (2000), 2017–2069.

[LSV92] J. Lakonishok, A. Shleifer, and R.W. Vishny, *The Impact of Institutional Trading on Stock Prices*, Journal of Financial Economics **32** (1992), no. 1, 23–43.

[LSV94] ______, *Contrarian Investment, Extrapolation, and Risk*, The Journal of Finance **49** (1994), no. 5, 1541–1578.

[Lux09] Thomas Lux, *Applications of Statistical Physics to Finance and Economics*, Handbook of Research on Complexity (B. Rosser, ed.), Cheltenham, 2009, pp. 213–258.

[Mal90] B.G. Malkiel, *A Random Walk down Wall Street*, Norton New York, 1990.

[Man63] B. Mandelbrot, *The Variation of Certain Speculative Prices*, The Journal of Business **36** (1963), no. 4, 394–419.

[Mar52] H.M. Markowitz, *Portfolio Selection*, Journal of Finance **7** (1952), no. 1, 77–91.

[Mar91] H.M. Markowitz, *Portfolio Selection: Efficient Diversification of Investments*, 2nd ed., Blackwell Publishers, 1991.

[Meh06] R. Mehra, *The Equity Premium in India*, National Bureau of Economic Research Cambridge, Mass., USA, 2006.

[Men34] K. Menger, *Das Unsicherheitsmoment in der Wertlehre*, Journal of Economics **51** (1934), 459–485.

[Mer72] R.C. Merton, *An Analytic Derivation of the Efficient Portfolio Frontier*, The Journal of Financial and Quantitative Analysis **7** (1972), no. 4, 1851–1872.

[Mer73] ______, *An Intertemporal Capital Asset Pricing Model*, Econometrica **41** (1973), no. 5, 867–887.

[MGU09] V. De Miguel, L. Garlappi, and R. Uppal, *Optimal versus Naive Diversification: How Inefficient Is the 1/N Portfolio Strategy?*, Review of Financial Studies **22** (2009), no. 5, 1915–1953.

[Mil27] F.C. Mills, *The Behavior of Prices*, New York: National Bureau of Economic Research, 1927.

[Mit15] W. C. Mitchell, *The Making and Using of Index Numbers*, no. 173, U.S. Bureau of Labor Statistics, 1915.

[MM58] F. Modigliani and M.H. Miller, *The Cost of Capital, Corporation Finance and the Theory of Investment*, American Economic Review **48** (1958), no. 3, 261–297.

[Mos98] W.S. Mossberg, *Making Book on the Buck*, Wall Street Journal (1998), 17.

[MQ02] M. Magill and M. Quinzii, *Theory of incomplete markets*, The MIT Press, 2002.

[MS92] R.C. Merton and P.A. Samuelson, *Continuous-Time Finance*, Blackwell Pub, 1992.

[nzz00] *Zahlen bitte – Untreue Ehemänner*, NZZ-Folio (12/2000).

[Oli26] M. Olivier, *Les Nombres Indices de la Variation des Prix*, 1926, Paris doctoral dissertation.

[Pap04] B.F. Papa, *Stock Market Volatility: A Puzzle? An investigation into the causes and consequences of asymmetric volatility*, Master's thesis, ETH Zurich and University Zurich, 2004.

[Pra64] J.W. Pratt, *Risk Aversion in the Small and in the Large*, Econometrica **32** (1964), no. 1/2, 122–136.

[Pre98] D. Prelec, *The Proability Weighting Function*, Econometrica **66** (1998), 497–527.

[Qui82] J. Quiggin, *A Theory of Anticipated Utility*, Journal of Economic Behavior and Organization **3** (1982), no. 4, 323–343.

[Rab00] Matthew Rabin, *Risk aversion and expected-utility theory: A calibration theorem*, Econometrica **68** (2000), 1281–1292.

[RH10] Y Ritov and W. Härdle, *From animal baits to investors preference: Estimating and demixing of the weight function in semiparametric models for biased samples*, Statistica Sinica (2010).

[Rie10] M.O. Rieger, *Co-Monotonicity of Optimal Investments and the Design of Structural Financial Products*, Finance and Stochastics **15** (2010), no. 1.

[Rie12] ______, *Why Do Investors Buy Bad Financial Products? Probability Misestimation and Preferences in Financial Investment Decision*, Journal of Behavioral Finance **13** (2012).

[Rou98] K.G. Rouwenhorst, *International Momentum Strategies*, The Journal of Finance **53** (1998), no. 1, 267–284.

[RR04] M. Renardy and R.C. Rogers, *An Introduction to Partial Differential Equations*, 2nd ed., Springer Verlag, 2004.

[RS76] M. Rothschild and J. Stiglitz, *Equilibrium in Competitive Insurance Markets: An Essay on the Economics of Imperfect Information*, The Quarterly Journal of Economics **90** (1976), no. 4, 629–649.

[RSW] M.O. Rieger, V. Stankovic, and P. Wöhrmann, *Normal Inverse Gaussian Distribution as a Model for Hedge Fund Returns*, in preparation.

[RT02] M. Rabin and R.H. Thaler, *Response from Matthew Rabin and Richard H. Thaler*, The Journal of Economic Perspectives **16** (2002), no. 2, 229–230.

[Rub74] M. Rubinstein, *An Aggregation Theorem for Securities Markets*, Journal of Financial Economics **1** (1974), no. 3, 225–44.

[Rub81] R.Y. Rubinstein, *Simulation and the Monte Carlo Method*, John Wiley & Sons, Inc. New York, NY, USA, 1981.

[Rub03] A. Rubinstein, *Economics and Psychology? The Case of Hyperbolic Discounting*, International Economic Review **44** (2003), no. 4, 1207–1216.

[RW06] M.O. Rieger and M. Wang, *Cumulative Prospect Theory and the St. Petersburg Paradox*, Economic Theory **28** (2006), no. 3, 665–679.

[RW08] ______, *Prospect Theory for Continuous Distributions*, Journal of Risk and Uncertainty **36** (2008), 83–102.

[S+05] K. Sydsæter et al., *Further Mathematics for Economic Analysis*, Prentice Hall, 2005.

[Sam64] P. Samuelson, *Theoretical Notes on Trade Problems*, The Review of Economics and Statistics **46** (1964), no. 2, 145–154.

[San00] A. Sandroni, *Do Markets Favor Agents Able to Make Accurate Predictions?*, Econometrica **68** (2000), no. 6, 1303–1341.

[Sch08] P. Schafheitlin, *Die Theorie der Besselschen Funktionen*, Teubner-Verlag, 1908.

[Sch09] *Handbook of Financial Markets: Dynamics and Evolution*, North-Holland, 2009.

[SDBW99] D. Schiereck, W. De Bondt, and M. Weber, *Contrarian and Momentum Strategies in Germany*, Financial Analysts Journal **55** (1999), no. 6, 104–116.

[She00] H. Shefrin, *Beyond Greed and Fear*, Harvard Business School Press, 2000.

[She08] H. Shefrin, *A Behavioral Approach to Asset Pricing (2nd edition)*, Elsevier, 2008.

[Shi81] R.J. Shiller, *Do Stock Prices Move too Much to Be Justified by Subsequent Changes in Dividents?*, American Economic Review **71** (1981), 421–436.

[Sin02] H.W. Sinn, *Weber's Law and the Biological Evolution of Risk Preferences: the Selective Dominance of the Logarithmic Utility Function*, CESifo, 2002.

[Sor98] G. Soros, *The Crisis of Global Capitalism: Open Society Endangered*, Public Affairs, 1998.

[SR09] L. Su and M.O. Rieger, *How Likely is it to Hit a Barrier? Theoretical and Empirical Estimates*, SSRN working paper (2009).

[TF95] A. Tversky and C.R. Fox, *Weighing Risk and Uncertainty*, Psychological Review **102** (1995), no. 2, 269–283.

[Tha93] R.H. Thaler, *Advances in Behavioral Finance*, Russell Sage Foundation Publications, 1993.

[TK92] A. Tversky and D. Kahneman, *Advances in Prospect Theory: Cumulative Representation of Uncertainty*, Journal of Risk and Uncertainty **5** (1992), 297–323.

[TR10] T. Talpsepp and M.O. Rieger, *Explaining Asymmetric Volatility Around the World*, Journal of Empirical Finance **17** (2010), 938–956.

[Var78] H.R. Varian, *Microeconomic Analysis, 2nd edition*, Norton New York (1978).

[Var93a] ______, *A Portfolio of Nobel Laureates: Markowitz, Miller and Sharpe*, The Journal of Economic Perspectives **7** (1993), no. 1, 159–169.

[Var93b] ______, *What Use is Economic Theory?*, Dept. of Economics, University of Michigan, 1993.

[vNM53] J. von Neumann and O. Morgenstern, *Theory of Games and Economic Behavior*, 3 ed., Princeton University Press, Princeton, 1953.

[Wak93] P.P. Wakker, *Unbounded Utility Functions for Savage's "Foundations of Statistics," and other Models*, Mathematics of Operationas Research **18** (1993), 446–485.

[WG96] G. Wu and R. Gonzalez, *Curvature of the Probability Weighting Function*, Management Science **42** (1996), 1676–1690.

[WRH16] M. Wang, M.O. Rieger, and T. Hens, *How Time Preferences Differ: Evidence from 53 Countries*, Journal of Economic Psychology **52** (2016), 115–135.

[ZK09] V. Zakalmouline and S. Koekebakker, *A Generalization of the Mean-Variance Analysis*, European Financial Management **15** (2009), no. 5, 934–970.

Index

T. Hens, M.O. Rieger, *Financial Economics*, Springer Texts in Business and Economics, DOI 10.1007/978-3-662-49688-6

MIX
Papier aus verantwortungsvollen Quellen
Paper from responsible sources
FSC® C105338
FSC
www.fsc.org

If you have any concerns about our products,
you can contact us on
ProductSafety@springernature.com

In case Publisher is established outside the EU,
the EU authorized representative is:
Springer Nature Customer Service Center GmbH
Europaplatz 3, 69115 Heidelberg, Germany

Printed by Libri Plureos GmbH
in Hamburg, Germany